D
521
.B64
1854

MAGNIFICENT WORK OF HISTORY.

A Whole Library in Itself!

Cost $11,000—1207 Pages—70 Maps—700 Engravings.

Eighth Thousand, now ready and for sale by Subscription:

A

HISTORY OF ALL NATIONS,

FROM THE EARLIEST PERIOD TO THE PRESENT TIME;

OR,

UNIVERSAL HISTORY;

IN WHICH THE

HISTORY OF EVERY NATION, ANCIENT AND MODERN,

IS SEPARATELY GIVEN.

BY S. G. GOODRICH,

Consul to Paris, and Author of several Works of History, Parley's Tales, etc.

It contains 1207 pages, royal octavo, and is illustrated by 70 Maps and 700 Engravings: bound in imitation Turkey morocco.

Invariable retail price, $6,00 in one volume; $7,00 in two volumes.
The same, full gilt edge and sides, $8,00 in one volume; $10,00 in two vols.

*** It is believed that the above work, by Mr. Goodrich, will be very acceptable to the American public. It is the result of years of toil and labor, assisted in his researches by several scholars of known ability, and has been gotten up at a great expense by the proprietors. No pains have been spared in the execution of the Illustrations and Maps, which are entirely new, and prepared by the distinguished author *expressly* for the work. Indeed, all the other historical writings of Mr. Goodrich sink into insignificance, when compared with this, the result of his riper and maturer years. It is admitted that *One Hundred Dollars* could not purchase the same matter in any other shape: and the publishers confidently expect that, in consideration of the great literary value of the work, the large sum expended in preparing it for the press, and the exceedingly moderate price at which it is offered, that it will be favorably received by every lover of good books.

DERBY & MILLER, *Sole Publishers, Auburn, N. Y.*

SUBSCRIPTION BOOKS:

Published by Derby & Miller, Auburn, N. Y.

BLAKE'S FARMER'S EVERY DAY BOOK;

Or, How a Farmer may Become Rich: Being Sketches of Life in the Country, with the Popular Elements of Practical and Theoretical Agriculture, and Twelve Hundred Laconics and Apothegms, relating to Morals, Regimen, and General Literature; also Five Hundred Receipts on Health, Cookery, and Domestic Economy; with ten fine Illustrations, representing the various scenes attendant upon Farming, etc. By JOHN L. BLAKE, D. D., author of "Biographical Dictionary," etc., etc.

☞ The publishers respectfully announce that they have undertaken the publication of this large and beautiful work, with a view to supply a *desideratum* that has long been felt—a book for *every Farmer's Library*—believing that the venerable author has produced a work that will be worth its weight in gold to every Farmer's Family that thoroughly peruse it. The work contains 664 pages, large octavo, with a motto surrounding each page. It is printed on fine paper, and bound in substantial imitation Turkey morocco, gilt back. Retail price $2,50.

FROST AND DRAKE'S BOOK OF INDIAN WARS AND CAPTIVITIES,

BEING A COMPLETE HISTORY OF THE INDIAN WARS OF THE UNITED STATES,

Embracing the early Indian Wars of the Colonies and the American Revolution, King Philip's War, the French and Indian Wars, the North-Western War, Black Hawk War, Seminole War, &c., &c., together with Indian Captivities—being true Narratives of Captives who have been carried away by the Indians, from the Frontier settlements of the United States, from the earliest period to the present time. By JOHN FROST, LL. D., and SAMUEL G. DRAKE, M. D. Well printed, on good paper, and bound in elegant red morocco binding, 684 pages, large octavo, and illustrated with 166 engravings, from original designs by W. Croome, and other distinguished artists. Retail price $2,50.

Prof. Frost's New Work!

GREAT MEN AND GREAT EVENTS IN HISTORY,

From the Earliest Period to the Present Time, beginning with Egypt under the first Pharaohs, and brought down to the Discovery of Gold in California. Illustrated with 300 beautiful wood Engravings, by Croome and other eminent artists. In one large octavo volume of 832 pages; elegantly bound in imitation Turkey morocco, gilt back and sides. Retail price $3,00.

Frost and Sidney's

HISTORY OF CALIFORNIA AND AUSTRALIA,

A History of the State of California, from the period of Conquest by Spain, to the year 1853, inclusive, containing an account of the discovery of the Gold Mines and Placers, the large number of Gold Seekers, the quantity of Gold already obtained, a description of her Mineral and Agricultural Resources, with thrilling accounts of Adventures among the Miners, with advice to Emigrants of best routes and preparations necessary to get there. By JOHN FROST, LL. D. Also,

A Complete History of the Colonies of Australia, New South Wales, Victoria and South Australia, from their first discovery until the year 1852, with a detailed account of their Pastures, Copper Mines and Gold Fields. By SAMUEL SIDNEY, Esq., author of "The Australian Hand Book." The whole making an octavo volume of 600 pages, illustrated by numerous Engravings, and bound in substantial scarlet cloth. Retail price, $2,50. (In press, and will be published during the Summer of 1853.)

WOMEN OF THE BIBLE,

Being historical and descriptive Sketches of the Women of the Bible, from Eve of the Old, to the Marys of the New Testament, by Rev. P. C. HEADLEY, illustrated, 16mo. muslin,........................ *Price.* 1 00

THE SAME—gilt edge and full gilt sides,.............................. 2 00

POETS AND POETRY OF THE BIBLE,

By GEORGE GILFILLIAN, 12mo. muslin,.......................... 1 00

THE SAME—gilt edge and full gilt sides,.............................. 2 00

GIFT BOOK FOR YOUNG MEN,

Or Familiar Letters on Self-knowledge, Self-education, Female Society, Marriage, &c., By Dr. WM. A. ALCOTT, 12mo. muslin......... 75

THE SAME—gilt edge, and full gilt sides,.............................. 1 50

GIFT BOOK FOR YOUNG LADIES.

Or Woman's Mission; being Familiar Letters to a Young Lady on her Amusements, Employments, Studies, Acquaintances, male and female, Friendships, &c., by Dr. WILLIAM A. ALCOTT, 12mo. muslin, 75

THE SAME—gilt edge and full gilt sides,.............................. 1 50

YOUNG MAN'S BOOK,

Or, Self-Education, by Rev. WM. HOSMER, 12mo. muslin........... 75

THE SAME—muslin, gilt edge and full gilt sides,...................... 1 50

YOUNG LADY'S BOOK,

Or, principles of Female Education, by Rev. WM. HOSMER, 12mo. muslin,.. 75

THE SAME—muslin, gilt edge and full gilt sides,...................... 1 50

WESLEY OFFERING,

By Rev. D. HOLMES, 16mo.. 75

THE SAME—muslin, gilt edge and full gilt sides, ... 1 50

SUMMERFIELD, OR LIFE ON THE FARM,

By Rev. DAY KELLOGG LEE, 16mo.................................. 1 00

THE SAME—muslin, gilt edge and full gilt sides, ... 2 00

GOLDEN STEPS FOR THE YOUNG,

To Usefulness, Respectabilty and Happiness, by JOHN MATHER AUSTIN, author of "Voice to Youth," &c., 12mo. muslin, steel frontispiece,.. 75

THE SAME—gilt edge and full gilt sides,.............................. 1 50

POEMS OF JOHN QUINCY ADAMS,

On Religion and Society, with notices of his life and character, 12mo.. muslin,.. 50

THE SAME—gilt edge,.. 75

SILVER CUP OF SPARKLING DROPS,

From many Fountains, by Miss C. B. PORTER, 12mo. muslin...... 75

THE SAME—gilt edge and full glit sides,.............................. 1 50

THE AMERICAN ORATOR'S OWN BOOK,

Compiled by JOHN AGAR, 12mo.................................... 1 00

FRESH LEAVES FROM WESTERN WOODS,

By METTA V. FULLER, 12mo. muslin,.............................. 1 00

LIFE OF BENJAMIN FRANKLIN,

Written by himself, fine portrait, 12mo. muslin, ... 1 25

LIFE OF GENERAL LAFAYETTE,
By P. C. HEADLEY, 12mo. with a portrait, uniform with Headley's Josephine, muslin,... *Price.* 1 25

LIFE OF JOHN QUINCY ADAMS,
Sixth President of the United States, by WILLIAM H. SEWARD, 12mo. muslin,.. 1 25

LIFE OF LOUIS KOSSUTH,
Governor of Hungary, by P. C. HEADLEY, with an Introduction by HORACE GREELEY, one vol. 12mo. muslin,.......................... 1 25

LIFE OF GENERAL ZACHARY TAYLOR,
Twelfth President of the United States, by H. MONTGOMERY, 12mo. muslin,... 1 00

LIFE OF WINFIELD SCOTT,
By E. D. MANSFIELD, illustrated, 12mo. muslin,.................... 1 25

LIFE OF GENERAL FRANK PIERCE,
Fourteenth President of the United States, by D. W. BARTLETT, 12mo. muslin,.. 75

GENERALS OF THE LAST WAR
With Great Britain, with portraits of Generals Brown, Macomb, Scott, Jackson, Harrison and Gaines, by JOHN S. JENKINS, muslin, 12mo.. 1 00

LIVES OF JAMES MADISON AND JAMES MONROE,
By J. Q. ADAMS, 12mo. muslin,.................................... 1 00

LIFE OF ANDREW JACKSON,
President of the United States, by JOHN S. JENKINS, muslin 12mo. 1 00

VOICE TO THE YOUNG,
Or, Lectures for the Times, by W. W. PATTON. 12mo. muslin, ... 63

MISSIONARY OFFERING,
A Memorial of Christ's Messengers in Heathen Lands, dedicated to Dr. Judson, 8 engravings, 12mo. muslin,.......................... 1 00

PURE GOLD,
Or, Truth in its Native Loveliness, by Rev. D. HOLMES, muslin, 1 00

METHODIST PREACHER,
Containing 28 Sermons upon Doctrinal Subjects, by Bishop Hedding, Dr. Fisk, Dr. Bangs, Dr. Durbin, and other eminent preachers of the M. E. Church, octavo,.................................. 1 00

EPISCOPAL METHODISM AS IT WAS AND IS,
Being a History of the M. E. Church in the United States, &c., by Rev. P. DOUGLASS GORRIE, 12mo. muslin,........................ 1 00

LIVES OF EMINENT METHODIST MINISTERS,
By Rev. P. DOUGLASS GORRIE, 12mo. muslin,...................... 1 25

INCIDENTS AND NARRATIVES,
For Christian Families, by Rev. A. RUSSEL BELDEN, 12mo. muslin, 1 00

HISTORY AND CONDITION OF OREGON,
Including a Voyage round the World, by Rev. G. HINES, of the Oregon Mission, 12mo. muslin,...................................... 1 25

HOME IN THE COUNTRY.

THIRD THOUSAND.

THE MODERN FARMER;

OR,

HOME IN THE COUNTRY:

DESIGNED FOR

INSTRUCTION AND AMUSEMENT

ON

RAINY DAYS AND WINTER EVENINGS.

BY REV. JOHN L. BLAKE, D. D.,

AUTHOR OF FARMERS' EVERY-DAY BOOK, THE FARMER AT HOME, THE FARM AND THE FIRESIDE, FAMILY ENCYCLOPÆDIA, AND A GENERAL BIOGRAPHICAL DICTIONARY.

AUBURN:
DERBY AND MILLER.
BUFFALO:
DERBY, ORTON AND MULLIGAN.
CINCINNATI:
HENRY W. DERBY.
1854.

Entered, according to the Act of Congress, in the year 1851, by
JOHN L. BLAKE, D. D.
In the Clerk's Office of the District Court of the United States for the
District of New Jersey.

Mrs. Eliz. Rathbone
9t.
1-5-1923

J. P. JONES & CO., STEREOTYPERS,
183 William-street, N. Y.

INTRODUCTION.

It has sometimes been affirmed that a majority of people in rural districts are wanting in literary taste, and consequently in the common elements of intellectual progress. It is undoubtedly a fact, that they do not spend so much time in reading the cheap publications of the day as do their accusers. This, however, furnishes no evidence that they are deficient in mental vigor, or in a disposition advantageously to spend their leisure time, whenever books adapted to their habits of thought, and to their circumstances, are presented to their consideration. If their minds have not received the tinsel embellishment, acquired by persons in some other occupations, it is no less true that they have escaped that moral contamination which has sometimes resulted from such embellishment. If they are without such flippancy and pertness it is equally true, that the simplicity of manners which attends this destitution is characteristic of their moral perceptions and their social affinities.

We have good opportunity of knowing the things whereof we make this affirmation. It is a fact that the population of rural districts are more exempt from vicious tendencies and crime than a corresponding number of individuals in a populous city. Of this there can be no doubt. To deny it is a mere assumption in opposition to the well authenticated records of crime. If any one doubts it, let him examine the annals of our criminal courts and of our penitentiaries. They will testify in favor of our position. They will show that an agriculturist will rarely be found chronicled with the hordes of those suffering the disreputable penalties of a violated law.

However, it is not denied, that there is occasion to incite the inhabitants of rural localities, and especially the younger portions of them, to more systematic habits of mental improvement. To make a successful effort for this, the means for doing it should be cast around them. These means, to be available, must be in accordance with the intellectual apprehensions originating in the previous culture enjoyed by those for whose benefit they are designed. They should be in accordance also with the natural scenery, the occupations, and the pecuniary considerations that exercise upon them an unvarying influence. It would be a perversion of the soundest maxims of philosophy to disregard the agency

414295

of such auxiliaries. If we would secure the attention of the sailor to any course of teaching, whether mental, moral, or physical, we must picture before him the scenes of ocean life, the dashing of the mountain wave, the howling of the midnight tempest, and the incessant rolling of the ship. With such scenes vividly appearing in the perspective, he readily becomes our pupil and yields a homage to our instructions. It is delightful to witness the flash of his eye, and the beaming joy rising upon his care and weather worn vizage. To such preliminaries the honest tar will never turn a deaf ear.

It is much the same in rural life. If we would awaken the mind of those who till the ground, we must show them that we are familiar with every thing pertaining to their mode of life. Not indeed that they be dosed inwardly and outwardly with the abstruse science of agriculture, without regard to their previous knowledge of it. The first lesson prescribed for them is ordinarily to be predicated on knowledge already attained. Then there will be a natural tendency to it from past effort. Each lesson prepares the way for an additional lesson. Each step taken generates an impulse that will quicken the succeeding one. Thus mental progress will be like accelerated motion on an inclined plane, constantly acquiring new power and velocity till it moves irresistibly onward from its own impetus. With such modes of calling into action the dormant energies of the human intellect, labor will never be spent in vain As much as has been said by those not acquainted with the facts to be considered in relation to their assumptions, of the want of congeniality between rural influences and mental culture, it will be found on more careful consideration, that such inferences are sadly deficient in logical deduction. In winning the ear and controlling the imagination of the farmer's family to a course of reading and study, we must spread before them graphic delineations of the husbandman's vocation; of the wonderful exhibitions in animal and vegetable physiology; of the most successful results from the improved modes of modern farming; and especially of the eminence to which the sons of farmers are not unwont to arrive both in agriculture and in the other pursuits to which they may be accidentally led. Well selected and well prepared topics of this character operate on the minds of persons living in the country similarly to the scenes of ocean life on the sailor.

Accordingly we have endeavored to put into a popular form—one which cannot be repulsive—some of the fundamental principles of agricultural science. Here will be found the analogies between animal and vegetable organization, and animal and vegetable development and subsistence. In them will be seen the well stored cabinet of nature's sublime and most curious mysteries. In them may be read the wisdom of nature's God. And in them is displayed a broad type of the marvelous perfection which pervades the wide creation—the earth and the heavens which encompass it. Connected with such principles will also be found the processes of ameliorating and fertilising the soil, the great secret of successful rural labor, whether on the farm, or in the garden. The latter cannot be made an enlightened service without an acquaintance with the former; and, if it ever becomes a well remunerated one, from accident, or habit, or an imitation of others, it is purely mechanical and physical, without one gleam of mental fervor or moral sympathy. The scientific farmer may become an enthusiast in his labor, while the unlearned and ignorant one will pursue his toils with the heartless and sullen monotony of the ox, which receives no impulse but from the voice or lash of its master.

Interspersed with the above are fragments and detached portions of rural literature, in prose and poetry. The country is full of the elements of poetry, and those who have no taste for it have drunk but moderately from her deep and pure fountains. But in this portion of the present volume the most interesting parts are selections from agricultural addresses, particularly at the annual fairs. These autumnal exhibitions are becoming more and more general, and are exerting on the agricultural community a most beneficial influence. In the addresses which characterize them, are found some of the finest specimens of a standard literature. The samples here furnished are designed to make the mass from which they are culled more generally known and appreciated. Here are facts more interesting, and oftentimes more stirring and absorbing, than the best of fiction; and not rarely in a style of philological drapery that would do credit to our most polished writers. It is a fact, whether generally understood or not, that on these occasions, there is, in bold relief, a display of real greatness rarely witnessed elsewhere. In this country, as it has long been in Great Britain, there is a grouping, at our agricultural fairs, of the highest grades of social and intellectual endowment, which even give type to our national character.

It is well known that our university commencements have ever been the occasions on which scholars are accustomed to mingle their sympathies and mutually to impart to each other the stimulants to an enduring fame. Here are formed those bonds of fellowship which bind in undying friendship the full grown sons of genius and erudition from every part of the land. Here burn the ethereal fires which give warmth and renewed vigor to all meeting within the circle of their genial radiations. It is good for scholars to the end of life thus annually to congregate and renew their early aspirations for mental renown. The annual fair of the agricultural society is to the farmer and to a whole rural community what the college commencement is to men of letters. It is the season of rest and of triumph to the thousands, and the tens of thousands, on whose wisdom and muscular power, are suspended the destiny and the happiness of mankind. As face answers to face in the water, and as the heart of man answers to man in the great realm of human brotherhood, so does the heart of the farmer swell with more vigorous pulsations when he beholds the face and when he grasps the hand of his unnumbered brethren as they gather round him. Here the feeble, too become strong, the unlearned become wise, and the wavering become resolute, both in toil and in loyalty to their honorable vocation. The fair of the agricultural society is the great jubilee of the husbandman! It should be a jubilee of the nation.

About one-third of this work consists of extracts from some of the best addresses delivered at recent agricultural fairs. No apology is necessary for this free use of them, either to the gentlemen who made them or to the public. The only difficulty experienced in making the extracts was in having so little space that could be devoted to such a purpose when the mass of materials from which they were taken was so ample Indeed, from some of the very best addresses that fell under the author's observation, nothing could be used, as the space was crowded before they were received. Nor can it be presumed that even a tenth part of those which have been called forth on such occasions, in different parts of the country, can have been seen by him. Still he became so well satisfied with the permanent value possessed by a large number of these popular productions, he resolved, that so far as able to collect them, on having them bound up for his own use. And, it is believed, a very large collection of such materials might advantageously be given to the public, were an enterprising publisher induced to engage in it.

Such a collection, judiciously made, would embody great merit. Rarely can a volume be found possessing an equal amount of interest, especially to persons for whose benefit it would be mainly designed. Nor would the general reader often find anything, on such subjects, more beautiful in the English language, than the addresses of Lewis Cass, Josiah Quincey, Jr., Edward Everett, H. G. O. Colby, Alex. H. H. Stuart, G. R. Russell, E. H. Derby, Marshall P. Wilder, Alex. H. Stevens, G. Bancroft, and several others. The present extracts are but a trifling sample of what may be procured; yet it is firmly believed that they in connection with the other portions of this volume, will do something at least, in giving edification and amusement, on rainy days and winter evenings, in the country more than in the city, calculated to engender the most gloomy and sluggish cogitations.

HOME IN THE COUNTRY.

MORAL DIGNITY OF AMERICAN LABOR.

It is not a condescending effort of the high to exalt the low, nor of the peculiarly cultivated to elevate and benefit the less refined and privileged of men. But it is a mutual agreement, to honor that imperishable element in man, which the power of the Creator has implanted within him; and to excite and cultivate to the highest possible degree, by an honorable competition, the skill and effort of man for the improvement and elevation of his present condition of being—not for the mere attainment of the means of luxurious indulgence, but for the widest disposal of benefits upon mankind; for the utmost melioration of the difficulties, and enlargement of the advantages, which the wisdom of the Creator has appended to the human station. No object beneath the effort to secure and bless the future immortality of men can be considered greater or of more importance.

The whole history of the prosperity of our country, whether general or sectional, will bear to a demonstration the assertion, that not to soil nor climate, nor sea nor land, nor zones nor temperatures, nor valleys, nor mountains, nor rivers, are we indebted for the wonderful display of genius and skill, and industry and resulting wealth, by which our nation has been marked, but to the elevating influence of Christian education upon youthful minds, and upon the society in which they have been trained, dignifying as the most honorable condition of man, free labor upon a free soil; making the cunning artificer a perfect equal to the eloquent orator; exalting the head that has humbly bent, through many a toilsome day, over the bench of industry, to preside with a dignity which commands united reverence upon the bench of judgment, and leading the feet that have followed through many a weary furrow in the field, to stand on a level with statesmen in the councils of the nation.

There is that, in all the influences and promises of this system of heavenly light, which is precisely adapted to excite man to stir up the gift that is in him—to make him feel that he was made to serve no master but God—to call him out to the utmost effort, in mental competition, for the improvement of his race—to make him deem himself inferior to no undertaking to which the line of his manifest right and duty shall lead him—to give him patience in effort, coolness in judgment, skill in discernment, and determination in execution—the elements of indubitable and certain success: and whether the wilderness blossom like the rose under his skill in agriculture, or the works of his hands seem almost to live, and speak, and act, in the beauty of his mechanical invention, Christianity honors his effort, and commands men to honor and protect the claims which it originates. It prepares a state of public mind, which smiles encouragingly upon his attainments and productions, and which confesses the honor that the whole community justly feels in having in its bosom, and cherishing as its own, individuals who have so distinguished themselves and their race.

This moral dignity of labor is purely an American scheme and thought. It has marked our country's history, from the earliest periods of its colonial establishment; not more arising from the first struggling condition of its original settlers, than from the very principles with which they emigrated, and upon which they determined to erect the empire which they founded. It is undoubtedly true that labor was at first the necessity of their being. Hands and arms that had never toiled before, were required to toil unceasingly upon the rugged shores which were selected as their future home. And, in this very fact, a dignity was given to human industry, which had never before been connected with it in modern times. The Winthrops, and Johnsons, and Endicotts of that day, would have dignified any station in life. And when they were seen, hewing out their future independence from the wilderness, and rearing their partial but honorable subsistence from a sterile and unwilling soil, never had the axe glittered with such light, nor the plough moved with such majesty before.

Within the recollection of our oldest citizens instances were not unfrequent, where our most eminent men considered it no degradation to discharge with their own hands, if occasion required it, services usually esteemed menial in the extreme; the grooming horses and blacking boots, not only for themselves, but for their guests. No station could exalt men who would voluntarily and cheerfully do this; and boot-blacking, in their hands, rose to a dignity which, in this country, luxurious idleness, though charioted in wealth, can never command. In the spirit which was thus cultivated, an honor was affixed to labor, and in the general feeling of the people, there was transmitted a moral dignity as connected with industry, even in the lowest shapes, in which the needs of man required it, which should be cherished by the present generation and made perpetual in the future.

The extreme difference between this general feeling, and the whole moral condition of the eastern continent, is a very remarkable fact. Throughout monarchical Europe the permanent distinctions of castes and classes make labor disreputable, and give no encouragement to the general enlargement of the human mind, nor to the innate ambition of individual thought. Agriculture in the hands of a peasantry who must live and die in the rude hamlet in which they were born; whose ignorance must never be enlightened beyond the clumsy implements of culture which their forefathers have used; who must feel themselves marked and distinguished, as the mere tolerated denizens of a soil which can never be their own; whose fare is of the coarsest and meanest provision which can sustain the life of man, and the average wages of whose labor is, in Austria, less than one seventh, in France less than one third, and even in England, less than one half of the average of agricultural wages among the freemen of America. Attempts to rise above this state, to attain a position in which man may have his honor as a man, and exercise a better influence upon the destiny of his own family, or his fellow-men, far from being considered a virtue which is to be encouraged, or a right which is to be acknowledged, is a crime for which men are to be shot.

One beneficent operation of the French Revolution, in the midst of all the horrors of its spirit, and its march, has been to break up this system of servile peasantry, and to multiply indefinitely the owners of the soil. But even in the agriculture of France, the mildew of the past is still thickly coated upon the efforts and hopes of the present; and the minds of men, cramped in infancy like the feet of Chinese women, by an unnatural and detestable pressure from without, are feeble and slow in all attempts to run into a new path, however attractive and promising.

In mechanical labor and skill, the absence of all honor as an habitual attendant, is in Europe equally manifest. It is known, that luxury purchases, often at a great price, the beautiful results of handicraft and skill. It is known, that individuals of boldness and energy—those irrepressible spirits whose elasticity no bounds can limit—have occasionally forced their way through all this downward pressure, and have compelled an acknowledgment of their greatness, and a respect for their mighty developments of mental and moral power, from those titled tribes who habitually fancy their interest to be in widening the gulf of their separation, and insulating their own condition as completely as possible. But what are these among so many? Their class are tradesmen and tradespeople still. And the habitual fact in their history is, not only no encouragement to rise, but great discouragement and jealousy of their possible ability to break the shell of caste, whose accumulated scales ages have riveted upon them.

I stood the other day by the bench of an English mechanic, whose remarkable skill I was admiring, and the genius of whose youthful son in his work I was noticing, when the father took from the drawer some beautiful crayon and pencil sketches, which this working boy had made. "Ah! sir," said the father, "this is America. My boy was taught all this for nothing, at your public school. Had I stayed at home, he would have lived and died unnoticed at the bench. Here he may take a stand, and be honored and encouraged." Yes, and this is but one of the multitude of illustrations, which a knowledge of facts would bring out, of the encouragement which American freedom gives to innate talent.

I knew a poor English carpenter, who with the utmost difficulty gathered the needful bread for his family. His children were in the public school of a neighboring city. His eldest son, having no chance of education before, laid hold of his opportunity greedily, passed with honor through all the stages of public education, at the public expense; and on his graduating at the summit of the career of the city's provision, was immediately appointed teacher, and is now (1848) a professor of ancient languages in one of the highest institutions, and honored the more for the industry which had made him, from neglected poverty, what he is. This is America. That boy might have lived and died a beggar in the streets of London, and no titled man have taken him by the hand, to bring out, in an elevating education, the noble powers his Creator had implanted within him.

Let a man make a tour of the single state of Connecticut, with no other knowledge or observation upon this subject than that which belongs to every intelligent American, he will never forget the impression of dignity, beauty, and power which will be made upon his mind. From the heading of a pin to the hammering of granite; from the polishing of the brass button to the beating of the brazen kettle; from the India-rubber suspender to the variegated and beautiful Brussels carpet; in every possible variety and shape and beauty of machinery; upon every flowing river, and upon every little rocky rivulet; from the immense brick or stone edifice of many stories, to the rude shed of pine boards in the woods, upon the margin of the hidden stream; he will see the effects of the American system, honoring, dignifying, prospering, and protecting American labor and American skill.

Human talent, industry, wisdom, and skill, under the favoring blessing of Heaven, must now go forth to sow and to gather in the harvest of the earth. We are teaching lessons of political economy which the world has never heard before. It is a noble dispensation for our country. Other nations may see us, but not with the vines or olives of Italy or France; nor with the oranges and grapes of Spain or Portugal; nor even the rich and glowing verdure, and the teeming harvests of England and lowland Scotland. The magnificence of their time-honored architecture we have not attained. And yet there are intelligence, prosperity, dignity, independence, and self-respect, marking the laboring classes of our population, which lift us far above all envy of the grandeur and glory of European display. They see that we have a people, flourishing and prosperous beyond comparison; but we have no rabble but that which their own degradation has thrown upon our shores. It is the province of America to build, not palaces, but men; to exalt, not titled stations, but general humanity; to dignify, not idle repose, but assiduous industry; to elevate, not the few, but the many; and to make herself known, not so much in individuals, as in herself; spreading to the highest possible level, but striving to keep it level still, universal education, prosperity and honor.

The great element of this whole plan of effort and instruction, is the moral, relative dignity of labor; an element which we are to exalt in public estimation in the highest possible degree, and transmit to our families and to posterity, as the true greatness of the country and the world. We are to look at this enlarging elevation of the working classes of men —a fact which may be considered the main index of our age —not as a difficulty to be limited, but as an attainment in which we greatly rejoice. And if our heraldry is in the hammer, and the axe, and the awl, and the needle, we are to feel it a far higher honor than, if in their place, we could have dragons and helmets, and cross-bones and skulls. Our country's greatness is to be the result, not of foreign war, but of domestic peace; not of the plunder of the weak, but of the fair and even principles of a just commerce, a thriving agriculture, and beautiful and industrious art.

Let us glory in every thing that indicates this fact, as an index also of our desire for renown. This great lesson—honor to the working classes, in the proportion of their industry and merit—the world will yet completely learn. And when the great exalting, leveling system of Christianity gains its universal reign, mountains will be brought down, and valleys will be filled; an highway shall be made for human prosperity and peace—for the elevation, and dignity, and security of man—over which no oppressor's foot shall pass; the poorest of the sons of Adam shall dwell unmolested and fearless beneath his own vine and fig-tree; the united families of earth shall all compete, to acquire and encourage the arts of peace; nation shall not rise up against nation, and men shall learn war no more.—*Address before Fair of American Institute*, 1848. *By the Rev.* STEPHEN H. TYNG, D. D.

THE HARBINGER OF SPRING.

The happiest bird of our spring, and one that rivals the European lark, in my estimation, is the Boblincon, or Boblink, as he is commonly called. He arrives at that choice portion of the year, which, in this latitude, answers to the description of the month of May, so often given by the poets.

With us, it begins about the middle of May, and lasts until nearly the middle of June. Earlier than this, winter is apt to return on its traces, and to blight the opening beauties of the year; and later than this, begin the parching, and panting, and dissolving heats of summer. But in this genial interval, Nature is in all her freshness and fragrance: "the rains are over and gone, the flowers appear upon the earth, the time of the singing of birds is come, and the voice of the turtle is heard in the land." The trees are now in their fullest foliage and brightest verdure; the woods are gay with the clustered flowers of the laurel; the air is perfumed by the sweet-briar and the wild rose; the meadows are enamelled with clover-blossoms; while the young apple, the peach, and the plum, begin to swell, and the cherry to glow, among the green leaves.

This is the chosen season of revelry of the Boblink. He comes amidst the pomp and fragrance of the season; his life seems all sensibility and enjoyment, all song and sunshine. He is to be found in the soft bosoms of the freshest and sweetest meadows; and is most in song when the clover is in blossom. He perches on the topmost twig of a tree, or on some long flaunting weed, and as he rises and sinks with the breeze, pours forth a succession of rich tinkling notes; crowding one upon another, like the outpouring melody of the skylark, and possessing the same rapturous character. Sometimes he pitches from the summit of a tree, begins his song as soon as he gets upon the wing, and flutters tremulously down to the earth, as if overcome with ecstacy at his own music. Sometimes he is in pursuit of his paramour; always in full song, as if he would win her by his melody; and always with the same appearance of intoxication and delight.

Of all the birds of our groves and meadows, the Boblink was the envy of my boyhood. He crossed my path in the sweetest weather, and the sweetest season of the year, when all Nature called to the fields, and the rural feeling throbbed in every bosom; but when I, luckless urchin! was doomed to be mewed up, during the livelong day, in that purgatory of boyhood, a school-room, it seemed as if the little varlet mocked at me, as he flew by in full song, and sought to taunt me with his happier lot. O how I envied him! No lessons, no tasks, no hateful school; nothing but holiday, frolic, green fields, and fine weather.

Farther observation and experience have given me a different idea of this little feathered voluptuary, which I will venture to impart, for the benefit of my school-boy readers, who may regard him with the same unqualified envy and admiration which I once indulged. I have shown him only as I saw him at first, in what I may call the poetical part of his career, when he in a manner devoted himself to elegant pursuits and enjoyments, and was a bird of music, and song, and taste, and sensibility, and refinement. While this lasted, he was sacred from injury; the very school-boy would not fling a stone at him, and the merest rustic would pause to listen to his strain. But mark the difference.

As the year advances, as the clover-blossoms disappear, and the spring fades into summer, his notes cease to vibrate on the ear. He gradually gives up his elegant tastes and habits, doffs his poetical and professional suit of black, assumes a russet or rather dusty garb, and enters into the gross enjoyments of common, vulgar birds. He becomes a bon vivant, a mere gourmand; thinking of nothing but good cheer, and gormandizing on the seeds of the long grasses on which he lately swung, and chanted so musically. He begins to think there is nothing like "the joys of the table," if I may be allowed to apply that convivial phrase to his indulgences. He now grows discontented with plain, every-day fare, and sets out on a gastronomical tour, in search of foreign luxuries. He is to be found in myriads among the reeds of the Delaware, banqueting on their seeds; grows corpulent with good feeding, and soon acquires the unlucky renown of the ortolan. Wherever he goes, pop! pop! pop! the rusty firelocks of the country are cracking on every side; he sees his companions falling by thousands around him; he is the *reed-bird,* the much-sought-for tit-bit of the Pennsylvanian epicure.

Does he take warning and reform? Not he! He wings his flight still farther south in search of other luxuries. We hear of him gorging himself in the rice swamps; filling himself with rice almost to bursting; he can hardly fly for corpulency. Last stage of his career, we hear of him spitted by dozens, and served up on the table of the gourmand, the most vaunted of southern dainties, the *rice-bird* of the Carolinas.

Such is the story of the once musical and admired, but finally sensual and persecuted Boblink. It contains a moral, worthy the attention of all little birds and little boys; warn-

ing them to keep to those refined and intellectual pursuits, which raised him to so high a pitch of popularity, during the early part of his career; but to eschew all tendency to that gross and dissipated indulgence, which brought this mistaken little bird to an untimely end.—Washington Irving.

THE OLD GRIST-MILL.

The grist-mill stands beside the stream,
With bending roof and leaning wall;
So old, that when the winds are wild,
The miller trembles lest it fall;
But moss and ivy, never sere,
Bedeck it o'er from year to year.

The dam is steep, and weeded green;
The gates are raised, the waters pour,
And tread the old wheel's slippery steps,
The lowest round for evermore;
Methinks they have a sound of ire,
Because they cannot climb it higher.

From morn till night, in autumn time,
When yellow harvests load the plains.
Up drive the farmers to the mill,
And back anon, with loaded wains;
They bring a wealth of golden grain,
And take it home in meal again.

The mill inside is dim and dark;
But peeping in the open door,
You see the miller flitting round,
And dusty bags along the floor;
And by the shaft, and down the spout,
The yellow meal comes pouring out.

And all day long the winnowed chaff
Floats round it on the sultry breeze,
And shineth like a settling swarm
Of golden winged and belted bees;
Or sparks around a blacksmith's door,
When bellows blow and forges roar.

I love our pleasant, quaint old mill!
It minds me of my early prime;
'T is changed since then, but not so much
As I am, by decay and time;
Its wrecks are mossed from year to year,
But mine all dark and bare appear.

I stand beside the stream of life;
The mighty current sweeps along,
Drifting the flood-gates of my heart,
It turns the magic wheel of song,
And grinds the ripened harvest brought
From out the golden field of Thought!

STODDARD.

THANKSGIVING DAY.

Thus shines the present, safe from war's alarms--
You till in peace your old ancestral farms;
Blithe with the Spring the busy task begin,
And feast at Autumn, when the harvest 's in.
Crowned is the board with all that man desires,
Bright blush the ceilings with your ruddy fires—
But brighter eyes are beaming round the board,
With mirth and fun, with love and frolic stored.
For who is sad when old Thanksgiving comes,
With all its wealth of sweetmeats, pies, and plums?
Behold the farm-house!—At the old farm gate
A merry group in high expectance wait—
The happy farmer, and the welcome guest,
The city cousin—very nicely dressed!
The village beauty, in her bran-new hood,
The happy children—most discreetly good;
The mother waiting for her eldest son,
Who brings the bride he has but lately won;
The village lovers, who have come to share
The evening revel and the generous fare;
The little boys, with collars white as snow,
Who all the good things in the larder know;
The little girls—their hair with ribbons tied,

Who wait to welcome the expected bride;
The trusty house-dog, with his knowing face,
Who seems to think that something will take place,
Though what that something is, he does not know,
Walks gravely round, with steps serenely slow.
But see, they come, the jingling bells are heard,
Forth flies to meet them, many a welcome word;
The mother holds within her warm embrace,
The new found daughter, with her smiling face;
The boys and girls around their brother crowd,
With eyes all welcome, and with greetings loud.
Oh! happy group! and oh! most happy day!
Ne'er shall New-England see its fame decay:
It still shall live—and all the future yet—
Shall never once Thanksgiving Day forget.

EUGENE BATCHELDER

SCIENTIFIC TERMS IN AGRICULTURE.

It is generally supposed, that in books on agriculture there is an array of scientific terms of which the common reader has no knowledge. It is indeed true, that these books necessarily have more or less of such terms, and without a knowledge of the terms the books cannot be read to the best advantage. This is too evident to need proof. Nevertheless, the labor requisite in the acquisition of this knowledge is far less than would be imagined, even if the reader had previously known nothing of chemistry. A very sensible man, who wrote letters to a young lady on botany, said, to encourage his pupil, it was possible to be a very good botanist without knowing one plant by name. So it is possible to be a very good agricultural chemist without knowing much beyond the names of a very few substances. Suppose an author were to introduce a succession of Latin terms or phrases into a work on any common subject, no matter what it is, and that some one or more of these words or phrases were to be found on every page, it is evident that the reader, not understanding them, would lose much of the meaning in what he

---

1. With what is it supposed that agricultural books abound? 2. What is said of the labor of obtaining a knowledge of them? 3. What is said about learning botany? 4. What is said of Latin terms?

read. But if there were a glossary of these terms and phrases, giving a translation of them, and he would spend a few hours in studying, it is clear he might read the entire book understandingly. It would be the same with agricultural books. If a very little time were spent in learning the few scientific terms as a preliminary exercise, the books would be read with interest and great profit.

A vast amount of knowledge may be gathered up by every person, on various subjects, with small effort, and in a manner altogether miscellaneous; that is, without any systematic habits of research or study. How easily do persons from frequently in the evening glancing an eye upon the heavens learn the names of the principal stars, though they had never studied astronomy? How readily do they learn the names of the common plants and trees, from seeing them so frequently, though they never studied botany? How soon do they become familiar with the names of new fabrics used in wearing apparel; or with new names in geographical science; or with new articles in domestic economy? In the same way we learn the names of the streets of a city, or the landmarks by the roadside in the country where we frequently sojourn; or the names of persons among whom we live, and think nothing of the labor of doing all this—indeed there is no labor in it. Yet this, if it were done at one time, as a matter of formal study, would require far more mental effort than to learn the chemical terms occurring in an ordinary book on agriculture.

A person uneducated in chemical science opening a book on agriculture—perhaps in fifty places—and discovering wherever he opened it half a dozen scientific terms, would not reflect but what they were all new terms from their to him peculiar and unwonted formation, whereas they were the same terms occurring over and over again, and perhaps not fifty different ones in the whole book, and half of these derivatives from the other half. Hence he is, as it were, appalled at the first aspect, by a merely imaginary or fictitious obstacle which has but little reality. Thus in looking at the stars in a clear winter evening, from their perpetually twinkling ap-

5. In what manner may a vast amount of knowledge be easily gained? 6. By what examples is this shown? 7. How is a person deceived in regard to the number of scientific terms in a book of chemistry? 8. How is a similar mistake made as to the number of the stars?

pearance, he apprehends the number of them an hundred or a thousand times greater than it is. So with these chemical terms. It is necessary, therefore, to divest the subject of this scare-crow incident in the study. Accordingly we say, that by making ourselves perfectly familiar with twelve or fifteen of these terms and their philological modifications—only one half the number of letters in our alphabet—and the task is made comparatively light. Who ever thought of being frightened to death at the idea of learning the letters of the alphabet; or, of being accused of cruelty in having his children taught to learn them? If children can learn these letters, adults might certainly learn a corresponding number of the terms of which we are speaking.

Again, suppose a person were to visit a menagerie and see there fifteen or twenty curious animals which he had never before seen, or of which he had not before heard, would he not in two or three hours learn the names of them all, and be able to give some general and appropriate description of each one of them? Or suppose he were to visit the Fair of the American Institute, or some similar Fair, and were to see a dozen or fifteen curious machines for the useful purposes of life, would he have any difficulty, after inspecting them, to tell pretty much all about them? Or suppose there were to be sent to him, by a friend, from some distant country, a basket containing fifteen kinds of ripe and rich fruit, of which he had never before heard, would he not speedily become familiar with the names and the distinctive attributes of each? Why should it then be thought a hardship to learn the names of the constituent elements of plants? The names and properties of these few elementary substances in plants might be as easily learned, if farmers and their sons would feel as much interested and apply themselves to it as in the other cases we have supposed. Besides, with most of the names of these fifteen substances most persons, without the aid of chemical knowledge, have some familiarity.

Let us now look at the list, and see if it is so formidable, as at first imagined. These fifteen terms which generally make the basis of scientific agriculture, or the names of the

---

9. What is said of learning the letters of the alphabet? 10. How is the subject illustrated by the menagerie? 11. How is the subject illustrated by the Fair of the American Institute? 12. Or by a basket of fruit?

elementary substances that enter into the composition of plants, are carbon, hydrogen, oxygen, nitrogen—potash, soda, magnesia, silex, lime, phosphorus, alumina, oxide of iron, oxide of manganese, chlorine, and sulphuric acid. No one surely ought to consider it a great task to make himself familiar with these fifteen terms, even if he felt no particular interest in the nature of the objects denoted by them—if they were simply names for objects in the moon; and especially, when he considers they are the names of substances found in plants he is annually cultivating, and that no plant exists without containing more or less of them. Out of these every plant is made. Every part of every plant that grows, from the rose and the dahlia that delight the eye with their delicate hues, to the stately oak that spreads its broad branches in Nature's wildness, contains some or all of these. How then can a person, especially in the country, fail to feel a most lively interest in such a subject? A man building a house needs more than twice the number of materials wanted in raising his crops on the farm; but does he not make out a list of the whole—become familiar with it, and ascertain where each one is to be had? And a mason or a carpenter would not deserve the names by which they are designated, if they could not tell the names and the qualities of the materials used by them in all departments of their labor. Or, what would be thought of a professed artisan who could not enumerate and describe every article required in the construction of a steam-boat, a carriage factory, or a cotton mill? Would such a person be deemed competent for such labors, or receive patronage? Yet, it might be asked, how such a case differs from that of the farmer who is ignorant of the elementary substances of plants, and of the ingredients composing soils and manures which produce vegetable growth?

Nor can it be said, in excuse for neglecting them, that they are all new and lock-jaw terms; or, in other words, terms of which the farmer has never heard; terms of great length and arbitrarily pronounced, like Russian and Polish

---

13. What are the scientific names of substances entering in vegetable composition? 14. Has the uneducated person no knowledge of them? 15. Why might it be supposed that every farmer would feel interested in them? 16. How is this explained by his building a house? 17. How is it explained by allusion to the professed artisan? 18. Are these terms particularly difficult to be remembered?

names of men. This objection does not apply. He is familiar with several of them, although he may yet have to learn the agency they have in vegetable composition. Who, without any formal scientific instruction, is not familiar with the names of iron, sulphur, potash, soda, and lime? No one; for they are of every-day occurrence. And, who does not know that carbon is coal; or rather that coal is carbon? And, who has never heard that oxygen and nitrogen form the air we breathe; or that water is composed of oxygen and hydrogen? With alumina or alum all are familiar. Silex or silica is but the name of portions of hard stones. Most persons too are no strangers to what is called ammonia, which is a combination of nitrogen and hydrogen, and is sometimes called sal volatile, and is what gives the sharp smell in the smelling-bottle or the manure heap. It is the ammonia in manure which is so efficacious in agriculture. And, if the reader is less familiar with the name of chlorine, it is easy to give him some hints of it. This is the substance used in bleaching cotton goods, and for a long time on opening them the smell is very pungent. And what is still more in point, chlorine united with soda makes our common salt; or, if united with ammonia, the product is what is called in the shops sal ammoniac. The circumstance that chlorine is a constituent of common salt is of itself sufficient to give it an interest in impressing it on the memory of every individual. Chlorine also is important from being used to disinfect the air when impregnated with foul vapors, as in the case of cholera and other diseases. Here then is the whole catalogue of these obnoxious terms, save manganese, which is simply a dark colored metal used in the manufacture of glass as well as in agriculture.

The list above, says Dr. Dana, may be divided as follows: First, the airy or volatile; secondly, the earths and metals; thirdly, the alkalies; fourthly, the inflammables. Only the third and fourth divisions require to be explained or defined.

---

19. Which are the five with which every farmer is familiar? 20. What does he probably know of carbon? 21. What does he probably know of the combination of oxygen and nitrogen? 22. And of oxygen and hydrogen? 23. What is said of alumina and silex? 24. Of ammonia? 25. And of chlorine? 26. What forms common salt? 27. What is said of chlorine as a disinfecting agent? 28. Into what four classes does Dr. Dana divide these substances?

The substances called potash and soda are termed alkalies They are said to have alkaline properties. Touch the tongue with a bit of quicklime, it has a hot, burning, bitter taste. These are called alkaline properties. Besides these, they have the power of combining with and taking the acid out of all sour liquids, that is, the acid and the alkali neutralize each other. This word alkali is of Arabic origin; its very name shows one of the properties of alkalies. *Kali* is the Arabic word for bitter, and *al* is like our word super, we say fine and superfine; so kali, is bitter; alkali, superlatively bitter, or truly alkali means, the *dregs of bitterness*. It is important that the reader should fix in his own mind what is here said about alkali and alkaline properties.

Further. Alkali is a general term. It includes all those substances which have an action like the ley of wood ashes, which is used for soap making. If this ley is boiled down dry, it forms potash, as all know. Now lime, fresh slaked, has the alkaline properties of potash, but weaker, and so has the calcined magnesia of the shops, but in a less degree than lime. Here we have two substances, earthy in their look, having alkaline properties. They are called, therefore, alkaline earths. But what we understand chiefly by the term alkalies, means potash, soda, and ammonia. Potash is the alkali of land plants; soda is the alkali of sea plants; and ammonia is the alkali of animal substances. Potash and soda are fixed; that is, not easily raised in vapor by fire. Ammonia always exists as vapor, unless fixed by something else. Hence there is a distinction among alkalies which is easily remembered. This distinction is founded on the source from which they are procured, and upon their nature, when heated. Potash is a vegetable alkali, derived from land plants; soda is a marine alkali, derived from sea plants; ammonia is animal alkali, derived from animal substances. Potash and soda are fixed alkalies; ammonia is a volatile alkali. Potash makes soft soap, with grease, and soda forms hard soap. Ammonia forms neither hard nor soft; it makes, with oil, a kind of ointment, used to rub a sore throat with,

---

29. Which two are alkalies? 30. How is it shown that they have alkaline properties? 31. Of what is the term alkali composed, and how is it explained? 32. How does lime as an alkali differ from potash? 33. What do we understand chiefly by alkalies? 34. From what are the three obtained? 35. What is the difference between fixed and volatile alkalies? 36. What is made from these three alkalies?

under the name of volatile liniment. But though there be three alkalies, and two alkaline earths, it should on no account be forgotten, that they all have common properties, called alkaline, and which will enable a person to understand their action, without any thing being said about their chemistry.

The inflammables, or the fourth division, are sulphur and phosphorus, both used in making friction matches. The phosphorus first takes fire by rubbing, and this sets the sulphur burning. Now, the smoke arising from these is only the sulphur and phosphorus united to the vital part of the common air. This compound of vital air, or oxygen, as it is called, and inflammables, forms acids, called sulphuric and phosphoric acids. So it is well known, that if coal or carbon is burned, carbonic acid or fixed air is produced. That is, by burning, the coal or carbon unites with the oxygen or vital part of common air, and forms carbonic acid. The heavy, deadly air, which arises from burning charcoal, has all the properties of an acid. Now let us see what these properties are. All acids unite or combine with the alkalies, alkaline earths, and the metals. When acids and alkalies do thus unite, they each lose their distinguishing properties. They form a new substance, called a salt. It is very important that this definition of a salt should be clearly understood, and kept in memory. The idea of a salt is not to be confined to common salt. Common salt is a capital example of the whole class. It is soda, an alkali, united to an acid, or chlorine, or, to speak in the most intelligible terms, to muriatic acid. So saltpetre is a salt. It is potash united to aquafortis. These have united, and their characters are neutralized by each other, so that in saltpetre one will not perceive either potash or aquafortis. They have formed a neutral salt.

By the above analysis, the list of substances to be found in plants—which is the same to be found in manures and soils on which plants grow—is reduced from things not known to things that are known. In this way, persons may feel

---

37. What makes the fourth division of Dr. Dana's classification? 38. How are friction matches made? 39. By what other name is fixed air called, and how is it produced? 40. How are salts formed? 41. What is common salt? 42. What is saltpetre? 43. What is said in the last paragraph of this chapter?

familiarized with them without further acquaintance with chemical science. And, by analogous processes, it will be seen that, in a multitude of cases, they understand the principles of the science, although ignorant of the terms which represent and explain these principles. Thus the house-wife proceeds in making bread on scientific principles, although she never saw a book on chemistry, or learned the meaning of a scientific term.

AGRICULTURAL CHEMISTRY.

What are we to understand by Agricultural Chemistry, as the term is now used by scientific farmers?

Before answering that question, it is expedient to explain the meaning of chemistry, without reference to agriculture, it being one of the fundamental sciences, the study of which is interwoven with every branch of Natural History or material philosophy, as well as with agriculture. Before we can fully appreciate its importance, when relating to the latter, we must understand its use in other departments of knowledge.

What then is the proper definition of chemistry as a science?

Perhaps no better definition can be given than that of Ure; to wit, that it investigates the composition of material substances, and the permanent changes of constitution which their mutual actions produce; or, that it treats of the mutual action of the integrant or constituent parts of a body. Brande gives a more copious definition, but the same in reality; to wit, that branch of natural knowledge which teaches us the properties of elementary substances, and of their mutual combinations; it inquires into the laws which effect, and into the powers which preside over, their union; it examines the proportions in which they combine, and the modes of separating them when combined; and endeavors to apply such knowledge to the explication of natural phenomena, and to useful purposes in the arts of life.

What are some of the most obvious examples of the science, as thus defined?

The detection of alloys in counterfeit coin; of poison mixed with other substances, either before being used, or when

"NEEDHAM'S WHITE BLACKBERRY. The above is a good illustration of this luxuriant fruit, raised by J. S. Needham, of Danvers, Massachusetts. The plant grows to the height of from six to ten feet. It yields plentifully. The fruit is large, amber-colored, and is very sweet and rich; and the demand for it exceeds the supply, although it is easily raised." 2

found in the stomach; of the component parts of any medicine, and hence its effect on the animal constitution or any particular disease; of the materials used in dye-stuffs, and hence the measure of their durability; of the elementary constituents forming any particular description of food, and hence its adaptation and power for sustaining animal life and growth. Indeed, there is seemingly no end to the benefits derived from chemical knowledge. By it we learn that grease or oil will prevent friction in machinery, or rust on metals; that the contact of the atmosphere on certain liquids will produce vinegar—on butter will cause it to become rancid, and on eggs will hasten putrefaction; and that yeast mixed with kneaded flour will occasion fermentation. Hence it may readily be seen how chemistry may be made beneficial to agriculture.

What are some of the cases that first occur to you of the beneficial effects of chemistry to agriculture?

To remedy defects in soil. If the soil is too light to preserve fertilizing agents, mix with it clay or other substances to increase its power of retaining them. If it is too cold, mix with it sand to render it warm. If clayey and impervious, mix with it substances that will cause it to be mellow and light. If sour, mix with it what will neutralize this property in it. Or, if wanting in some particular agent needful to the growth of a particular crop, cast upon it that description of manure which contains this agent. So, if the soil is found to be naturally too moist, chemistry will admonish the farmer to resort to draining; or, if naturally too dry, to resort to irrigation.

But most of this is probably known by most farmers, who never studied chemistry; how then do the cases named apply to the subject?

True, some of these things may have been learned from one's own experience in agriculture; some from observing the practice of others, or from the exercise of mere common sense. Nevertheless, they involve chemical principles; and if portions of a science are thus learned, as it were, from necessity, it shows the importance of pursuing the study more systematically in other matters that will be found equally valuable, although less obvious, to the uneducated farmer. It is by no means the fact that chemistry is learned from books only at school; but the fact that portions of it are learned under such

2

disadvantageous circumstances, and by a slow process, shows the advantage, to farmers especially, of making it a part of elementary or common school education to their sons.

How can the necessity for a rotation of crops be explained on principles of agricultural chemistry?

It is easily done. Here the knowledge imparted by the science is of immense value. It has ever been known that, in raising a succession of the same crop for a long course of years, there would be gradually and regularly a diminution in the amount of the product. The fact was apparent, because it was constantly observed by the most intelligent of uneducated farmers; but the reason of it was a perfect mystery. Now chemistry explains the mystery. It tells them that one crop requires mainly in its growth one particular fertilizing agent; that another crop in the same way requires mainly another fertilizing agent; and so of a third, and a fourth. Thus, for instance, potatoes being planted one year; then Indian corn; then oats; then wheat; and then grass —or any analogous rotation—there will be no lack or diminution of production: when the same, or a similar alternation of culture, may be followed to any indefinite extent, and with similar results. All this is made plain by chemistry, as the most familiar process with which one can be acquainted.

How is it that chemistry is able to comprehend these facts, unless by a succession of experiments, like those of the common farmer?

The chemist, in the first place, ascertains of what particular substance each particular vegetable to be raised is composed, or what material from the soil enters into its growth. He then ascertains of what elementary substances any particular portion of soil is composed. This being done, if the soil is deficient or destitute of any one elementary substance required in the growth of a particular vegetable, that vegetable cannot be produced on it. Thus four different vegetables being examined and found to require each mainly for itself a particular elementary substance; and that the soil contains these four substances, but only enough of each for one crop, it follows that either one of them planted more than one year in succession would fail of receiving nourishment, but if each one were planted one year only and in succession or rotation, there would be nourishment enough in the soil for all four of them, and there would be a good crop of each.

But is it a part of the theory in this supposition, that each of these vegetables in its own year and for its own crop extracts from the soil all of the elementary substance mainly required in its growth?

By no means. It is literally so in theory but not in fact. Portions of this substance may not be brought into contact with the roots of the vegetable, and hence not be absorbed. Consequently such portions of it will remain in the soil, and might be available for the same vegetable another year, but yielding a diminished crop; and, not only for one year but for a succession of years, but each one with diminished results.

How is it that the chemist can ascertain the elementary substances entering into the composition of vegetables, and the component parts of the soil?

An answer to this inquiry comes not properly into view in the present place. It would occupy too much space. It will receive attention elsewhere. Here it is assumed as a fact, to show the importance of chemical knowledge. All to be said at present is, that the chemist as much has the means at control for doing it, and doing it with accuracy, as the boy, on finding a bag of medals,—some gold, some silver, some copper; or a bag containing peas, beans, kernels of corn, and grains of wheat, is able to separate the mixture, putting each kind into a parcel by itself, and telling with accuracy how many of each kind there are. The one is as simple to the chemist, as the other is to the boy of ten years old.

Of what benefit to the farmer in the preparation and use of manures is agricultural chemistry?

It enables him to judge of the degrees of efficacy as fertilizing agents of the different kinds of manure; of the best modes of preserving and applying it; of the different substances that may advantageously be converted into it; and especially the most expeditious and certain process for the operation. The great secret of successful farming lies in the ability to provide manure of the best quality and in all needful measure; and the agricultural chemist has an advantage for this over all other persons.

In what way can the farmer be materially aided in feeding his stock by a knowledge of this science?

The fact is too palpable to need reiteration, that different kinds of vegetables used for the feed of horses, cattle, swine, sheep, and poultry contain different degrees or different proportions according to their bulk of nutritive aliment. When, therefore, it is known how much it costs to produce these several kinds, it is easily ascertained which kind in the feeding of stock is most economical. If one farmer in this way is enabled to raise pork at four dollars per hundred, while it costs another five dollars per hundred ; or if one is enabled to produce milk on his farm at one and a half cents per quart, while it costs another two cents per quart, it is evident that the one in agricultural profit has a material advantage over the other.

Does a knowledge of agricultural chemistry enable the farmer the better to distribute the crops upon his farm ?

It truly does. What was said on the rotation of crops is applicable to the present question. And, in addition to that to the common observer it might be apparent that all soils are not alike adapted to every species of vegetable production. A deep soil is required by some, and a more light one by others. Some require great moisture and others less. Some admit of a more cool and shaded position, and others as much of a direct exposure to the sun as possible. The individual who would consider a piece of land that answers for buckwheat well adapted to the growth of carrots is an agricultural simpleton; and the same might be said in reference to many other similar ill-judged distributions of crops. Much on these matters is to be learnt by experience and observation; but, if experience is based on scientific knowledge much time and money are saved.

What are some of the cases that illustrate this position ?

On very light lands, rye, of all grains, grows best; and of all food for cattle, says Professor Johnston, spurry grows best on light sandy soils. On loamy and gravelly soils, barley is a kind that grows best; and turnips and Indian corn will grow well on such soils. And he says barley could not grow on a stiff clay; while on heavy, clay lands, wheat, clover, and grass grow most luxuriantly.

THE CROP OF ACORNS.

There came a man in days of old,
To hire a piece of land for gold,

And urged his suit in accents meek,—
"One crop alone, is all I seek;
That harvest o'er, my claim I yield,
And to its lord resign the field."

The owner some misgivings felt,
And coldly with the stranger dealt,
And found his last objection fail,
And honied eloquence prevail,
So took the proffered price in hand,
And for one crop leased out the land.

The wily tenant sneer'd with pride,
And sowed the spot with acorns wide;
At first, like tiny shoots they grew,
Then broad and wide their branches threw,
But long before those oaks sublime
Aspiring reach'd their forest prime,
The cheated landlord mouldering lay
Forsaken with his kindred clay.

Oh ye, whose years unfolding fair,
Are fresh with youth, and free from care,
Should Vice or Indolence desire,
The garden of your soul to hire,
No parley hold, eject the suit,
Nor let one seed the soil pollute.

My child, their first approach beware,
With firmness break the insidious snare,
Lest as the acorns grew and throve
Into a sun-excluding grove,
Thy sins, a dark o'ershadowing tree,
Shut out the light of Heaven from thee.

MRS. SIGOURNEY.

THE AMERICAN PLOUGHMAN.

CLEAR the brown path to meet the coulter's gleam!
Lo! on he comes behind his smoking team,
With toil's bright dew-drops on his sun-burnt brow,
The lord of earth, the hero of the plough!

First in the field before the reddening sun,
Last in the shadows when the day is done.
Line after line along the burning sod
Marks the broad acres where his feet have trod;
Still where he treads the stubborn clods divide,
The smooth, fresh furrow opens deep and wide;
Matted and dense the tangled turf upheaves,
Mellow and dark the ridgy cornfield cleaves;
Up the steep hill-side where the laboring train
Slants the long track that scores the level plain;
Through the moist valley clogged with oozing clay,
The patient convoy breaks its destined way;
At every turn the loosening chains resound,
The swinging ploughshare circles glistening round,
Till the wide field one billowy waste appears,
And wearied hands unbind the panting steers.

These are the hands whose sturdy labor brings
The peasant's food, the golden pomp of kings;
This is the page whose letters shall be seen
Changed by the sun to words of living green;
This is the scholar whose immortal pen
Spells the first lesson hunger taught to men;
These are the lines, O Heaven-commanded toil,
That fill thy deed—the charter of the soil!

O gracious Mother, whose benignant breast
Wakes us to life, and lulls us all to rest,
How sweet thy features, kind to every clime,
Mock with their smile the wrinkled front of time!
We stain thy flowers—they blossom o'er the dead;
We rend thy bosom, and it gives us bread;
O'er the red field that trampling strife has torn,
Waves the green plumage of thy tasseled corn;
Our maddening conflicts scar thy fairest plain,
Still thy soft answer is the growing grain,
Yet, O! our mother, while uncounted charms
Round the fresh clasp of these embracing arms,
Let not our virtues in thy love decay,
And thy fond weakness waste our strength away

No by these hills, whose banners, now displayed,
In blazing cohorts Autumn has arrayed;

By yon twin crest, amid the sinking sphere,
Last to dissolve, and first to reappear;
By these fair plains the mountain circle screens,
And feeds in silence from its dark ravines;
True to their home, these faithful arms shall toil,
To crown with peace their own untainted soil;
And true to God, to Freedom, to Mankind,
If her chained bandogs Faction shall unbind,
These stately forms, that bending even now,
Bowed their strong manhood to the humble plough,
Shall rise erect, the guardians of the land,
The same stern iron in the same right hand,
Till Greylock thunders to the parting sun
The sword has rescued what the ploughshare won!

DR. HOLMES.

PHYSIOLOGICAL REFLECTIONS ON WATER.

Absolutely pure water, fresh drawn of the chemist's still, or formed from its elements by burning a gallon of hydrogen gas in half a gallon of oxygen, seems as simple and inert a substance as one can well conceive—devoid as it is of color, taste, and smell. Yet in the whole range of material substances there is perhaps not one whose transformations are more surprisingly Protean, or whose relations are more extensive and intricate. A solid body, stone-hard, falls from the sky and breaks your window. You pick it up, and find it a dense angular crystal; which, while you examine it in the palm of your hand, changes to a transparent fluid; which again, dwindling gradually as you gaze at it, becomes invisible, and vanishes into thin air. If the weather be frosty, the vanished substance soon re-appears in dew drops, softly deposited on the cold window—which just before its momentum had power to break; and these drops, while you examine them, shoot into delicate ramifications, and resume their previous crystalline solidity.

Nor is the hail stone less soluble on earth than in air. Placed under a glass bell with thrice its weight of lime, it gradually melts and disappears; and there remain four parts, instead of three, of perfectly dry earth under the glass. Of a plaster of Paris statue weighing five pounds, more than one pound is solidified water. Even the iridescent opal is but a mass of fluid and water, combined in the proportion of nine grains of the earthy ingredient to one of the fluid. Of an acre of clay land a foot deep, weighing twelve hundred tons, at least four hundred tons are water; and, even of the great mountain chains with which the globe is ribbed, many millions of tons are water solidified in earth.

Water, indeed, exists around us to an extent and under conditions which escape the notice of cursory observers. When the dyer buys of the drysalter one hundred pounds of alum, carbonate of soda, and soap, he obtains in exchange for his money, no less than forty-five pounds of water in the first lot, sixty-four pounds in the second, and a variable quantity, sometimes amounting to seventy-three and a half pounds, in the third. Even the transparent air we breathe contains in ordinary weather about five grains of water diffused through each cubic foot of its bulk, and this rarified water no more wets the air than the solidified water wets the lime or opal in which it is absorbed. But while water is thus capable of incorporating itself with earth and air, and of assuming alternately their respective conditions, it can, on the other hand, in its turn, dissolve both air and earth; giving to invisible gases its own palpable form, and liquifying, without chemically changing, the densest constituents of the crust of the globe.

Of the absorptive power of water for gases we have a practical example in the frequent contamination of London water by the coal gas, which leaks from the gas pipes into the soil, and is sucked into the water pipes by the vacuum which the water creates in its secession towards the mains when turned off. This pollution takes place to so great an extent in certain streets, where the ground is so saturated with escaped gas that the fire-plug boxes if covered over at night collect enough to take fire the next morning. So abundantly is this gas drawn into the service pipes that it has frequently been known to ignite at the water-taps; to the great consternation of those who, coming with their pitchers, have seen fire issue where water was wont to flow. Drain air and grave-yard gas must in some situations be pumped by this vacuum process into the pipes, and contribute to pollute the water.

As for the solvent power of water on solids, the phenomenon is as familiar to us as it is profoundly marvellous. Every one has seen salt vanish in water; the particles, just now opaque and fixed, strangely acquiring mobility and translucence. Every one, however, is not aware how extensive the range of this power of water is. The glass we drink from seems insoluble; yet Lavoisier found that glass retorts used in distilling water lost weight, the water at the same time acquiring an equivalent impregnation of the elements (fluid and alkali) of glass. This erosive action of water and the gases it contains on glass, takes place also, though more slowly, at the ordinary temperature of the air; and its results become apparent in the lapse of time. The old stained glass windows at Westminster Abbey are honey-combed on the outside by the rain, and in many parts nearly eaten through.

Comminution quickens the effect; if a common drinking-tumbler be pounded and moistened, enough of the powder will be dissolved to give the water a powerful reaction on turmeric paper. Pure flint, which, as opal, we have seen solidifying water, may, in its turn, be converted by combination with water into transparent tremulous jelly—or even, in minuter portions, be taken up as clear aqueous solution of flint. Thus granite rock, of which silicates, such as form glass, are a main ingredient, is gradually disintegrated by water; and the hot springs of Iceland bring up from the deep Plutonic strata so much siliceous matter in solution that objects dipped in them become coated with a flinty deposit.

The salubrity of earthy, alkaline, and metallic salts in water used as beverage, is strenuously asserted and denied by authorities of equal eminence. Some physiologists contend that lime, magnesia, iron, and the alkalies in combination with carbonic, sulphuric, phosphoric, and other acids, are essential constituents of the animal body, their presence in water is not only harmless, but positively beneficial, and their elimination from our beverage would, in particular, according to these writers, deprive our bones of the material necessary to their growth. In opposition to this view, the cogent fact is alleged that the citizens of Aberdeen, who drink the purest water in Great Britain, have also fully developed bones; whence it is inferred that the earthy and alkaline salts supplied to us in our solid food furnish the organism with a due proportion of mineral constituents.

This position is still more indisputably established by the fact that the ejected residue of the solid food contains a large proportion of superfluous mineral salts. On the whole, the weight of scientific evidence seems in favor of the salubrity of water free from earth; towards which, at all events, the instinct of mankind manifestly inclines.

Such is our own experience, and such is the evidence furnished by observation from facts all around us. As for the inferiority of soft water, in point of freshness and sapidity, to the hard water drawn from wells and springs, this difference depends not on any pleasantness of savor inherent in the earthy salts, but on the superior coolness and more abundant æration of newly drawn spring water. Let the pure water be cooled to forty-five degrees and the spring water warmed to sixty-five degrees, and deprived of its carbonic acid, and the former will be chosen in preference to the latter. Alexander knew this; who, at the siege of Petra, had thirty pits filled with snow to cool his water; and this also Mahomet knew, who describes, as one of the principal tortures of the damned, a quenchless thirst, with nothing to slake it but *warm*, filthy water. The subterranean tanks of Madrid, and the colossal cisterns of Constantinople, protected from the sunshine by groined coverings, argue the acquaintance of their ancient constructors with the value of *coolness* in water, and put to shame the London reservoirs—exposed, as they are, not only to the solar heat and light, with all the growths which they encourage, but also to the impure exhalations of two millions of people, and to the filthy droppings of the London air.

But whatever differences of opinion may exist as to the palatability of hard or earthy water, its inferiority for detergent, culinary, and manufacturing purposes is admitted on all hands. Lime and magnesia in water spoil alkaline soaps, by combining with the fatty acids which give them their lubricity, and so reducing them to the state of insoluble earth-soaps, which are unpleasant to the skin in the bath, and injurious to linen in the wash-tub. The tannin of tea—its astringent part—is thrown down by the lime of hard water as a tannate, along with coloring, extractive, and aromatic matter; so that of the tea infused in spring water of average hardness, at least one third is wasted. Hard water, used for boiling meat and vegetables, extracts their juices less thoroughly than soft, and toughens their fibres, shriveling greens and peas, giving spinage and asparagus a yellow tinge, and seriously impairing the flavor of soups. Hard water is equally prejudicial, for like reasons, in many manufactures. The tannin of oak bark, like that of tea, is precipitated from solution by lime, to the great injury of leather. The valuable juices of certain dye-woods, of the brewer's malt and hops, and of apothecary's drugs, are, like those of meat and vegetables, less readily yielded to hard water than to soft; and, as the extra dose of carbonic acid, by which chalk is upheld in water, is driven off by heat, steam-engine boilers, in which hard water is used, become rapidly encrusted with an earthy deposit, which hinders the transmission of heat to the water; and thus not only occasions waste of fuel, but exposes the over-heated iron to burn and burst.

Of organic bodies, whether vegetable or animal, water is also a large constituent during life, and a powerful solvent after death. Potatoes for example, contain seventy-five per cent.—by weight—and turnips no less than ninety per cent. of water;—which explains, by the way, the small inclination of turnip-fed cattle and sheep for drink. A beef-steak strongly pressed between blotting-paper yields nearly four-fifths of its weight of water. Of the human frame, bones included, only about four-fifths is solid matter—chiefly carbon and nitrogen; —the rest is water. If a man weighing 160 lbs. were squeezed flat under a hydraulic press, 120 lbs. of water would run out, and only 40 lbs. of dry residue would remain. A man is therefore, chemically speaking, a little more than fifty pounds of carbon and nitrogen diffused through six pailfuls of water. Berzelius, indeed, in recording the fact, justly remarks, "the living organism is to be regarded as a mass diffused in water;" and Dalton, by a series of experiments tried in his own person, found that of the food with which we daily repair this water-built fabric, five-sixths is also water. Thus amply does science confirm the popular saying that water is the "first necessary of life."

Nor of life only. Of death, considered as the final predominance of chemical over vital forces, water is also the indispensable minister; taking as it does, an active part in the processes of fermentation, putrefaction, and decay—through which organized bodies pass in their gradual relapse to the inorganic condition. These changes deserve our particular attention, for they go on in our ordinary rivers; and at a certain degree of activity they turn water into a deadly poison.

Some years since the putrescent residuum of a starch factory at Nottingham was suffered to contaminate a brook containing fish and frogs, and resorted to by cattle for drink. The fish and frogs disappeared from the water, and the cattle suffered a series of symptoms of disease. Their muscles, their blood, and all the more putrefiable tissues of their bodies wasted; their coats became rough and staring; their yield of milk fell off rapidly; a bloody purging ensued; and they died in a state of extreme emaciation. After twenty-four cows and nine calves had thus miserably perished, the contamination of the water was stopped by an action at law upon which the fish and the frogs began to reappear, and the mortality of the cattle ceased.

Advancing to a higher point of view, we note properties and functions of water, as it operates in the organism of plants and animals, and in the still wider laboratory of the world at large, we shall find it still the great solvent; the principal carrier of circulating substances and forces, and the universal medium of physical and vital transformations. The sap of plants is a solution of nutrient matters, saline and organic, in water, which distributes them so rapidly, that its upward course through the minute vessels looks like the rushing of a swift stream. A pailful of water suitably impregnated with salts is speedily sucked up by the root of a growing tree immersed in it; the salts are assimilated, as also is part of the water, the remainder being evaporated from the leaves. Food or poison may be thus artfully administered to plants, and timber is thus hardened in France, and even stained, while living, of divers different hues. As for the evaporation from foliage, it is so abundant that a sunflower perspires five gills per day, and a cabbage nearly as much; and, it appears from experiments, that a wheat-plant, during the period of its growth, 172 days, exhales 100,000 grains of water; so that, taking the ultimate weight of the mature plant at one hundred grains, and its main weight at fifty grains, which is a full estimate, its mean daily transpiration actually exceeds ten times its own weight. At this rate an acre of growing wheat, weighing at least two tons at maturity, should exhale on an average fully ten tons of water per day. However, it is calculated that the rain daily falling on an acre of land is not equal to this quantity. Hence the exhalations from the plant with which the experiment was made must have been above the average of those from plants growing in a wheat field.—ABSTRACT FROM LONDON QUARTERLY REVIEW.

THE SUPERIORITY OF EDUCATED LABOR.

The most abundant proof exists, derived from all departments of human industry, that uneducated labor is comparatively unprofitable. I have before me the statements of a number of the most intelligent gentlemen in Massachusetts, affirming this fact as the result of an experience extending over many years. In Massachusetts we have no native born child wholly without school instruction; but the degrees of attainment in mental development are various. Half a dozen years ago, the Massachusetts Board of Education obtained statements from large numbers of our master manufacturers, authenticated from the books of their respective establishments, and covering a series of years, the result of which was, that increased wages were found in connection with increased intelligence, just as certainly as increased heat raises the mercury in the thermometer.

Foreigners, and those coming from other states, who made their marks when they receipted their bills, earned the least; those who had a moderate or limited education occupied a middle ground on the pay-roll; while the intelligent young women who worked in the mills in winter, and taught schools in summer crowned the list. The larger capital in the form of intelligence yielded the larger interest in the form of wages. This inquiry was not confined to manufacturers, but was extended to other departments of business, where the results of labor could be made the subject of exact measurement.

This is universally so. The mechanic sees it, when he compares the work of a stupid with that of an awakened mind. The traveller sees it, when he passes from an educated into an uneducated nation. There are countries in Europe, lying side by side. where, without compass or chart, without bound or land-marks, I could run the line of demarcation between the two, by the broad, legible characters which ignorance has written on roads, fields, houses, and the persons of men, women, and children on one side, and which knowledge has inscribed on the other.

This difference is most striking in the mechanic arts; but is clearly visible also in husbandry. Not the most fertile soil, not mines of silver and gold, can make a nation rich without intelligence. Who ever had a more fertile soil than the Egyptians? Who have handled more silver and gold than the Spaniards? The universal cultivation of the mind and heart is the only true source of opulence;—the cultivation of the mind, by which to lay hold on the treasures of nature; the cultivation of the heart, by which to devote those treasures to beneficent uses.

Where this cultivation exists, no matter how barren the soil or ungenial the clime, there comfort and competence will abound; for it is the intellectual and moral condition of the cultivator that impoverishes the soil, or makes it teem with abundance. He who disobeys the law of God in regard to the culture of the intellectual and spiritual nature, may live in the valley of the Nile, but he can rear only the "lean kine" of Pharoah; but he who obeys the highest law, may dwell in the cold and inhospitable regions of Scotland or of New England, and "well-formed and fat-fleshed kine" shall feed on all his meadows.—HORACE MANN.

TRIBUTE TO GENIUS AND LABOR.

The camp has had its day of song;
The sword, the bayonet, the plume
Have crowded out of rhyme too long
The plough, the anvil and the loom!
O, not upon our tented fields
Are Freedom's heroes bred alone;
The training of the workshop yields
More heroes true than War has known!

Who drives the bolt, who shapes the steel,
May, with the heart as valiant, smite,
As he who sees a foeman reel
In blood before his blow of might!

The skill that conquers space and time,
That graces life, that lightens toil,
May spring from courage more sublime
Than that which makes a realm its spoil.

Let Labor, then, look up and see
His craft no path of honor lacks;
The soldier's rifle yet shall be
Less honored than the woodman's axe?
Let Art his own appointment prize,
Nor deem that gold or outward height
Can compensate the worth that lies
In tastes that breed their own delight.

And may the time draw nearer still
When men this sacred truth shall heed,
That from the thought and from the will
Must all that raises man proceed!
Though Pride should hold our calling low,
For us shall duty make it good;
And we from truth to truth shall go,
Till life and death are understood.

EPES SARGENT.

A PLEA FOR OUR PHYSICAL LIFE.

We do our nature wrong,
Neglecting over long
The bodily joys that help to make us wise;
The ramble up the slope
Of the high mountain cope—
The long day's walk, the vigorous exercise,
The fresh luxuriant bath,
Far from the trodden path,
Or 'mid the ocean waves, dashing with harmless roar,
Lifting us off our feet upon the sandy shore.

Kind Heaven! there is no end
Of pleasures as we wend

Our pilgrimage in life's undevious way,
If we but know the laws
Of the Eternal Cause,
And for His glory and our good obey,
But intellectual pride
Sets half these joys aside,
And our perennial care absorbs the soul so much,
That life burns cold and dim beneath its deadening touch.

Welcome, ye plump green meads,
Ye streams and sighing reeds!
Welcome, ye corn-fields, waving like a sea!
Welcome, the leafy bowers,
And children gathering flowers?
And farewell, for a while, sage drudgery!
What! though we're growing old,
Our blood is not yet cold!
Come with me to the fields, thou man of many ills,
And give thy limbs a chance among the daffodils!

Come with me to the woods,
And let their solitudes
Re-écho to our voices as we go,
Upon thy weary brain
Let nature come again,
Spite of thy wealth, thy learning, or thy woe!
Stretch forth thy limbs and leap—
Thy life has been asleep;
And though the wrinkles deep may furrow thy pale brow,
Show me, if thou art wise, how like a child art thou!

CHARLES MACKEY.

ORGANS AND STRUCTURE OF PLANTS.

What is the first step to be taken in understanding what may be termed vegetable physiology or the processes employed by nature in the growth of plants?

It is to learn the organs possessed by each plant, and the functions for which they are designed in the development of the different parts of the vegetable structure.

What will be the consequence, if this is neglected?

Without this knowledge, it will be impossible to understand how plants derive their nutriment from the air and the soil, and indeed it is and must be the basis of all our studies in relation to the subject.

To what are the organs of the plants analagous?

They are analagous to the organs of an animal. The one as much needs a specific structure, through the agency of which the aliments of vegetable life are maintained, as the other. An animal might as well live and grow without food; and might as well receive and assimilate the food when furnished, without mouth and teeth and stomach, and without veins and arteries, as a plant can flourish without corresponding organs. The organs of a plant answer precisely the end that these organs do in the animal economy.

What other similarity is there between the growth of animals and vegetables?

A supply of fresh air is alike needed by both. Food and drink of some kind or other—that is, solid and fluid or gaseous substances, also. Without them animals and vegetables would both sicken and die. This fact should never be forgotten.

What are the most obvious organs of plants?

It is a general law in vegetable economy, that all plants consist of a root, a stem, and leaves; and these organs consist of distinct parts, each designed to answer a specific purpose, and is essentially necessary to the perfection of the whole.

To what have the organs of a plant distinct reference?

The production of a fruit, which being accomplished, the plant either dies entirely, or lies torpid for a season, until a succession of the same circumstances which gave it life in the first instance, shall again call its productive organs into action.

What are the offices of the root?

In the first place, the root keeps the plant fixed in a proper position; and, in this respect is therefore analagous to the limbs of an animal. In the second place, the plant receives most of its nourishment through the root; which is therefore analagous to the mouth of an animal.

What is said of the form of the root?

The form of the root is diversified to an unlimited extent; sometimes it is large; sometimes it is small; sometimes straight; sometimes crooked; sometimes rough; sometimes smooth. The roots of plants are as much diversified as the tops or branches.

What is said of the instincts of the root?

The instincts of roots are truly wonderful. If there is in the vicinity a little spot of soil more rich than the rest, they are sure to find it as a mouse is sure to find a little crumb of cheese sunk into the crevice between the edges of two boards. Or, if there is a seam in a large stone obstructing their progress, they will be as sure to find and penetrate this avenue, as a man with a good lantern, in a dark night, would search out and pursue a small foot-path in a grass field.

Of how many parts does a root consist?

The body, the crown or collar, the branches, and the fibres; each of essential importance in its place—the fibres particularly; so also is the crown, being the portion of the plant between the stem or leaves and the body of the root.

What is said of the importance of the crown?

In many plants, of a hardy nature, nearly the whole of the body of the root may be cut away, and yet, if the crown be uninjured, still the plant will flourish; but, in the generality of plants, if the crown be injured, no matter how perfect so ever the body may be, the plant is usually destroyed.

If the crown is slender what will be the consequence?

If the crown of plants is slender, they dry up as the seeds ripen, and the plants soon die. Such plants are termed annuals, including wheat, barley, oats, and a multitude of others—but when the crown from any cause, such as the soil, climate, or culture, is rendered strong, such annuals are brought to grow two years, and then are called biennials; or for a succession of years, and then are called perennials.

What is said of the fibres of a root?

The fibres are an essential part of each root, yet in most cases they may be removed without injury to the plant, provided the crown is sufficiently healthy and vigorous to push out new ones.

What are the spongelets of a root and what is said of them?

The spongelets are the tips or the extreme ends of the fibres; and they absorb the pabulum from the soil. In case the spongelet, from any cause, is removed from the point of the fibre, two lateral shoots are immediately thrown out, provided the plant is sufficiently vigorous to bear the temporary loss it thus sustains,—each furnished with its spongelet, and thus the destruction of the one becomes a source of strength to the plant.

How does tillage operate to make this power of the plant in reproducing spongelets advantageous?

In ploughing between the rows of corn, if the small roots are cut, the number of spongelets will be increased and consequently the vigor of the plant will be promoted. The same may occur with other crops, particularly in garden culture.

What is said of the duration of fibres on the root?

They are produced annually like leaves, in some cases, the old ones having fallen off. The Dahlia is an instance of it. In other cases the fibres are constantly increasing in size, and becoming harder like the parent of the root, and subsequently throwing out new fibres themselves, as is the case with large trees.

What general description is given of the stem of plants?

The stem of all plants rises immediately from the crown of the root, and is, consequently, always above the ground. The same variety prevails in this part of the plant as in the root; for instance, the stems of wheat, barley, and the grasses, rise to some height, and are termed the straw; the stems of mushrooms, fungi, and the like, are termed the stalk; and the stem of the strawberry is termed the runner; all of them being appropriately described from the appearance each presents.

Of what several parts is the stem composed?

The pith, the wood, and the bark. This will appear by cutting across a young twig of a common tree or shrub. This is readily seen in the Ash and Elder, because in them and some others the parts are more distinctly seen, yet where less apparent to the eye they exist.

What is said of the pith?

It is a soft spongy substance occupying the centre. If a slice of it be cut, either across or vertically, and magnified, it is seen to consist entirely of cellular tissue, the cells of which are mostly of a regular form. When young it contains a good deal of fluid; when the branch is old it becomes white and hard; and in an old stem or branch, it is often found to have shriveled up and almost entirely disappeared.

What curious product is obtained from the pith?

Generally the pith of trees is applied to no important use; but one curious product is obtained from the large pith which constitutes nearly the entire stem of a herbaceous plant. This is the substance known as *Rice Paper*, which is made by cutting the soft portion of the stem with a sharp knife in a spiral manner, so as to cause it to spread out, as if a sheet of paper were being unrolled from a round roller. It is then flattened out and made smooth by pressure.

What is said of the formation of the wood?

It surrounds the pith in rings or layers, the number and thickness of which depend upon the age of the branch or stem. These rings or layers occupy the entire space between the pith and the bark. The number of them is easily reckoned on cutting the stem or branch across; and they correspond exactly in this climate with the number of years which the part has existed; or in other words, a distinct ring or layer is formed each year, so that in this way the age of any tree is easily ascertained.

How is it supposed that this takes place?

There is reason to suppose, that on the falling of the leaves, which is annual in temperate climates, or in correspondence with the formation and decay of each set of leaves, an additional layer or ring of wood is also produced.

What exception is there to this?

In tropical climates there are many kinds of trees having two or three successions of leaves in each year, and the presumption is a corresponding number of layers of wood is also formed. Unless this be the fact, the Boabab trees of Senegal, in many instances, have reached the age of 5000 years, their wood being composed of that number of layers. It is not credible that they have reached that age; and the appearance of supposed evidence for such a hypothesis, is explained by their having an additional layer with each succession of leaves.

What changes are constantly taking place in the wood of trees, as they increase in age?

In most trees, of which the wood is used as timber, the inner and older portion is much harder and dryer than the exterior. Sometimes there is an evident line of demarcation between the *heart-wood* or *duramen*, as it is called, and the *sap-wood*, or *alburnum*. This is seen in the lignum vitæ, and cocoa-wood. But in most cases, the exchange of character is more gradual, and the lines of demarcation are less distinct.

To what is this change owing?

It is attributed to the consolidation of the interior wood, by the deposition in its tubes of resinous and other matter secreted by the plant. The portion of the stem in which this has taken place thus acquires great toughness and durability, but it is no longer fit to perform any office in the living system, save that of mechanically supporting the rest; since no fluid can now pass in any way through the now-filled-up channels. It is through the new layers, or sap-wood, therefore, that the sap entirely ascends; and these, in their turn, become enclosed by others, and are at last consolidated, like the more aged ones, into duramen. The heart-wood alone is used by the artisan; for the sap-wood soon splits and decays.

How does it appear that sap does not ascend to nourish the plant through the heart-wood?

It is well known, that the *heart* or *duramen* of an aged tree with some portions of the exterior stem sometimes entirely decays, leaving the remainder a mere shell or hollow trunk. Yet, vegetable life does not become extinct. The sap continues to ascend as before. New layers of alburnum are annually made. Annually there is a new succession of leaves; and the process of throwing out new branches continues without change.

What account is given of the formation of bark?

It is obvious to the sight that the wood is enclosed by the bark; and the latter, like the former, is composed of regular layers, although they cannot be so plainly seen. The layers of the bark are formed from the *interior*. So that the older are on the outside. These are generally lost, either by decay, or falling off; so that it is very seldom that the same number can be traced in the bark as in the wood, although an additional one is formed in each at the same time. As the new layer of wood is formed on the *outside* of the previous one,—at the point, therefore, at which it is in contact with the bark, and as the new layer of bark is added to the *inside* of the previous one; at the point, therefore, at which it was in contact with the wood, it is obvious they are produced at the same spot, and that the newest layers of both will be always in contact with each other.

What is said of the bark from which cork is obtained?

It contains a great quantity of cellular tissues, and is therefore thick and spongy, and may be regarded as a sort of external pith. The tree producing this bark is a species of the oak, found in the southern part of Europe, and mostly in Spain and Portugal.

Why is the inner coat of some bark called LIBER ?

Liber, in the Latin language, is the name of a book ; and before the invention of paper, the Romans used these inner coats of certain bark, which were very fine and delicate, for the purposes of writing. Thus the inner coat of bark and a book had the same name.

For what other purposes has the inner coat of bark sometimes been used ?

In the Polynesian Islands especially it is wrought into cloth, mats, sails, and cordage. It is said that a very beautiful kind of this substance is obtained from the Vegetable-Lace tree of Jamaica. When its layers are unfolded, it has the appearance of a delicate lace.

What is the most obvious agency of leaves in the growth of vegetation ?

Leaves perform in the vegetable kingdom, the same offices as the lungs in the animal kingdom. Through them, from the pores covering their surface, the respiration of the plant is carried on ; and, more than this, for at the same time that the respiration is going on, through the pores, a constant assimilation of one of the gases of the atmosphere is taking place. From this source to a considerable extent, the plant derives its nourishment.

What other agency is performed by the leaves ?

A constant chemical action is always in operation in the leaves, in the formation of the resinous, and oleaginous, and acid matters they contain. These processes of the leaves are constantly in operation, from the first formation of the leaf until the seed is perfected, and they only cease when from the ripening of the fruit, their assistance is no longer required.

How is it known that plants do absorb water, and other nourishment from the ground in their growth ?

It may be known from their being dried and burnt, and found to contain the same elementary substances existing in the ground. It may be known also from placing certain plants in a vessel of water, the water disappearing at a much more rapid rate, than it would evaporate if there were no plants immersed in it.

What experiments of this have been made?

An experiment has been made with four plants of spearmint. They had been placed with the roots in water for the period of fifty-six days. During this time, they had taken up 54,000 grains of water, or about seven pints of the fluid, although their own united weight was but 403 grains, showing also that the remainder had been evaporated, or cast off by the leaves.

How can it be proved that the sap ascends through the alburnum or the new layers of wood?

By coloring the water provided for the nourishment of the plant, it will be found that this part of the stem is tinged with the coloring matter. It may also be known by making an incision or boring a hole into this wood, from which the sap will ooze out. In this way the sap is procured for maple sugar.

What singular fact has been related to show that every succeeding new layer of wood is on the outside?

Adamson relates, that, in visiting Cape Verd in the year 1748, he was struck by the venerable appearance of a tree, 50 feet in circumference. He recollected having read in some old voyages an account of an inscription made on a tree thus situated. No traces of such an inscription remained, but the position having been accurately described, Adamson was induced to search for it by cutting into the tree, when, to his great satisfaction, he discovered the inscription entire under no less a covering than three hundred layers of wood, proving that each new layer is formed on the outside of the last preceding one, and that the inscription had been made 300 years.

If the sap in its ascent meets with any obstruction what will be the consequence?

When this happens, there is an accumulation of the sap at this point which causes a protuberance or wart on the tree, and sometimes a new shoot for the formation of an additional branch.

What is said of the size to which leaves sometimes attain?

Leaves are of every imaginable form, and their size varies almost as much as their forms. In the mosses which abound in cold climates, they are extremely minute; and the forest trees of the North are adorned with leaves which appear very diminutive, when compared or rather when contrasted with the foliage of equatorial plants. There we find the leaves of the Banana, perhaps the same which were employed by our first parents, to supply the want of more artificial dress; they being in the opinion of many writers, the "Fig leaves" of sacred history. In Ceylon, a country alternately exposed, for many months in succession, to the rays of the vertical sun, and the inclemencies of an unceasing storm, is found the singular Talipot, a single leaf of which is sufficiently large to shelter twenty men from the vicissitudes of the climate in which they dwell. This tree is venerated by those who find beneath its branches so kind a shelter, and travellers consider it the greatest blessing which Heaven has bestowed on the country. And when we regard its subserviency, to the wants of the human race, it is not surprising that by the ancients, the wide spreading tree, decorated with leaves, and occasionally beautified with flowers, should have been held sacred as the very temple of the Deities they worshipped.

PATRIOT FARMERS OF NEW-YORK.

Farmers of New-York—the hour of separation for this dazzling array of beauty, this vast multitude of men, is at hand. Fruits richer than ever graced the gardens of Pomona —a paradise of flowers—needle-work the most exact, delicate and even—ingenious farming implements and manufactures of all sorts, cloths of the finest quality, from your own looms, and from looms in Massachusetts—horses fit to win prizes at Olympia—cattle such as never fell in a Hecatomb to Jove, and never were dreamed of by the highest of the Dutch painters—all these and more have arrested our gaze and filled us with wonder and delight. And now I am commissioned to summon you, and through you the population of this mighty commonwealth, to come up and join us, as, under the auspices of the State, honor and distinction are awarded to agricultural industry and genius.

But the farmers of New-York are not content with improvements in the material world alone. From their generous impulses spring your system of free schools. They have proved themselves the liberal benefactors of academies and colleges. They too have been careful for the means of their own special culture, and have founded and nurtured societies for promoting agriculture. For an example of the virtues of private life, I name to you the farmer of Westchester county, the pure and spotless Jay, who assisted to frame our first treaty of peace, which added Ohio and the lovely West to our agriculture. Side by side with him, I name the friend of his youth, Robert R. Livingston, the younger, the enlightened statesman of our Revolution, whose expansive mind succeeded in negotiating for our country a world beyond the Mississippi, and gained access for our flag to the gulf of Mexico. Here, on the banks of the Hudson he is celebrated as it were by every steamboat, and remembered on your farms through his experimental zeal. On this day be remembered the virtues of Stephen Van Rensselaer, who first brought Durham cattle to this State, and liberally diffused the breed.

Join with me also in a tribute to Mitchill, the faithful advocate, and perhaps institutor, of one of the earliest agricultural societies; to Jesse Buel, who connected science with fact, taught how the most barren soil may be made vastly productive, diffused his acquisitions by the press, and by life and by precept was the farmer's friend; to Willis Gaylord, whose agricultural essays are standard authorities, honorable to the man and the State; to Le Ray de Chaumont, who kept alive an agricultural society in Jefferson county, when all others had expired, and gave the impulse to the formation of the State Society, of which he was the first president; to James Wadsworth, for his skill as a cultivator, and still more for his liberal exertions, pouring out thousands after thousands, at the impulse of a generous mind, as if from a well-spring of good will, to promote agricultural science in primary schools.

And I should be wanting on this occasion, did I not tender the expression of your regard to the present president of the State Society, to the influence of that institution of which he is the honored head; to its Journal of Agriculture, to its annual fairs. But let me also entreat its friendly wishes to its purpose of establishing an agricultural school; and to that other more diffusive design of introducing through its secretary, scientific works on agriculture into school libraries. I am happy also to announce that efforts are now making to constitute agriculture, as it deserves to be, a branch of instruction in one, at least, of your universities.

I have named to you some of the benefactors of agriculture in New-York. Their benefits endure. The pursuits of the farmer bind him to home. Others may cross continents and vex oceans; the farmer must dwell near the soil which he subdues and fertilizes. His fortunes are fixed and immovable. The scene of his youthful labors is the scene of his declining years; he enjoys his own plantations, and takes his rest beneath his cotemporary trees.

But the farmer is not limited to the narrow circumference of his own domain; he stands in relation with all ages and all times. Your Society has done wisely to urge on those who bear the Gospel to untaught nations, to study their agriculture, and report for comparison every variety of tillage. All ages and all climes contribute to your improvement. For you are gathered the fruits and seeds which centuries of the existence of the human race have discovered and rendered useful. Tell me, if you can, in what age and in what land the cereal grasses were first found to produce bread? Who taught to employ the useful cow to furnish food for man? When was the horse first tamed to proud obedience? The pear, the apple, the cherry, where were these first improved from their wildness in the original fruit?

And whose efforts led the way in changing the rough skin of the almond to the luscious sweetness of the peach? All ages have paid their tribute to your pursuit. And for you the sons of science are now scouring every heath and prairie and wilderness, to see if some new grass lies hidden in an unexplored glade; if some rude stock of the forests can offer a new fruit to the hand of culture. For you the earth reveals the innumerable beds of marl; its mineral wealth, the gypsum and the lime, have remained in store for your use from the days of creation. For you Africa and the isles of the Pacific open their magazines of guano; for you old Ocean heaves up its fertilizing weeds.

And as the farmer receives aid from every part of the material world, so also his door is open to all intelligence. What truth is not welcomed as an inmate under his roof? To what pure and generous feeling does he fail to give a home? The great poets and authors of all times are cherished as his guests. Milton and Shakspeare, and their noble peers, cross his threshhold to keep his company. For him, too, the harp of Israel's minstrel monarch was strung; for him the lips of Isaiah still move, all touched with fire; and the apostles of the new covenant are his daily teachers. No occupation is nearer heaven. The social angel, when he descended to converse with men, broke bread with the husbandman beneath the tree.

Thus the farmer's mind is exalted; his principles stand as firm as your own Highlands; his good seeds flow like self-moving waters. Yet in his connection with the human race, the farmer never loses his patriotism. He loves America—is the depository of her glory and the guardian of her freedom. He builds monuments to greatness, and when destiny permits, he also achieves heroic deeds in the eyes of his race. The soil of New-York, which he has beautified by his culture, is consecrated by the victories in which he has shared. Earth! I bow in reverence, for my eyes behold the ground wet with the blood of rustic martyrs, and hallowed by the tombs of former heroes! Where is the land to which their fame has not been borne? Who does not know the tale of the hundred battle fields of New-York? Not a rock juts out from the Highlands, but the mind's eye sees inscribed upon it a record of deeds of glory. Not a blade of grass springs at Saratoga, but takes to itself a tongue to proclaim the successful valor of patriot husbandmen.

Here the name of Schuyler, the brave, the generous, the unshaken patriot, shall be remembered; the zealous, reliable George Clinton, a man of soundest heart, a soul of honesty and honor, a dear lover of his country and of freedom. Nor do we forget him—the gallant Montgomery—twin martyr with Warren—who left his farm on the Hudson, not, as it proved, to conquer Quebec, but to win a mightier victory over death.

I renew the theme once more, to recount how the farmers of New-York have served their country and mankind. They were invested with sovereignty, and abdicated. Glorious example! Highest triumph of disinterested justice! They themselves peacefully and publicly renounced their exclusive authority, and transferred power in this republic from its territory to its men. May your institutions, under the spirit of improvement, be perpetual. May every pure influence gather round your legislation. May your illustrious example show to the world the dignity of labor; the shame that lights on idleness; the honor that belongs to toil. To the end of time, be happiness the companion of your busy homes, and the plough ever be found in the hands of its owner.

The farmer is independent. With the mechanic and manufacturer as his allies, he makes our country safe against foreign foes, for it becomes perfect by its own resources. But why do I say this? To foster a spirit of defiance? Far otherwise. Let us rejoice in our strength, but temper it with gentleness and the spirit of love for all mankind—a love that shall perpetuate tranquility, and leave the boundless and rapidly increasing resources of the country at liberty for its further development.

And has it occurred that this great commonwealth—the most numerous people ever united under a popular form of government—is emphatically a commonwealth of the living? Go to the Old World, and your daily walk is over catacombs, your travel among the tombs. Here the living of the present day outnumber the dead of all the generations since your land was discovered. All, all who sleep beneath the soil of New-York, are fewer in number than you who move above their graves. Look about you and see what the men of the past have accomplished.

Concentrate in your mind all that they have achieved; the beauty of their farms, the length and grandeur of their canals and railroads, the countless fleets of canal boats they have constructed; their ships that have visited every continent and discovered a new one; their towns enlivening the public plains; their villages that gem the valleys; the imperial magnificence of their cities; and when you have collected all these things in your thoughts, then hear me when I say to you, that you of this living generation, as you outnumber all the dead—are bound before your eyes are sealed in death, to accomplish for New-York more than has been accomplished for New-York thus far in all time. Mighty commonwealth! lift up your heart; let your sun ascend with increasing splendor towards its zenith. You shall be a light to humanity; a joy to the nations—the glory of the world.—George Bancroft, LL. D., *Address, New-York State Fair*, 1844.

GOOD IN ALL SEASONS.

I love to see a city street,
With all its living swarm;
Men, women, coaches, carts, to meet,

Children with bright eyes and quick words,
And infants that, like little birds,
Sit perch'd upon the arm.

Yet not the less I love to go,
Leaving all these behind,
To where the brooks and rivers flow,
And where the wide sky may be seen,
Where flowers are sweet and leaves are green,
And stirring in the wind.

'T is merry, merry in the spring,
And merry in the summer time,
And merry when the great winds sing
Through autumn's woodlands brown—
When from the tall trees scatter down,
Ripe acorns fringed with rime.

And in the winter, wild and cold,
'T is merry, merry too;
Then man and boy are blithe and bold,
Then rings the skate upon the ice;
Then comes the hoar-frost in a trice,
And everything is new.

Free are the woods and hills! There dwell
Creatures that serve not man—
Glad things, that neither buy nor sell,
That want not aught we have to give,
That ask us not for leave to live,
But live just as they can.

To God who made and loves them all
They hymn their praise serene;
Things great, things wondrous, things so small
Their very forms escape your sight
Two worlds of beauty and delight—
The hidden and the seen.

When first leaves cluster on the trees,
And spring flowers star the ground,
And birds come o'er the southern seas,

And build their nests and sing aloud;
And insects, a gay, shining crowd,
Glitter and hum around.

When winter comes, and beasts and men,
Retreating from the field,
Seek fire-lit house, and winter den;
In town or country still the same,
God's love all living things proclaim,
Their good all seasons yield.

Therefore, for us let seasons change;
Let the sun shine, or tempests rage;
Through street or forest still we'll range,
And find God present in each spot,
His guiding hand in every lot;
His grace from age to age.

WM. HOWITT.

THE COUNTRY BURIAL-GROUND.

I like that ancient Saxon phrase, which calls
The burial-ground God's Acre! It is just;
It consecrates each grave within its walls,
And breathes a benison o'er the sleeping dust.

God's Acre! Yes, that blessed name imparts
Comfort to those, who in the grave have sown
The seed, that they have garnered in their hearts,
Their bread of life, alas! no more their own.

Into its furrow shall we all be cast,
In the sure faith, that we shall rise again
At the great harvest, when the archangel's blast
Shall winnow, like a fan, the chaff and grain.

Then shall the good stand in immortal bloom,
In the fair garden of the second birth;
And each bright blossom, mingle its perfume
With that of flowers, which never bloomed on earth.

With thy rude ploughshare, Death, turn up the sod,
And spread the furrow for the seed we sow;
This is the field and Acre of our God,
This is the place, where human harvests grow!

LONGFELLOW.

THE RAINBOW.

I sometimes have thoughts in my loneliest hours,
That lie on my heart like the dew on the flowers,
Of a ramble I took one bright afternoon
When my heart was as light as a blossom in June;
The green earth was moist with the late fallen showers,
The breeze fluttered down and blew open the flowers,
While a single white cloud, to its haven of rest
On the white wing of peace, floated off in the west.

As I threw back my tresses to catch the cool breeze,
That scattered the rain drops and dimpled the seas,
Far up the blue sky a fair rainbow unrolled
Its soft tinted pinions of purple and gold.
'T was born in a moment, yet, quick as its birth
It was stretched to the uttermost ends of the earth,
And, fair as an angel, it floated as free,
With a wing on the earth and a wing on the sea.

How calm was the ocean! how gentle its swell!
Like a woman's soft bosom it rose and it fell;
While its light sparkling waves, stealing laughingly o'er,
When they saw the fair rainbow knelt down on the shore,
No sweet hymn ascended, no murmur of prayer,
Yet I felt that the spirit of worship was there,
And I bent my young head, in devotion and love,
'Neath the form of the angel, that floated above.

How wide was the sweep of its beautiful wings!
How boundless its circle! how radiant its rings!
If I looked on the sky, 't was suspended in air;
If I looked on the ocean, the rainbow was there;
Thus forming a girdle, as brilliant and whole,
As the thoughts of the rainbow, that circled my soul.

Like the wings of the Deity, calmly unfurled,
It bent from the cloud and encircled the world.

There are moments, I think, when the spirit receives
Whole volumes of thought on its unwritten leaves;
When the folds of the heart in a moment unclose,
Like the innermost leaves from the heart of the rose.
And thus, when the rainbow had passed from the sky,
The thoughts it awoke were too deep to pass by;
It left my full soul, like the wing of a dove,
All fluttering with pleasure, and fluttering with love.

I know that each moment of rapture or pain
But shortens the links in life's mystical chain;
I know that my form, like that bow from the wave,
Must pass from the earth and lie cold in the grave;
Yet Oh! when death's shadows my bosom uncloud,
When I shrink at the thought of the coffin and shroud,
May hope, like the rainbow, my spirit unfold
In her beautiful pinions of purple and gold.

AMELIA B. WELBY.

ELEMENTARY CONSTITUENTS OF PLANTS.

What is meant by an elementary constituent or substance entering into the composition of vegetables?

It is a simple substance; that is, a substance not composed, or compounded, or made, of two or more other substances; or, which we have as yet no means of reducing to two or more other substances.

How can the difference between a compound and a simple substance be explained?

A measure or a bag of grain having portions of wheat, rye, barley, oats, rice, and maize, would be neither a simple nor a compound, but a mixture of different compound substances. But if these were reduced to flour and then kneaded into one mass of paste they would form a compound substance. Silver and gold, or any other metals, melted into a solid consistence, would be a compound substance. This is a process comprehended by the eye alone. The parts in a mixed body can be separated by the hand or machinery; but the simple

SHORT HORNED BULL, "LAMERTINE," owned by LEWIS G. MORRIS, of Fordham, N. Y., and which obtained the First Premium at the Fair of the American Institute, 1849.

substances in a compound one can be separated from the others only by the aid of chemical agents.

But are the substances specified, different kinds of grain and metals, simple substances?

By no means. Each of them is a compound substance; and, they were selected simply to show in what manner a compound substance may be formed; that persons with the eye and without the aid of chemistry may understand the nature of the difference between the two things.

What other illustration may be given of the difference between a compound and a simple substance?

One of the most familiar examples is water. Formerly it was supposed that water is a simple substance; yet chemistry has demonstrated that it is a compound; completely reducing any quantity of water into two other substances—called oxygen and hydrogen—two of the gases in our atmosphere. This analysis or separation of the constituents of water is rendered perfect and clear as would be the separation of gold and silver which had been melted together and made into a spurious coin.

What inference should be drawn from such facts?

It might be supposed that every person would desire a knowledge of chemistry; at least to understand the nature of the objects with which we are surrounded, and with which we are constantly coming in contact. Without this knowledge we may unconsciously use for food the most destructive poisons; or may cast away as worthless objects of the greatest value.

What opinion formerly prevailed on this subject?

It was the doctrine of ancient philosophers, that the vegetable and the animal kingdom were composed, in various combinations and proportions, of four elementary principles, called earth, air, fire, and water. Later investigations show that neither of these is a simple substance; and, hence, that each of them is a compound of elementary principles or essences, which are now known to exist, and to form the basis of the whole mass of matter.

What are the elementary constituents of vegetables?

They are, according to some writers, ten in number—others say fifteen—carbon, hydrogen, oxygen, and nitrogen, which are denominated the *organic;* and, potash, soda, magnesia, silex, lime, phosphorus, alumina, oxide of iron, oxide of manganese, chlorine, and sulphuric acid, which are denominated the *inorganic*. As one or more of these is a component part of every plant that exists, it may thence be fairly inferred that its presence is absolutely necessary.

What further reason is there to affirm that all of these substances are indispensable to the perfection of vegetable life?

Plants growing in situations where these substances cannot be had, are diminutive in size, their organs are not fully developed, and in many instances, they perish altogether, after having attained a certain growth. Besides, soda, lime, magnesia, and phosphorus enter largely into the structure of bones and teeth. Animals are depending on vegetation for their support, either directly or indirectly, and as no other means naturally exist for supplying these necessaries to the animal frame, it is not too much, perhaps, to say that this property of plants was designed for that purpose.

What is the foundation for the distinction between the organic and the inorganic parts of vegetables?

The organic portions of a plant are gases, as will appear from the fact that in the combustion of wood or any other vegetable substance nothing of it remains, having passed off in the atmosphere; whereas the inorganic portions of wood or other vegetables, on being decomposed by combustion, will remain in the form of ashes.

What is the relative proportion of the organic and the inorganic parts of a plant?

It varies in different plants, from one-tenth to one-hundreth part of the weight. This may be ascertained by weighing a piece of wood, and of other vegetable formations, before combustion, and the ashes that afterwards remain.

What account is given of carbon?

Till reduced, by combustion, to gas, when it is called *carbonic acid gas*, it is a solid substance, usually of black color, without taste or smell, and burns more or less readily in the fire. Wood-charcoal, coke, black-lead, and the diamond, are varieties of carbon. The gas or air that is produced from their combustion, is likewise exhaled from the lungs of animals in breathing, and being unfit for respiration, is destructive of animal life.

Where else is this gas to be found?

It exists in coal-mines, and sometimes in caverns and deep

wells, and is there called fixed air; and, it is used in cities and large towns for lighting houses and streets. When thus used, it is a combination of carbon and hydrogen.

How does carbon effect the quality of wood?

It is chiefly to the carbon which it contains that the hardness and solidity of wood is owing; and, in proportion as the tissues of the plant are deficient in carbon, do we find them deficient in firmness of structure.

How is this great amount of carbon, needed in vegetable productions, furnished?

The quantity of coal used, especially in cities, is immense. The gas proceeding from it, all returns in process of time, to its previous condition; that is, to the formation of new plants. In addition to this, no one can estimate how much gas is extricated by the respiration of the countless millions of animals on the globe.

What is said of hydrogen?

It is a kind of gas or air which burns as coal gas does, but in which a candle will not burn, nor an animal live, and which, after being mixed with common air, explodes when brought near the flame of a candle. Hydrogen is contained largely in plants; and in most substances into whose composition it enters, it is combined with oxygen nearly in the same proportion as in water. Being an important constituent of water, it passes readily by their roots into the substance of plants.

What is said of oxygen?

This gas enters largely into the organization of vegetable substances. It makes one-fifth part of the air we breathe; and without it, plants will no more flourish, than animals can live. A portion of this air is combined with water; and it is in this manner that fishes and other aquatic animals, as well as plants, are supplied with it. Without oxygen, combustion will not take place; and it is estimated that it forms the basis of one-third part of all solid substances.

What are some of the most common effects of the combination of oxygen with other substances?

By the aid of this gas, iron or steel may be burned. It is this gas, as existing in the atmosphere, which principally causes rust on metals, and hence this rust is called oxidation. It is the principal agent in producing acids; and hence, when an acid is to be produced, the substance with which, as for instance in the making of vinegar, it is to be combined, should receive the free and constant contact with the atmos phere. And it also causes rancidity in butter, lard, and oil. Hence they should be kept secluded from the atmosphere.

What may be said of nitrogen?

Nitrogen forms an important part in the growth of both animals and vegetables. It constitutes about four-fifths of the atmosphere. In vegetables it is the substance called *gluten*, which exists so largely in the seeds of various kinds of corn; and most of all in wheat. On this account in part, wheaten bread is the most nutritious of all vegetable substances ordinarily used as food. The value of wheat, and all grain, depends on the quantity of gluten it contains. If persons breathe for a few minutes the gas, they will be affected much the same as if they had drank an excess of fermented liquors.

What becomes of the organic portions of a plant, on being liberated by combustion?

It has been said that they pass off in the atmosphere, and are invisible. The carbon uniting with the oxygen, and some additional oxygen from the air, becomes carbonic acid. Hydrogen, uniting in the same manner with oxygen, becomes a watery vapor. Thus in connexion with the carbon and the hydrogen, the oxygen is entirely disengaged. And the nitrogen combining with what remains of the hydrogen, becomes ammonia.

What is said of potash?

The substance found in the shops, and used as merchandize, called potash, is prepared from ashes deposited in the burning of vegetables. As the article is manufactured in iron pots, the name is a compound of the word pot, and the first syllable of the word ashes. The taste is extremely acrid; and it is so corrosive, it destroys the texture of the skin the moment they touch each other. On this account it is employed in surgery, for the purpose of opening abscesses, or for destroying excrescences, and is called a caustic.

What is soda?

This is an alkaline salt, much resembling potash, but is more fusible; and when it comes into the air, it crumbles into a fine powder. It is manufactured from marine plants and plants growing upon the sea shore, as well as from sea water; and it may be found occasionally in its natural state. Common salt is the muriate of soda.

What is said of lime?

Lime or quick lime, is obtained by exposing limestone and shells to a very strong heat—usually in some kind of a furnace. After being subjected to this heat, if water be poured upon it, heat is evolved, and it slakes, or becomes a fine white powder. Quicklime is called calcareous earth, and the soils which abound in this substance are called calcareous soils. Most soft stones contain more or less of lime.

What is the use of lime in agriculture?

One of its most obvious purposes is to give stability and substance to those parts of the plant which otherwise would not, perhaps, have sufficient strength to perform the functions alloted them; but that it answers other purposes there is little doubt, as it is found not only in the stalk, but in the leaves fruit, and indeed in almost all parts of the plant.

What is said of magnesia?

Magnesia is called a primitive earth, having for its base a metallic substance, denominated magnesium. The name is said to be derived from the place where it was first found. It may be obtained from sea water, and from some kinds of limestone rock, known as *magnesian* rock. It is used for medicinal purposes, having the power to correct the acidity of the stomach. As an object of merchandise or medicine, it exists in the form of a white powder.

What is said of silex?

Silex and silica denote the same substance; or rather the latter is the scientific name of the former. It is one of the primitive earths, and is a constituent of all stones, being found in greatest abundance in agates, jasper, flints, quartz, and rock crystal. In the latter it exists nearly in a state of purity. When pure it is perfectly white or colorless, and without taste or smell.

What is said of phosphorus?

It is a combustible substance, not as yet decomposed, and classed among those that are elementary. It is of a yellowish color, and semitransparent, resembling fine wax. It burns in common air with great rapidity; and in oxygen gas, with the greatest vehemence. Even at the common temperature, it combines with oxygen, undergoing a slow combustion and emitting a luminous vapor. It was originally obtained from urine; but it is now manufactured from bones, which consist of the phosphate of lime.

What is the oxide of iron?

The existence of iron is too abundant, and its use too common, not to be well known by all. When polished and exposed to the air, it becomes gradually covered with rust. This rust, or *oxide of iron*, as it is called, is formed by portions of the metal combining with the oxygen of the atmosphere. Of course between the iron and the oxygen there is an affinity.

What is said of the oxide of manganese?

Manganese is a metal of a dusky white, or whitish gray color, very hard and difficult to be fused. It never occurs as a natural product in a metallic state. The substance usually called manganese is an oxide of the metal, but not pure. It occurs in soils and vegetables only in very small quantities.

What is the substance called chlorine?

Chlorine is a new name given to what was formerly called oxymuriatic gas. It is a kind of air which has a greenish yellow color, and a strong suffocating smell, existing largely in common salt. Chlorine means green, and hence its new name. Hitherto, it has resisted all efforts to decompose it, and is now reputed to be a simple substance.

What is said of the substance called sulphuric acid?

It is sometimes called oil of vitriol. It derives its name from sulphur and a liquid, formed by sulphur saturated with oxygen. It exists in common gypsum or plaster of paris, in alum, and in glauber and epsom salts.

Does every vegetable contain a portion of each of the four organic elementary substances?

All vegetables do not contain the four; most plants contain only three of them—carbon, hydrogen, and oxygen. The more common ones which contain these three only, are oils, fats, starch, gum, sugar, and the fibre of wood.

Are all the inorganic parts of plants above named to be found in the ashes of all plants?

Generally in all the cultivated plants, but not in the same proportion, nor in the same amount. Some leave a larger quantity of ashes than others; and, there is also a manifest unequal proportion in different ones. For instance, in a ton of hay there are 180 lbs. of ashes, while in a ton of wheat there are but 40 lbs. of ashes. Then, the ashes of wheat contain more phosphoric acid than the ashes of hay; and in return the ashes of hay contain more lime than the ashes of wheat. Similar inequalities and disproportions will be found in the ashes of other plants.

THE SOURCES OF NATIONAL WEALTH.

What is wealth? In what does it consist? Wealth is every thing that supplies human wants, natural or artificial. Here is, of course, an end to its multiplication. The artificial wants of mankind have no limits, of course wealth has no bounds, but the productiveness of nature, and the capacities of human industry. And what are human wants? The first is food. This can be procured only from the soil. Hence, the first and most universal of human pursuits is agriculture. The first item in a nation's wealth is cultivated land. Before this, every other species of property dwindles into insignificance, and strange as it may seem, the greatest investment in this country, the most costly production of human industry is the common fences which divide the fields from the highlands, and separate them from each other. No man dreams, that when compared with the outlay of these unpretending monuments of human art, our cities and our towns, with all their wealth are left far behind. You will scarce believe me, when I say that the fences of this country have cost more than twenty times the specie there is in it.

In many of the counties in the northern states, the fences have cost more than the farms and the fences are worth. It is this enormous burden, there can be no doubt, which keeps down the agricultural interest of this country, and it is freedom from it which enables the north of Europe, with a worse climate and an indifferent system of cultivation, to undersell us in the markets of England. There, travellers tell us, fences are almost unknown. The herds and flocks are under the care of herdsmen and shepherds, and thus an untold expenditure is saved, besides the loss of the land which the fences occupy and the accumulation of soil, that, with the most careful management, is apt to be thrown up around them by the plough.

The farmer contributes to the weath of a country by his perpetual toil. Every thing begins with him. Every day in the year has its various and its continuous operations; all directed, however, to this one point—to bring the greatest quantity of produce from a given number of acres. Such is the nature of his work, that little can be done to expedite or shorten the process. Every foot of every field must be passed over by the plough. There are no fire-horses yet invented to do this at the rate of twenty miles an hour. The ploughman, therefore must rise early and work late. His labors, too, must be generally confined to the hours when the sun is above the horizon. In autumn and winter these are few. He must work the harder during that part of the year when the days are long. Every industrious farmer is continually adding to the substantial and permanent wealth of a nation. He is continually adding to the productive power, which is the best species of wealth. His savings, if any he makes, go back into the soil, to increase its fertility, or they go into fixtures, which add comfort or diminish the labors of all coming years. The savings of the farmer, and he cannot make any thing only by the most assiduous industry, increase the fund that is most wanting, especially in such a country as this, i. e., agricultural capital. The farmers of this country can do nothing, they say, for the want of money. How are they ever to get it, but by the improvement of their farms ? as things have been managed in this country hitherto, there has been a tendency to deterioration.

The radical mistake has been committed, of supposing that the best investment for the farmer is the purchase of more land, whereas, in most instances the latter policy would have been the better cultivation of that which he already had. The plan has been to exhaust the soil of one field, and then turn to another. Such a plan can result in nothing but ruin. Nothing has been more neglected in this country than agriculture. The soil of the United States is capable of sustaining two hundred millions of inhabitants better than it sustains seventeen.

Eighty years ago the population of England and Wales was only six millions, and a most miserable living did they get, black bread, barley cakes, and oatmeal porridge, were the main food of the rural population. Since that time, the population has more than doubled, and, in ordinary times, fare better than half the number did then. Their annual agricultural productions have increased more than two hundred millions of dollars, and yet the productive powers of the whole island are scarcely as great as those of the single state of Illinois.

But agriculture, to flourish, must have a market for its surplus productions. And what is a market? Does that magic word reside in any place? most people seem to think so. A market is every where. It is people, not a place—people not engaged in agriculture, but employed in the production of something which supplies human wants. And the nearer it is found to the farmer's door the better, the less of his productions are spent in getting them to market. Agriculture can flourish, then, only where there is a large population engaged in manufactures and commerce.

The second source of national wealth is manufacturing industry. No nation ever became wealthy by raising the raw material, and then exchanging it for the manufactured article. The manufacturing people always have the advantage. They may work day and night, summer and winter, in fair and stormy weather. An agricultural population work only in the day time, when the earth is free from frosts, and when the clouds are not disburdening themselves upon the earth. A manufacturing population can avail themselves to any extent, of the aid of machinery. The fall of water in the town of Lowell is made to do the work of a million of human beings. Every thing the farmer raises must be brought out of the earth by main force, by hard work. The farmer's productions are bulky, and are often almost consumed in getting them to market. The manufactured article is usually comparatively light in proportion to its value.

The farmer, moreover, is obliged to take the chances of unpropitious seasons, and occasionally a short crop. But no variation of the seasons has ever been known to produce a short crop of boots and shoes, and drought has never been so great as to blight the labors of the loom. With these advantages, a manufacturing people will always continue to keep an agricultural people in debt. Towns and cities will spring up among them, and the very fact of a condensed population gives them great advantages. An exclusively agricultural people, in the present age of the world, will always be poor. They want a home market. They want cities and towns, they want a diversity of employment. They want that enterprise and activity, which is engendered merely by bringing masses of people to act upon each other by mutual stimulation and excitement. Why is the balance of trade continually in favor of the North? Because our labor is not sufficiently diversified, because the raw material goes from this very city to the North to be manufactured, and then comes back to be worn by our citizens, while we have among us thousands and thousands who might work it up, but who are lying here idle, and many of them supported by public charity!

One of the postulates of national wealth, is education, universally diffused. It is this alone that can give skill to the hand, and wisdom in the general conduct of affairs. Without that, the physical power of a nation is like the strength of the sightless Cyclops, working in the dark. Physical strength is generally available in proportion to the intelligence by which it is guided. Most of our readers have heard of the Lowell Offering, a periodical written exclusively by the girls, who are engaged every day in carding, spinning, and weaving. Mr. Dickens tells us that he carried home a number of that work, as one of the most wonderful phenomena of the western world. I was told myself, at that place, by one of the superintendants, that the principal writers in that publication were the most profitable operators in the several establishments, obtained the highest wages, and made the best use of their money. So, after all the sneers which are cast on literary ladies,—to them blue stockings is no disqualification for the most common employments of life. So it is, all the world over. The schoolmaster's wages is an investment which yields, in an economical point of view, the highest per centum.

It is to enlightened education that we must look for the extinction of that false sentiment, so adverse to the true prosperity of a nation—the degradation which sometimes attaches to personal toil. No community can ever grow rich, when it is thought to be more respectable to be a genteel loafer, than to get an honest living by the labor of the hands.

No nation can be prosperous and rich without a good government. And what is a good government? It is one which protects, instead of making war upon property. It is one which hallows the marriage between capital and labor, two things, which God's providence has joined together, and nothing but human folly will ever put asunder,—a union from which proceeds the fair family of industry, wealth, contentment, harmony, peace. Once divide them, and the whole structure of society is broken up.—REV. MR. BURNAP, *of Baltimore.*

VALUE OF THE COCOA-NUT TREE.

The cocoa-nut, one of the most profitable fruits that the earth produces, is turned to no account whatever by the Jamaicans, though it grows as luxuriantly there as in any quarter of the globe. I was told by a gentleman who had a large number of these trees growing, that he would esteem it the best property on his estate, if he could get one dollar a hundred for the nuts, but that there was but a very limited market for them at any price. And yet there is no part of this fruit that is not valuable. It thrives in a sandy soil, and bears in Jamaica within three or four years after it is planted. From its flowers the finest arrack in the world may be distilled, and the best of vinegar. A coarse brown sugar may also be prepared from the flower. The green fruit yields a nutritious and delightful drink, and a more substantial food in the pulp which contains the liquid. When ripe, the fruit is popular as an article of diet in all parts of the world. From that fruit a pure oil may be extracted, which may be manufactured into candles, soap, and used in a variety of other ways, in which vegetable oils are available, while the refuse, or oil cake, as it is called, is a most excellent food for cattle.

A medicinal oil is extracted from the bark, which is used, I understand, in Ceylon as an efficacious remedy in cutaneous diseases; the root is also used for medicinal purposes; its elastic fibres are sometimes woven into strainers for liquids, while the timber may be used in building, or converted into beautiful articles of furniture. The husk consists of a tough fibre, from which cordage and rigging of the best quality may be manufactured, and which furnishes the finest stuffing for mattresses that is used, not excepting hair. I saw some of this fibre manufactured at the Penitentiary in Kingston, for mattress stuffing. I satisfied myself that if its value was known in America, it would bring a higher price than any commodity now in use for bedding. The specimens that I saw were manufactured by the convicts, at a cost, I was told, of six cents a pound. Hair costs with us, I believe, about twenty-five cents. The process of manufacturing it is very simple—the husk shells are soaked till perfectly soft, and then are pounded out until the fibres are all separated. This was done in the prison by hand labor, and without the use of machinery, and yet the article could be produced by them for six cents a pound. By the aid of a very simple machine, something, for instance, like that to which rags in a paper mill are first subjected, it is very apparent that the cost of manufacturing it might be reduced at least one half. When I asked why machinery was not employed in this department of the prison, I was told that they had not work enough to occupy the convicts if machinery was employed. Of course I had nothing to say to a reason so conclusive as that.

The supply of these husks would be almost inexhaustible. They have no more use or value than walnut shells have with us, and may be had by the ship-load for the mere expense of cartage. A cargo of a thousand tons could be manufactured for a thousand dollars, and be worth in the port of New-York not less than $4,000, as soon as the usefulness of the article became generally known.—SELECTED.

THE FAMILY MEETING.

We are all here!
Father, mother,
Sister, brother—
All who hold each other dear.
Each chair is filled, we 're all *at home;*
To-night let no cold stranger come;
It is not often thus around
Our old familiar hearth we 're found.
Bless then the meeting and the spot,
For once be every care forgot;
Let gentle peace assert her power,
And kind affection rule the hour—
We 're all—all here.

We 're *not* all here!
Some are away—the dead ones dear,
Who thronged with us this ancient hearth,
And gave the hour of guiltless mirth.
Fate, with a stern, relentless hand,
Looked in and thinned our little band.
Some like a night-flash passed away,
And some sank lingering day by day.

The quiet graveyard—some lie there,
And cruel ocean has his share—
We're *not* all here.

We *are* all here!
Even they—the dead—though dead, so dear.
Fond memory, to her duty true,
Brings back their faded forms to view.
How life-like, through the mist of years,
Each well-remembered face appears;
We see them as in times long past,
From each to each kind looks are cast;
We hear their words, their smiles behold,
They're round us as they were of old—
We *are* all here.

We are all here!
Father, mother,
Sister, brother—
You that I love with love so dear.
This may not long of us be said:
Soon must we join the gathered dead,
And by the hearth we now sit round,
Some other circle will be found.
Oh, then, that wisdom may we know,
That yields a life of peace below;
So, in the world to follow this,
May each repeat in words of bliss,
We're all—all *here*.

CHARLES SPRAGUE.

THE OLD ARM CHAIR.

I love it, I love it; and who shall dare
To chide me for loving that old arm-chair?
I have treasured it long as a sainted prize,
I've bedewed it with tears, and embalmed it with sighs;
'Tis bound by a thousand bands to my heart:
Not a tie will break, not a link will start.
Would ye learn the spell? a mother sat there,
And a sacred thing is that old arm-chair.

4

In childhood's hour I lingered near
The hallowed seat with listening ear;
And gentle words that mother would give,
To fit me to die and teach me to live.
She told me shame would never betide,
With truth for my creed and God for my guide;
She taught me to lisp my earliest prayer,
As I knelt beside that old arm-chair.

I sat and watched her many a day,
When her eye grew dim, and her locks were gray;
And I almost worshipped her when she smiled
And turned from her Bible to bless her child.
Years rolled on, but the last one sped—
My idol was shattered, my earth-star fled;
I learned how much the heart can bear,
When I saw her die in that old arm-chair.

'Tis past! 'tis past! but I gaze on it now
With quivering breath and throbbing brow:
'Twas there she nursed me, 'twas there she died;
And memory flows with a lava tide.
Say it was folly, and deem me weak,
While the scalding drops start down my cheek,
But I love it, I love it, and cannot tear
My soul from a mother's old arm-chair.

ELIZA COOK.

THE FOOD OF VEGETABLES.

What is to be understood by the food of plants?

All those substances which are derived from the soil and the atmosphere by plants, and are assimilated or changed into the several component parts of the vegetable structure, are called food. These substances are the elements or simple bodies, of which all vegetables are composed.

What is the propriety of calling them food?

Because they subserve the same end in the vegetable kingdom that food does in the animal kingdom. Food is something that is used for nourishment and the support of life; and, this is as much needed by plants as by animals

McCORMICK'S REAPING MACHINE. A good Reaping Machine, by large farmers, is of the first importance; several have been constructed possessing different degrees of merit. HUSSEY'S is, from common fame, of great excellence; still, from the favor, particularly at the late World's Fair, bestowed on McCORMICK'S, popular feeling now seems to verge towards the latter. It is drawn by two horses, to be relieved in due time by two others, and the four are thus to work alternately through the day. A boy, of sixteen years, can drive them, and a man is required to rake the grain from the Machine into parcels on the ground, as it passes on, of a suitable size for sheaves. It will take six or seven hands to bind the sheaves and put them into shocks, as fast as they can be made ready. It is also affirmed, that in every acre of land, a bushel of wheat that would be lost from being trodden down, or shelled out by the use of the cradle, is saved by this Machine, which is equal to about three-quarters of the cost of operating it.

We commonly say that the latter eat their food; and that the former assimilate theirs. This is the only difference. The food of plants is sometimes called pabulum. So is the food of animals.

In what form should the food of plants be furnished?

It is not very common to speak of cooking the food of plants; but it should undergo a process that is similar to cooking; it should be reduced to a state that renders it susceptible of being taken up by the organs through which it is to be transmitted, and of being digested or assimilated to the different materials found in the full grown plant.

How can the necessity for this preparation of vegetable food be further explained?

Animal food must be divided into small parcels so as to be admitted into the mouth; then it must be masticated and converted into a kind of pulp, before being conveyed to the stomach. Otherwise the gastric juice cannot act upon the different particles, a process necessary to cause the support and growth of the animal. It is so with vegetable food. In the one case food might as well be forced into the stomach in half pound lumps, as to place in the ground for their nourishment the food of vegetables in a compact mass. The latter must be placed, not only within the reach of the spongelets of the plant, which are to it the same as the mouth is to the animal; but it must be made fine and pulpish, and hence easily received and converted into vegetable matter.

Into what vegetable matter is vegetable food converted?

Plants consist chiefly of woody fibre, starch, and gluten; and, of course these are the substances for the production of which the food or nourishment of vegetables is mainly designed, and into which it is converted or assimilated.

What description is given of woody fibre?

A woody fibre is defined to be a filament or slender thread in plants; or the slender root of a plant; and is the substance which forms the principal portion of trees, shrubs, the shells of nuts, bark, hay, flax, hemp, cotton; the straw of wheat, rye, oats, and barley; the vines of peas and beans; the stems of flowers, and all analagous vegetable productions.

What description is given of starch?

It is that portion of the flour of wheat and other grain, sometimes called fecula, which, when the flour is mixed with water, settles to the bottom. It is then, on being dried, a white powder, and in weight about half the flour from which it was taken. It is also obtained from potatoes, of which it forms nearly the entire substance.

What is said of the gluten of vegetables?

It is the viscid or ropy and tenacious substance found in most plants, and particularly in the flour of wheat and other grain. It contributes much to the nutritive quality of flour, and may be obtained in a separate state, by washing the flour wrapped in a coarse cloth, placed under a stream of water so as to carry off the starch and soluble matter.

In what proportions are the different kinds of vegetable food required in the production of woody fibre, starch, sugar, humic acid, and gluten?

Woody fibre consists of equal portions of carbon and water; dry starch consists of four-fifths of carbon and one-fifth of water; sugar consists of about three-sevenths of carbon and four-sevenths of water; and humic acid consists of four-sevenths of carbon and three-sevenths of water; and as water is composed of oxygen and hydrogen, these vegetable substances are composed of carbon, oxygen and hydrogen. But in the composition of gluten may be found carbon, oxygen, hydrogen, and nitrogen.

What are the constituents of air and water?

Common air, or the air we breathe, consists of one-fifth oxygen or thereabouts and four-fifths nitrogen. And water consists of oxygen and hydrogen in the proportion of four parts of the former and one of the latter.

Through what medium are these different kinds of food conveyed to the plants they are to nourish?

The carbon, oxygen, and hydrogen are in part absorbed by the leaves of the plant from the air, and in part by the roots from the soil; but the nitrogen found in the gluten is supposed to be obtained directly from the soil by the roots of the plant.

How is it that leaves are able to obtain so much of their food from the air?

Leaves are furnished with vessels adapted to the office of this absorption. They are not visible to the naked eye, but are no less real. When we consider the countless number of leaves on a tree, and that each one has two surfaces exposed to the air, it is not surprising that they are able to take in an adequate amount of nourishment.

What are some of the most incomprehensible facts in the science of vegetable physiology?

We could not believe it did we not know it, that the brilliant diamond and the black charcoal are of the same nature; that water will extinguish fire, when the substances of which it is composed, oxygen and hydrogen, are both combustible; that the only ingredients of pure white starch are water and charcoal; and that gum and sugar are composed of the same substances that enter into the composition of starch and woody fibre.

To what is the size of a plant owing?

It has been affirmed that the size is in proportion to the surface of the organs destined to convey food to it. When the food is more abundant than the existing organs require, the superfluous nutriment is employed in the formation of new organs; so that at the side of a cell, a twig or a leaf, arises another.

To what is the amount of food from the air proportioned?

The nutriment received by plants from the air is evidently in proportion to the extent of the surface of the leaves; and new developments correspond with this amount. When new products are no longer employed, the nutriment they imbibe goes to the formation of woody fibre and other solid parts.

What are the particular products of the food received through the agency of the leaves?

After contributing to the woody fibre and other solid parts of the plant the nourishment received in this way from the air goes to the production of sugar, starch, and the acids; and also to the production of the blossoms and the maturing of the fruit.

What is said of the amount of leaves on the globe?

It has been estimated, that the superficies of leaves and other green parts of plants which absorb carbonic acid gas, are more than double the whole surface of the globe, yet are furnished adequately with all the carbon necessary for the support and growth of the vegetable creation. Such estimates must indeed be quite indefinite and uncertain: nevertheless they serve to impress the mind with something like an approximation to the reality.

What calculation has been made of the entire quantity of carbon in reserve for vegetable production?

The quantity is evidently sufficient for the support of the whole vegetable world. It is easily ascertained, that a column of air of about 2400 lbs. rests on every square foot of the earth's surface, the one thousandth part being carbonic acid, 27 per cent. of which is carbon. The whole atmosphere, then, contains 3,306 billion lbs. of carbon, a weight more than equal to all the plants and mineral surface of the earth.

What is said of ammonia, as food for vegetables?

Every part of vegetables contains ammonia; the *root*, as beet—the tree *stem*, as maple—and in all blossoms and unripe fruit. It forms the red and blue coloring matter of flowers. Ammonia is a gas, one part being nitrogen and three parts hydrogen. It has a peculiarly pungent smell, and is the same article known under the name of hartshorn used for smelling bottles. It forms the gas that rises from the vaults of privies, heaps of manure, and fermented urine.

What curious fact exists in relation to the extraction of nourishment from the food of vegetables by the different organs?

Each organ extracts from its food that which is necessary for its own sustenance, and other parts, not assimilated, are separated, as excrements. This coming in contact with another organ in its circulation, affords nutriment to it, and so on with a third, and a fourth; and, when incapable of further transformation, it is separated from the system by appropriate organs. Each part or organ, therefore, is fitted for special functions and one may receive very different substances from another.

To what is this analagous in the animal economy?

Man may receive carbonic acid into the stomach with impunity, and even with advantage; but, to receive it into the lungs, might, as it often does, produce death. So also with other transformations in the animal economy; the kidneys, for example, separate from the body substances containing a large proportion of nitrogen, the liver those with an excess of carbon, and the lungs principally those composed of oxygen and hydrogen. Volatile oils and alcohol, which are incapable of being assimilated, are exhaled through the lungs. Superabundant nitrogen is excreted as a liquid excrement from the body and passes through the urinary ducts; all gaseous matter passing through the lungs and all incapable of further transformation, through the intestinal canal.

What particular cases of this in the vegetable function are among the most prominent?

Transformations of the compounds of plants are constantly taking place during their life; and, as a consequence, gaseous substances are eliminated by the leaves and blossoms. Solid excrement is deposited in the bark. Soluble substances containing carbon, are excreted by the roots and are absorbed by the soil, where they decay or putrify and become nutriment, as humus, for another generation of plants.

Generally, what food in the soil is necessary for the growth of plants?

Such substances as contain carbon or nitrogen, or which are capable of yielding these two elements for its organization; and, also water, or its two elements, oxygen and hydrogen; and finally a soil is required which will furnish the plant with the metallic oxides, or inorganic bases.

How are the different kinds of food to be adapted to the different kinds of plants?

Each genus of plants requires special conditions for their life; and individuals require many conditions. They cannot be brought to maturity, even if but one of these be wanting. Their organs, like those of animals, contain substances of very different kinds; and in all are found metallic salts For the production of all their organs, therefore, the soil must contain all their elements. These may be united in one substance, or they may exist in several.

What reciprocation, in the way of food, is there between vegetables and animals?

The life of plants is evidently connected very closely with that of animals. Vegetation may exist without animal life, but the existence of animals depends on the life and growth of plants. These, therefore, afford both nutriment for the existence and growth of animals, and the essential gaseous element, oxygen, for their respiration. Animals expire carbon and plants inspire it; plants expire oxygen and animals inspire it. Thus, by this wonderful economy of nature, both are enabled to exist, and the due composition of the air is uniformly maintained.

What besides food, in its more obvious signification, has an influence on the growth of plants?

They have habits strikingly illustrative of the harmony of nature. We see them adapted to the peculiarities of their situations. If indigenous to the tropical climate, they cannot live in our temperate zone without the aid of art; if inhabitants of the valley, they cannot dwell on the mountain summit; nor, if the rugged tenants of the bleak and frosty mountain, can they endure the enervating dalliance of the luxurious vale; nor can either dwell with the aquatic plant immersed in a liquid element.

How do plants become objects of peculiar interest when we administer to their growth?

They almost seem emulous to administer to our pleasure as a reward for what we do for them. We call them inanimate; yet, it is no great effort of the imagination to endow them with attributes that will respond to every kind administration to their wants. How are their numberless shoots and branches urging into life and action millions of buds that are expanded into light and being by the genial sun, rivalling one another in their efforts to produce the fairest flower and the choicest fruit!

THE PRACTICAL USE OF LEAVES.

There are two facts in the function of the leaf which are worth consideration on account of their practical bearings. The food of plants is, for the most part, taken in solution through the roots. Various minerals—silex, lime, alumina, magnesia, potash,—are passed into the tree in a dissolved state. The sap passes to the leaf, the superfluous water is given off, but not the substances which are held in solution. These, in part, are distributed through the plant, and, in part, remain a deposit in the cells of the leaf. Gradually the leaf chokes up, its functions are impeded, and finally entirely stopped. When the leaf drops it contains a large per cent. of mineral matter.

An autumnal or old leaf yields, upon analysis, a very much larger proportion of earthy matter than a vernal leaf, which, being yet young, has not received within its cells any considerable deposit. It will be found, also, that the leaves contain a very much higher per cent of mineral matter than the wood of the trunk. The dried leaves of the elm contain eleven per cent. of ashes, (earthy matter,) while the wood contains less than two per cent.; the leaves of the willow eighteen times as much as the wood; the leaves of beech an excess over the wood a small fraction less; the leaves of the European oak nineteen times as much as the wood; and those of the pitch pine twelve times as much as the wood.

It is very plain from these facts, that, in forests, the mineral ingredients of the soil perform a sort of circulation; entering the root, they are deposited in the leaf; then, with its fall to the earth, and by its decay, they are restored to the soil, again to travel their circuit. Forest soils, therefore, instead of being impoverished by the growth of trees, receive back annually the greatest proportion of those mineral elements necessary to the tree, and, besides, much organised matter received into the plant from the atmosphere; soils therefore are gaining instead of losing. If the owners of parks or groves, for the sake of neatness, or to obtain leaves for other purposes, gather the autumnal harvest of leaves, they will in time take away great quantities of mineral matter, by which the soil ultimately will be impoverished, unless it is restored by manures.

Leaf manure has always been held in high estimation by gardeners. But many regard it as a purely vegetable substance; whereas, it is the best mineral manure that can be applied to the soil. What are called vegetable loams, (not peat soils, made up principally of decomposed roots,) contain large quantities of earthy matter, being mineral-vegetable rather than vegetable soils. Every gardener should know that the best manure for any plant is the decomposed leaves and substances of its own species. This fact will suggest the proper course with reference to the leaves, tops, vines, haulm, and other vegetable of the garden.

The other fact connected with the leaf, is its function of exhalation. The great proportion of crude sap which ascends the trunk, upon reaching the leaf is given forth again to the atmosphere by means of a particularly beautiful economy. The quantity of moisture produced by a plant is hardly dreamed of by those who have not specially informed themselves. The experiments of Hales have been often quoted. A sunflower, three and a half feet high, presenting a surface of 5,616 square inches exposed to the sun, was found to perspire at the rate of twenty to thirty ounces avordupois every twelve hours, or seven times more than a man. A vine, with twelve square feet, exhaled at the rate of five or six ounces a day. A seedling apple tree, with twelve square feet of foliage, lost nine ounces a day.

These are experiments upon very small plants. The vast amount of surface presented by a large tree must give off immense quantities of moisture. The practical bearings of this fact of vegetable exhalation are not a few. Wet forest lands, by being cleared of timber, become dry, and streams fed from such sources become almost extinct as civilization approaches on wild woods. The excessive dampness of crowded gardens is not singular, and still less is it strange that dwellings covered with vines, whose windows are choked with shrubs, and whose roof is overhung with branches of trees, should be intolerably damp, and when the good house wife is scrubbing and scouring, and nevertheless, marvelling that her house is so infested with mould, she hardly suspects that her troubles would be more easily removed by the axe or saw than by all her cloths and brushes.

A house should never be surrounded closely with shrubs. A free circulation of air should be maintained all about it, and shade trees so disposed as to leave large openings for the light and sun to enter. The unusual rains that, some seasons produce great dampness in our residences, cannot but be noticed by all, both on account of the effect on the health of the occupants and upon the beauty and good condition of their household substance. Such facts should always be kept in mind, when locating houses, and when planting trees and shrubs about them.—Rev. Henry Ward Beecher.

THE DELIGHTS OF AUTUMN.

With what a glory comes and goes the year !
The buds of Spring, those beautiful harbingers
Of sunny skies and cloudless times, enjoy
Life's newness, and earth's garniture spread out;
And when the silver habit of the clouds
Comes down upon the autumn sun, and with
A sober gladness the old year takes up

His bright inheritance of golden fruits,
A pomp and pageant fill the splendid scene.

There is a beautiful spirit breathing now
Its mellow richness on the clustered trees,
And, from a beaker full of richest dyes,
Pouring new glories on the Autumn woods,
And dipping in warm light the pillared clouds.
Morn on the mountains, like a summer bird,
Lifts up her purple wing, and in the vales
The gentle wind, a sweet and passionate wooer,
Kisses the blushing leaf, and stirs up life
Within the solemn woods of ash deep-crimsoned,
And silver beech, and maple yellow-leaved,
Where Autumn, like a faint old man, sits down
By the way-side a-weary. Through the trees
The golden robin moves. The purple finch
That on wild cherry and red cedar feeds,
A Winter bird, comes with its plaintive whistle,
And pecks by the witch-hazel, whilst aloud
From cottage roofs the warbling blue bird sings,
And merrily, with oft-repeated stroke,
Sounds from the threshing-floor the busy flail.

Oh what a glory doth this world put on
For him who, with a fervent heart, goes forth
Under the bright and glorious sky, and looks
On duties well performed, and days well spent!
For him the wind, ay, and the yellow leaves
Shall have a voice, and give him eloquent teachings.
He shall so hear that solemn hymn, that Death
Has lifted up for all, that he shall go
To his long resting-place without a tear.

LONGFELLOW

THE RAINY DAY.

The day is cold, and dark, and dreary;
It rains, and the wind is never weary;
The vine still clings to the mouldering wall,
But at every gust the dead leaves fall,
And the day is dark and dreary.

My life is cold, and dark, and dreary;
It rains, and the wind is never weary;
My thoughts still cling to the mouldering Past,
But the hopes of growth fall thick in the blast
And the days are dark and dreary.

Be still, sad heart! and cease repining;
Behind the clouds the sun is shining;
Thy fate is the common fate of all,
In each life some rain must fall
Some days must be dark and dreary.

LONGFELLOW.

A DEFENCE OF AGRICULTURE.

When harvests are exuberant, joy and health follow in their train; but let delusive prosperity draw industry from agriculture; let an insiduous disease attack one of its important products; let an insect, or a parasite, fasten on a single esculent, and mark the effect upon commerce and human life. Upon such an event all business is deranged; the commercial marine of the world proves itself unequal to the crisis; sloops of war and frigates become carriers of grain; warehouses, canals, railroads, and ports, prove insufficient for the exigency; masses of specie flow from the guarded treasuries of the old world to the rude cabins of the prairies; manufactures and public improvements stop in their course, famine and pestilence invade provinces and states; and the pale survivors, reckless of those ties which bind man to his birth-place, brave storms and shipwreck, sickness and death, on the route to new and untried regions.

Agriculture has become essential to life. The forest, the lake, and the ocean, cannot sustain the increasing family of man. Population declines with a declining cultivation, and nations have ceased to be with the extinction of their agriculture. In ancient times, agriculture was esteemed and honored. In classic Greece and Rome it was the theme of the popular poets of the age, and was not deemed unworthy of distinguished warriors and statesmen. We read of Cicero at his Tusculan villa, of Cato at his farm, of Cincinnatus leaving his plough to command the armies of the republic;

while the great naturalist, Pliny, in his beautiful letters, prides himself on his vineyards.

The overflow of the Nile, the fertilizer of Egypt, has been celebrated for centuries as the great festival of the country; and in that "central flowery land," which claims such remote antiquity, the sovereign of three hundred millions, "the son of heaven, whose person is too sacred to be seen, whose imperial despatch is received amid burning incense and prostration, and in whose presence no one dares speak but in a whisper," annually exhibits himself to his subjects, holding a plough in honor of agriculture.

In England, too, whose nobles shrink from all connection with trade, agriculture is highly honored. Earls, dukes, and princes, preside at agricultural festivals, compete for prizes, and do not disdain to write treatises on the culture of roots, the rotation of crops, and manufacture of composts. Sir Robert Peel, the great statesman of the age, is one day bearing down by his eloquence the opposition of Parliament to his vigorous and enlightened policy, and another discussing the prospects of agriculture among the farmers of Tamworth.

It is, too, with mingled pleasure and pride that we recur to the fact, that the hero and statesman, who led the armies of our Revolution, was himself a practical farmer. Amid all the excitement, harrassing duties, and embarrassments of a protracted war, he directed by letters the operations of his farm, and finally retired from the highest position to which talent and patriotism could aspire, followed by the love of his countrymen, to devote to agriculture the close of his life; and it is a little remarkable his example has been followed by nearly all who have succeeded to the office of President.

Our fathers did not enjoy, as farmers, the privileges which we possess. The country, emerging from a long war, was deficient in capital. Implements and buildings were rude and defective; a few small seaports and fishing towns formed their principal markets; and access to these was by no means easy; for the bridle-path blazed through the forest, the ford and the ferry, were but a poor substitute for the country road, the turnpike, canal, and railroad.

At a period, too, when the wars of Europe made us carriers of the world, it is not astonishing that talent and enterprise should have been drawn from the secluded home of the farmer to the perilous "march upon the deep;" to the uncertain pursuits of trade, or to the sharp competition of professional life; growing with the growth of commerce, or be tempted to exchange the rudeness of the country for the enervating refinements of the city. Besides, temperance, taste, and progressive art, education and the weekly press, had not yet gilded the home of the farmer ; judicious enterprise had not yet drawn the daughters of New England from the distaff to the water-falls, and enlivened the adjacent districts by the creation of valuable markets.

Contrast Massachusetts to-day (1847,) with Massachusetts half a century since. Counties checkered with factory villages, tied together by a fast spreading net-work of railroads, sparkling with school-houses, churches, and tasteful residences, and improving farms, and peopled by an intelligent and energetic race—compare these with all that preceded them, and we shall find much to cheer us in the contrast, without detracting in any degree from the courage and patriotism of our progenitors. If in addition to the progress of the country, we take into account the vast increase of wealth, the advance in the mechanic arts, the discoveries of chemistry, shall we not arrive at the conclusion that agriculture now presents a new aspect, assumes a new importance, and offers new attractions to all who engage in it?

In estimating the importance of science and capital to agriculture, we learn, from the lessons of experience, that a fertile soil alone does not carry agriculture to perfection. Should we seek the spots where agriculture gives the largest and most remunerative returns for a given space, we should find them not on the fertile banks of the Nile or the Ganges, the rich plains and valleys of Sicily, or the prairies of the West, where a virgin soil and low prices attract so many youthful cultivators. Far otherwise. You must look to Flanders and Holland. There, science and capital combined, in a harsh climate, have rescued vast wastes from the ocean, and converted sterile marshes and barren sands into productive fields, the very garden of Europe ; or look at England, our parent land, where the same powerful combination has transformed the sandy plains of Norfolk, for centuries abandoned to the rabbit, into luxuriant fields of wheat, clover, and turnips ; and changed the fens of Lincolnshire, which encircle the old town of Boston—fens, for centuries, the resort of wild ducks, geese, and other birds of passage, into the granary of England.

The soil of Belgium was originally sand and clay alone. It has been enriched by ashes and composts, until it has become rich, black, loamy mould. Tanks are provided on the farms for liquids, and each cow is estimated to produce ten tons of solid, and twelve of liquid manures. Every expedient is resorted to, both to increase their quantity, and to improve their quality. Rotations of crops are followed; and the result of these efforts is, that Belgium sustains a population of 350 people, 67 cattle, and 17 horses, to the square mile; usually raises her own bread stuffs, and exports wheat, madder, flax, wool, and bark, to other parts of Europe. In Holland, where the dike, steam engine, and wind mill are employed to prevent the incursion of the sea upon the land gained from its bosom, a population of 214 to the mile is sustained, and large exports are made of butter, cheese, and other agricultural products. The average value of land is nearly $300 per acre, although it is burdened with oppressive taxes.

Does the farmer aim at a life useful and beneficial to his race?—let him remember that every acre that he reclaims, every blade of grass that he bids to grow where none grew before, ameliorates the condition of his fellows. Does he aspire to wealth?—let him reflect that his gains, if less brilliant and striking than those of trade and the professions, are more certain and uniform; and that gradual improvement of his estate, and the silent but continued rise of property, promise eventual prosperity. Is he tasteful?—he will here find a theatre for taste in woods, orchards, and flowers, and the design of his buildings. Is he ambitious?—here are obstacles to be surmounted, objects to be controlled, races to be improved, a kingdom in miniature to be governed by wise and wholesome regulations.

Or would the farmer make conquests and achieve victories?—here weeds and water are his enemies: here uncultivated plains are his Mexico, and deep fens and morasses his Texas and California; and no philanthropist or casuist, will complain of his conquests, should he subdue them. Let him guard against the ambush of the crow, the wire-worm, the squirrel, and the fox; and repel the invasion of the blight, the white weed, and the sorrel. He shall see his battle-fields not stained with blood, but blossoming with clover; and when, in his green old age, he points out to his children his

Palo Alto, Buena Vista, Cerro Cordo, and Cherubusco, and recounts his bloodless achievements, he shall feel greater satisfaction than if his victories had been saddened by the sacrifices and tears of thousands !—*From the address before the Middlesex County Agricultural Society, Massachusetts,* 1847, *by* E. H. Derby, Esq.

THE THEORY OF MANURES.

What does the agriculturist understand by manures?

Manures consist of the remains of organized bodies of every description, whether animal or vegetable, in a state of decomposition; that is to say, resolving themselves into those principles and elements which can re-enter into the vegetable system. Hence, they may be composed of animal or vegetable substances; or they may consist of mineral matter; or they may be derived partly from mineral and partly from animal and vegetable substances.

What is the design of manures?

It is to furnish the soil with those elementary substances that enter into the formation of vegetable structures; for if the soil do not contain an adequate quantity of these substances, the farmer would obtain only a feeble crop. Without this supply a vigorous vegetation cannot be produced. This is seemingly self-evident. In human art, if a machine is to be formed there must be a supply of the materials of which it is to be constructed. A man might be considered insane were he to talk of constructing a fence, or a plough, or a wagon, without the wood and the iron needed for them. And not less so in talking of raising plants without the requisite elements for them.

Whence arises the necessity for this artificial supply of the soil with the substances in manure?

The soil may be compared to a rain-water cistern which is successively filled with the falling liquid collected on the roofs of houses or otherwise, and from which, so long as thus replenished, the water may be daily drawn for the purposes of domestic economy. But if it ceases to rain, or if the fixtures for saving the water be destroyed, the supply will soon be exhausted, and the cistern will be of no daily use for household purposes. It is much so in obtaining plants from the ground. They can no more be produced from a soil

SOUTHERN HOUSE for a large family, copied from WHEELER'S RURAL HOMES.

destitute of the elements which compose them than water can be drawn from an empty cistern.

How does this appear?

The fertilizing substances in manure are mostly the same ones that in previous years had been drawn or taken from the soil in producing a crop of vegetables. The vegetables either decayed or were used for food by animals, from which in the form of excrements they became manures. To be able to produce again a similar crop, the soil must be made by artificial means, or by manuring, to contain a sufficient amount of the same substances. If it does not contain them a new crop of vegetables cannot be raised ; or if obtained, it will be a diminished one.

Is it true that nothing can be produced from the soil without manure?

So long as there is water in the cistern one may draw water from it though it be not replenished by the falling rain. So likewise a soil already supplied in ample quantity with the constituent elements for plants may yield a good crop without manure. The theory is, that to keep his crops from diminution the farmer must put as much into his soil as his crops take from it. And, if he desire to increase his crops, he must put more into it, than they take out of it.

What other illustration of this can be given?

A man who takes money out of his pocket faster than he puts it in will soon have none to be taken out and will be a bankrupt. It is so with the farmer who takes away from his soil more than he returns to it. The woman who is constantly cooking up her provisions without replenishing the larder will soon have nothing remaining to be cooked. So it is with the husbandman who is all the time using up the very stamina of his land, and does nothing to fill the consequent vacuum. And the cloth dyer who is constantly extracting from his dye stuffs all the coloring agents for new fabrics without replenishing his vats, will soon run down in his occupation. So it will be with the agriculturist who allows his fields to become impoverished from not causing them to be invigorated with successive new supplies of fertilizing substances.

Must then the soil be manured in every case before a crop can be obtained?

If the crop is permitted to remain and rot upon the soil;

or if it be ploughed in, as it is occasionally done, this becomes a manure, and none beside is requisite. In this way forest lands for ages retain and perhaps increase their powers for vegetable production by the annual decay of the fallen leaves. The necessity for other manures is only where the crops are removed from the soil.

In what condition should be the substances used for manures before being cast into the ground?

They should be reduced to their original state; not simply reduced to fine lumps or particles as meat is chopped up for sausages; but, be so decomposed that they may be assimilated and combined with all the other vegetable elements in the formation of new vegetable compounds. Till thus reduced, they cannot become fertilizing agents any more than would be a stone or a piece of solid metal. Let a stone, or a piece of iron, or a piece of hard wood be converted into small dust and they will readily mix with the soil and become subject to that chemical action connected with the formation of new compounds.

How are the substances used for manures to be reduced to an elementary condition?

The process employed by Nature in this work is fermentation. By this process the elementary parts of the substance fermented assume new forms of combination and become fitted to supply the matter for nutrition to plants in that form in which it can be received by the pores of the roots. This may be done by first mixing the matters to be fermented with the soil; as the process in this way is very slow, it is usual to collect them, previous to being mixed with the soil, in heaps or masses, where heat will be generated, which in connection with the moisture will hasten the process of fermentation.

What are some other results from fermentation, besides the decomposition of the primary elements?

In addition to the separation of these elements in vegetable substances from fermentation, the whole mass is reduced to fine particles, so as to mix readily with water and thus be speedily conveyed to the roots of the plants. It is mainly by being well mixed with water as it exists in the soil, that the fertilizing properties of manure are conveyed to the growing plant.

To what has the process of fermenting decaying vegetables been compared?

To a slow mouldering fire. It is well known that fire reduces compound bodies to simple ones. For instance the inorganic portions of wood are reduced to ashes, while the organic or gaseous portions of it are at the same time disengaged and disappear. The result is the same as that which comes from the decay of bodies in the process of fermentation.

What are the relative advantages and disadvantages between fermenting manures before being cast into the soil, or at first mixing them with the soil?

If at first mixed with the soil prior to fermentation and the disintegration of its parts, there may be and usually will be much delay before its effects will be fully manifest in the augmentation of the crops; but, then as an offset for the delay there will be less loss from the escape of the fertilizing gases when exposed to metereological influences. On the other hand, the beneficial effects will be sooner felt; but in the end the soil will be less enriched from it. The presumption is, that a field of clover or buckwheat ploughed in and completely buried up till well rotted will enrich the ground far more than it would if it were removed and used for feed or were reduced to putrefaction in a mass, and then cast upon the soil, for in the latter mode of preparing it much of its fertilizing properties would be lost.

Do animal substances, particularly the solid excrements as much require an accelerated process of fermentation as vegetable substances?

They do not. Ordinarily the dung of animals will soon, if at first mixed with or buried up with the soil, combine in the formation of the vegetable structure. But if there is connected with it straw, which is usual in the barn-yard, it should previously be subjected to fermentation when in masses. The urine also especially, which is an important part of animal manure, should, before being applied to the soil, be well fermented.

What injury results to manure, if exposed to the sun and air?

The fertilizing gases are liable to escape and be lost. Hence manures should always be under cover to shield them from the action of the sun's rays, the rain, and the atmosphere. Provided a farm have no shelter for them thus to shield them from the metereological influences, they should be carefully arranged in regular and solid heaps, and then be well covered with a thick coat of mould. This, in a measure, will prevent the escape of the gases.

What amount of water is needed in fermentation?

Only a moderate amount; barely sufficient for a slight moisture, similar to a sweat. If immersed in water the generation of heat is prevented; and, without heat fermentation is impracticable. It is known that wood, buried in water, will continue in an undecayed state for an indefinite length of time. Hence, a barn-yard should be so constructed as to avoid an excess as well as a deficiency of moisture.

How long should vegetable substances, particularly in the barn-yard, in connection with animal manure, be continued in a state of fermentation?

Simply long enough to be made tender, so that one can take up a wisp of hay or straw of the size of his arm, and with his hands easily break it off; or so that a spade will readily pass through it lying in a mass, or that it will offer no adhesive resistance when removing it with a fork. When it has reached this period of decay, it will rather receive injury than benefit from protracted fermentation.

What description of manures is ordinarily preferable?

None for general use are equal to those of the barn-yard; for these are supposed to embrace, and do embrace, if there has been no waste of materials susceptible of conversion into fertilizing agents, every element of vegetable structure that had been removed from the soil the previous year; a far greater number of such elements than can be found in any other fertilizer. For particular crops indeed, other fertilizing agents may be applied with augmented efficacy.

What should be the great aim of every farmer in relation to the subject of manures?

To understand the nature of soils, the physiology of vegetables, the theory of mauures; and, then by every possible means to increase his stock of manure to the greatest extent, letting no substance which can be converted into it be lost, and searching out wherever to be found every hitherto unobserved substance which can be discovered and reached. In this way a farmer will soon renovate his grounds and render them highly productive. In this way he will thrive and become independent.

What has the poor farmer to do who has only scanty means for the creation of manures?

If a farmer is poor, there is so much more need of his having them. Without them, he is not half paid for his labor in tillage, and is constantly declining in his circumstances; with them he becomes thrifty by obtaining good remuneration for his labor. A prudent poor farmer should ordinarily shudder at the idea of running into debt, especially for victuals, clothing, or furniture; but, it is wise in him to do it for manures, if he cannot obtain them otherwise, because they will not only enable him to raise enough more on his farm to pay for them, but he gets doubly paid for his tillage, and greatly improves his farm for future use and profit.

LIGHTNING RODS.

In erecting rods for the protection of buildings from the effects of lightning, a few things must always be kept in mind, in order to ensure efficiency and consequent safety. As to the point of erection, the most exposed and elevated part of the building should be chosen. If a dwelling house, the chimney is the point that will require protection; if there are several of them, the most elevated one, or the one most exposed to the general course of storms—as the westerly one; or if but one has a fire in it, that one will be found most liable to be struck by lightning, and will of course more require protection. Numberless recorded cases prove that all heated currents of vapor, whether rising from a chimney, or the masses of hay and grain in a barn, are excellent conductors of the electric fluid, and are to be considered as such in all arrangements for protection.

The materials to be used as the conducting medium or rod, is another point that should be attended to. Iron is the most generally used for this purpose, but copper is preferable, as it possesses greater conducting powers, is not liable to rust or fusion, and, being tougher, is not broken to fragments by an electric discharge, as iron sometimes is. The greater cheapness of iron, however, will probably continue its use, and when well put up, it gives all reasonable security. No iron rod should be used of a less diameter than three-fourths of an inch, and an inch rod is still better; as it must be remembered that the surface only has any effect in conducting electricity, and, therefore, the larger the surface over which it passes, the less intense its action, and the less danger of breaking or fusion. A small quantity of metal, if of the right kind, and a large surface given to it, will make a better conductor, than a larger quantity in an improper form.

Thus a copper ribbon, two inches wide, and of the proper length, will be superior to a copper wire of the same weight, as the process of rolling and flattening it, gives a much larger surface; and the same remark will be true in regard to iron. A number of small iron or copper wires, twisted into a rod, is better than a solid rod of the same weight, for the same reason; that is, a greater surface is exposed by the small wire than by the solid one, and this would be the best manner of constructing rods, were it not to be apprehended, that heavy discharges passing over wires would fuse and destroy them. Such have been recommended for ship conductors, as they would be entire and flexible, and perhaps it would be found that the distribution of the fluid over such an extent of surface as such a rod would afford, would prevent the danger of fusion in any case.

In preparing the rod, the most essential thing is the making of the elevated points. These should be several in number, slightly diverging from the main rod, and the sharp points gilded or tipped with silver, so as to prevent their rusting, and losing, in a great degree, their conducting power. Perhaps the easiest mode of pointing them, is to make points of large silver wire an inch in length, turning a screw on one end, and insert this into an opening drilled in the tapered end of each branch of the conductor, to receive it. If wire of the proper size is not convenient, they may be made by cutting them from a half-dollar, and hammering them into the right form. If the rod is made in pieces, they should never be put up by turning hooks on the ends, and connecting them in that way; as interruption to the fluid in its descent is frequently attended with bad consequences. The several pieces should be put together with screws, the connecting-piece receiving the ends of two rods, and being as near the size of the conductor as strength and security will admit. The rod should never be secured to the building by metal staples or fastenings; or if such are necessary, the connection between the rod and these should be broken by pieces of glass, which is a non-conductor. Wood is the best for fastenings, and should always be used, except from necessity. It is not enough that the rod attract and receive the discharge; it must also conduct it to the earth, or no adequate security is afforded.

It has been estimated that a rod properly made, affords protection to a distance of five or six times its height above the roof; that is, a rod standing six feet above a building, will protect the building for a distance of thirty feet. Instances have been known, however, in which a chimney having a column of heated vapor rising from it, has been struck, when within the limits usually considered safe, in a protected building. In such cases, the result must be ascribed to the height and conducting power of the vapor. The foot of the conductor should gradually recede from the walls of the building, and enter the earth to such a depth as to reach moist earth, and if the bottom of the rod is pointed, or split and parted different ways in the earth, the passing off of the fluid will be facilitated. No paint should ever be allowed on a conductor.—ALBANY CULTIVATOR.

MOUNTAIN SCENERY.

Mountains! how one's heart leaps up at the very word! There is a charm connected with mountains so powerful, that the merest mention of them, the merest sketch of their magnificent features, kindles the imagination and carries the spirit at once into the bosom of their enchanted regions. How the mind is filled with their vast solitude! how the inward eye is fixed on their silent, their sublime, their everlasting peaks! How our heart bounds to the music of their solitary cries—to the tinkle of their gushing rills, to the sound of their cataracts. How inspiriting are the odors that breathe from the upland turf, from the rock-hung flower, from the hoary and solemn pine; how beautiful are those lights and shadows thrown abroad, and that fine, transparent haze which is diffused over the valleys and lower slopes as over a vast, inimitable picture.

Whoever has not ascended mountains, knows little of the beauties of nature. Whoever has not climbed their long and healthy ascents, and seen the trembling mountain-flowers, the glowing moss, the richly-tinted lichens at his feet; and scented the fresh aroma of the uncultivated sod, and of the spicy shrubs; and heard the bleat of the flock across their solitary expanses, and the wild cry of the mountain plover, the raven, or the eagle ; and seen the rich and russet hues of distant slopes and eminences, the livid gashes of ravines and precipices, the white glittering line of falling waters, and the cloud tumultuously whirling round the lofty summit; and then stood panting on that summit, and beheld the clouds alternately gather and break over a thousand giant peaks and ridges of every varied hue,—but all silent as images of eternity ; and cast his gaze over lakes and forests, and smoking towns, and wide lands to the very ocean, in all their gleaming and reposing beauty, knows nothing of the treasures of pictorial wealth in rural scenes.

We delight to think of the people of mountainous regions; we please our imaginations with their picturesque and quiet abodes; with their peaceful secluded lives, striking and unvarying costumes, and primitive manners. We involuntarily give to the mountaineer heroic and elevated qualities. He lives amongst noble objects, and must imbibe some of their nobility ; he lives amongst the elements of poetry, and must be poetical ; he lives where his fellow-beings are far, far separated from their kind, and surrounded by the sternness and the perils of savage nature ; his social affections must therefore be proportionally concentrated, his home-ties lively and strong ; but, more than all, he lives within the barriers, the strong-holds, the very last refuge which Nature herself has reared to preserve alive liberty in the earth, to preserve to man his highest hopes, his noblest emotions, his dearest treasures, his faith, his freedom, his hearth, and home.

How glorious do those mountain ridges appear when we look upon them as the unconquerable abodes of free hearts; as the stern, heaven-built walls from which the few, the feeble, the persecuted, the despised, the helpless child, the delicate woman, have from age to age, in their last perils, in all their weaknesses and emergencies, when power and cruelty were ready to swallow them up, looked down, and beheld the million waves of despotism break at their feet :—have seen the rage of murderous armies, and tyrants, the blasting spirit of ambition, fanaticism, and crushing domination, recoil from their bases in despair.

"Thanks be to God for mountains!" is often the exclamation of my heart as I trace the history of the world. From age to age, they have been the last friends of man. In a thousand extremities they have saved him. What great hearts have throbbed in their defiles, from the days of Leonidas to those of Andreas Hofer! What lofty souls, what tender hearts, what poor and persecuted creatures, have they sheltered in their stony bosoms from the weapons and tortures of their fellow men.

"Avenge, O Lord, thy slaughtered saints whose bones
Lie scattered on the Alpine mountains cold!"

was the burning exclamation of Milton's agonized and indignant spirit, as he beheld those sacred bulwarks of freedom for once violated by the disturbing demons of the earth; and the sound of his fiery and lamenting appeal to Heaven will be echoed in every generous soul to the end of time.

Thanks be to God for mountains! The variety which they impart to the glorious bosom of our planet were no small advantage; the beauty which they spread out to our vision in their woods and waters; their crags and slopes, their clouds and atmospheric hues were a splendid gift; the sublimity which they pour into our deepest souls from their majestic aspects; the poetry which breathes from their streams, and dells, and airy heights, from the sweet abodes, the garbs and manners of their inhabitants, the songs and legends which have awoke in them, were a proud heritage to imaginative minds; but what are all these when the thought comes, that without mountains the spirit of man must have bowed to the brutal and the base, and probably have sunk to the monotonous level of the unvaried plain.

When I turn my eyes upon the map of the world, and behold how wonderfully the countries where our faith was nurtured, where our liberties were generated, where our philosophy and literature, the fountains of our intellectual grace and beauty, sprang up, were as distinctly walled out by God's hand with mountain ramparts from the eruptions and interruptions of barbarism, as if at the especial prayer of the early fathers of man's destinies, I am lost in an exulting admiration. Look at the bold barriers of Palestine! see how the infant liberties of Greece were sheltered from the vast tribes of the uncivilized north by the heights of Hæmus and Rhodope! behold how the Alps describe their magnificent crescent, inclining their opposite extremities to the Adriatic and Tyrrhine Seas, locking up Italy from the Gallic and Teutonic hordes till the power and spirit of Rome had reached their maturity, and she had opened the wide forest of Europe to the light, spread far her laws and language, and planted the seeds of many mighty nations!

"Thanks to God for mountains! Their colossal firmness seems almost to break the current of time itself; the geologist in them searches for traces of the earlier world, and it is there too that man, resisting the revolutions of lower regions retains through innumerable years his habits and his rights. While a multitude of changes has remoulded the people of Europe, while languages, and laws, and dynasties, and creeds, have passed over it like shadows over the landscape, the children of the Celt and the Goth, who fled to the mountains a thousand years ago, are found there now, and show us in face and figure, in language and garb, what their fathers were; show us a fine contrast with the modern tribes dwelling below and around them; and show us, moreover, how adverse is the spirit of the mountain to mutability, and that there the fiery heart of Freedom is found for ever.

WILLIAM HOWITT.

SUNRISE ON THE HILLS.

I stood upon the hills, when heaven's wide arch
Was glorious with the sun's returning march,
And woods were brightened, and soft gales
Went forth to kiss the sun-clad vales.
The clouds were far beneath me;—bathed in light,
They gathered mid-way round the wooded height,
And, in their fading glory, shone
Like hosts in battle overthrown,
As many a pinnacle, with shifting glance,
Through the gray mist thrust up its shattered lance
And rocking on the cliff was left
The dark pine blasted, bare, and cleft.
The veil of cloud was lifted, and below
Glowed the rich valley, and the river's flow
Was darkened by the forest's shade,
Or glistened in the white cascade;
Where upward, in the mellow blush of day,
The noisy bittern wheeled his spiral way.

I heard the distant waters dash,
I saw the current whirl and flash,—

And richly, by the blue lake's silver beach,
The woods were bending with a silent reach.
Then o'er the vale, with gentle swell,
The music of the village bell
Came sweetly to the echo-giving hills ;
And the wild horn, whose voice the woodland fills,
Was singing to the merry shout,
That faint and far the glen sent out,
Where, answering to the sudden shot, thin smoke,
Through thick-leaved branches, from the dingle broke.

If thou art worn and hard beset
With sorrows, that thou would'st forget,
If thou wouldst read a lesson, that will keep
Thy heart from fainting and thy soul from sleep,
Go to the woods and hills !—No tears
Dim the sweet look that Nature wears.

LONGFELLOW.

RELIGIOUS INFLUENCES IN THE COUNTRY.

No situation in life is so favorable to established habits of virtue, and to powerful sentiments of devotion, as a residence in the country, and rural occupations. I am not speaking of a condition of peasantry, of which, in this country, we know little, who are mere vassals of an absent lord, or the hired laborers of an intendant, and who are, therefore, interested in nothing but the regular receipt of their daily wages; but I refer to the honorable character of an owner of the soil, whose comforts, whose weight in the community, and whose very existence depend upon his personal labors, and the regular returns of abundance from the soil which he cultivates. No man, one would think, would feel so sensibly his immediate dependence upon God, as the husbandman. For all his peculiar blessings, he is invited to look immediately to the bounty of heaven. No secondary cause stands between him and his maker. To him are essential the regular succession of the seasons, and the timely fall of the rain, the genial warmth of the sun, the sure productiveness of the soil, and the certain operations of those laws of nature, which must appear to him nothing less than the varied exertions of omnipresent energy.

In the country we seem to stand in the midst of the great theatre of God's power, and we feel an unusual proximity to our Creator. His blue and tranquil sky spreads itself over our heads, and we acknowledge the intrusion of no secondary agent in unfolding this vast expanse. Nothing but omnipotence can work up the dark horrors of the tempest, dart the flashes of the lightning, and roll the long-resounding rumor of the thunder. The breeze wafts to his senses the odors of God's beneficence; the voice of God's power is heard in the rustling of the forest; and the varied forms of life, activity and pleasure, which he observes at every step in the fields, lead him irresistibly, one would think, to the source of being, and beauty, and joy. How auspicious such a life to the noble sentiments of devotion!

Besides, the situation of the husbandman is peculiarly favorable, it should seem, to purity and simplicity of moral sentiment. He is brought acquainted, chiefly, with the real and native wants of mankind. Employed solely in bringing food out of the earth, he is not liable to be fascinated with the fictitious pleasures, the unnatural wants, the fashionable follies and tyrannical vices of a more busy and splendid life.

Still more favorable to the religious character of the husbandman is the circumstance, that, from the nature of agricultural pursuits, they do not so completely engross the attention, as other occupations. They leave much time for contemplation, for reading and intellectual pleasures; and these are peculiarly grateful to the resident of the country. Especially does the institution of the Sabbath discover all its value to the tiller of the earth, whose fatigue it solaces, whose hard labor it interrupts, and who feels, on that day, the worth of his moral nature, which cannot be understood by the busy man, who considers the repose of this day as interfering with his hopes of gain, or professional employments

If, then, this institution is of any moral and religious value, it is to the country we must look for the continuance of that respect and observance which it merits. My friends, those of you, especially, who retire annually into the country, let these periodical retreats from business or dissipation bring you nearer to your God; let them restore the clearness of your judgment on the objects of human pursuits, invigorate your moral perceptions, exalt your sentiments, and regulate your habits of devotion; and if there be any virtue, or sim-

"HOUGHTON'S SEEDLING GOOSEBERRY. This is a very superior gooseberry for the dessert, and it is also excellent for cooking. The skin is thin. reddish brown; and the flesh is fine, very tender, sweet, and of a fine delicious flavor. It is pronounced by competent judges the best known variety of that highly esteemed fruit."

plicity remaining in rural life, let them never be impaired by the influence of your presence and example.—REV. JOS. S. BUCKMINSTER.

NATURE AND VARIETY OF SOILS.

What in agriculture is to be understood by soil?

The soil is the portion of the ground in which plants are produced. It forms a stratum varying from a few inches to a foot or more in depth. It is usually somewhat dark in color, arising from its intermixture with decomposed stems, leaves, and other parts of plants, which have grown upon it, and, in part, often by the presence of animal substances.

What else enters into the formation of the soil?

Primary rocks, on their surface, are constantly undergoing a kind of decomposition, by the crumbling from them of an infinite number of fine particles. These particles combine with the vegetable substances in the formation of the soil in an endless variety of proportions; and the entire process of disintegration and combination is effected by the mechanical and chemical action of the air and water, and the different degrees of temperature to which they are subjected.

By what other names is soil sometimes called?

The finer portions of it are called mold. It is also called loam. Both terms are derived from the fine and smooth quality of the substance denoted.

What is marl?

It is defined by Webster, to be a species of calcareous earth, of different composition, being united with clay and fuller's earth—that is, a species of earth which absorbs grease?

What is the subsoil?

The subsoil is the bed or stratum of earth lying between the surface soil and the base on which it rests. The name is derived from the prefix to the word soil, signifying *under;* and the definition is the same as *under-soil.* It is in part composed of the same materials that form the soil; but ordinarily only in a small degree.

What is the use of the sub-soil?

When loosened by the sub-soil plough or otherwise it becomes a broad reservoir or fountain for the rain not retained by the surface soil. Hence here it remains to fertilize vegetation in seasons of drought. And the subsoil is also valuable, when thus loosened, for the roots of such plants as require more space than is found in the surface soil.

What is the use of the soil?

It affords a place for vegetable life, by enabling the seed and plant there to become fixed, to receive nourishment in all stages of its development till reaching maturity, and thus furnishing a supply of food for man and animals. The aliment required for this nourishment is in the soil, to be imparted to the plant as wanted.

What are the fertilizing agents in the soil produced by the decomposition of rocks?

Among others are potash, soda, phosphoric acid, magnesia, and silex. It is evident that the particles thus furnished to the soil possess a fertilizing power from the fact that even lava from burning mountains is, after a while, covered with a rich loam that could come from no other source.

To what is this fertility owing?

Undoubtedly to the alkali contained in the lava, and which by exposure to the combined action of the air and moisture are reduced to a state capable of being absorbed by plants. As the lava is formed of stony substances, it is apparent, that when reduced to small particles, whether before or after having been melted, they enrich the soil and become important agents in vegetable development.

By what means does the soil become impoverished or exhausted?

Ordinarily by removing from it the vegetables that grew upon it. They, it is obvious, are composed mostly of materials existing in the soil. If permitted to remain upon it, and rot where they grew, the particles would again mix with the soil, leaving it much as it was before. But, if they are carried away, the soil is as much exhausted, as though the materials composing them had been dug up by a spade and removed in a waggon to some other place.

How is this exhaustion of soil to be prevented?

In agriculture the crops are ordinarily of course removed to the barn or granary of the farmer. Therefore to prevent the exhaustion thus occasioned, he must return to the soil in the form of manure the amount of fertilizing agents absorbed by the growth of the crop taken away. If this be done, it may continue for any indefinite period capable of producing good crops.

What besides this manure is needful in producing good crops?

Good, thorough tillage. The soil is to be frequently loosened by the plough, spade, or some other instrument, so that all the different particles shall be well mixed together, and the whole rendered mellow by an exposure to the action of the sun and the atmosphere. Crops are ordinarily abundant as much in proportion to the degree of tillage as the amount of manure used in the culture. Without either, the benefit of the other will be comparatively small.

How does it appear that such tillage is needful?

In the first place, without it, the rain instead of sinking into the soil, will pass from it as if falling upon a board or stone, and of course there would be no moisture in it to nourish plants. In the second place, if in a hard uncultivated soil the seed were to germinate, the roots could not extend themselves to be brought into contact with fertilizing agents. It would cause a smile, to drill a hole into a stone to receive seed with a view to germination and vegetable development. But this is no more absurd than to cast seed into a hard and compact soil, like a public highway, for a like purpose. Yet, reduce the stone to powder, mix with it fertilizing agents, cast seed upon it, and a crop may be expected. The principle is the same for thorough tillage.

Why is it that gardens are more productive than other cultivated lands?

Simply because of better tillage and more ample manuring. If fields were in these respects made equal to well managed gardens, we should rarely hear of unproductive farms, or lost labor in agriculture.

How are the different kinds of soil denominated?

Some of the more obvious and common distinctions of soil are the following—rich and poor soils—stiff and free or light soils—wet and dry soils—clayey and sandy soils—calcareous soils—peaty soils—gravelly soils, and some others. Though soils are thus distinguished by external characters, they pass into each other by such minute gradations, that it is difficult saying to what class they belong. And these intermediate soils are the most numerous in all countries. It may be said, therefore, that the greater portion of soils consists of intermediate classes, and that it is often difficult to bring them under any division, derived from their texture or any external features.

What is said of rich and poor soils?

This classification would include all others of every variety, having reference to their powers of producing useful plants. Those that have this power in a good measure are called rich soils, and those which have it not are called poor, the distinction being based solely on their productive qualities. Their fertility, all other things being equal, is indicated by the greater or smaller proportion of mold which enters into their composition.

What is said of stiff and free, or light soils?

The stiff soils are those which are tenacious and cohesive in their parts; the light or free soils are those which are of looser texture, and whose parts are easily separated. All soils which possess this tenacious or cohesive property in a considerable degree, are termed clayey; while all the looser ones are termed free or light soils. And all soils are more or less clayey, or more or less light, as they possess more or less of this tenacious or cohesive property, or of this looser texture. The transition from the one to the other is by such imperceptible gradations, that it is often difficult to tell to which class a particular soil belongs.

What is said of wet and dry soils?

When water, from any particular cause, is generally abundant, the soils may be termed wet; and when there is habitual deficiency of water, they may be termed dry. The cultivator may very easily learn how to adapt his crops, and the season for their tillage, to these states of the soil. It will usually be found advantageous to have each on a farm.

What is said of clayey and sandy soils?

In the former of these, clay preponderates or forms the principal element, while the same may be said in reference to the latter of the preponderance of sand. The one is tenacious, stiff, very retentive of moisture, and can be worked only in favorable periods, and requires extra labor in its tillage. The other is loose, easily worked, but is not retentive of manure or moisture, owing to its porous texture. The former when well prepared, yields a heavy crop, and owing to the expense of tillage is usually appropriated to meadow and pasture. The latter is better adapted to tap root plants, as carrots, turnips, clover, Indian corn, and alternate hus bandry.

What is said of rich sandy soils?

They are early in maturing the cultivated plants, and are fit for the production of every kind of herbage and grain. They yield to the richer clays in the power of producing wheat, but they surpass them in the production of rye and barley. They are well suited to the growth of the cultivated grasses; and, when left in perennial pasture, they are quickly covered with the natural plants of the soil. But their distinguishing character is their peculiar adaptation to the raising of the plants cultivated for their roots and tubers.

What is said of gravelly soils?

The composition of a gravelly soil is indicated by its very name. It is even more porous than a sandy soil, and is easily exhausted, for the animal and vegetable matters which it receives not being attracted by earthy constituent parts of it, rarely sufficiently abundant for that purpose, are more liable to be decomposed by the action of the atmosphere, and carried off by the water. With gravelly soils should be mixed clay, chalk, marl, peat, and other earthy substances; and especially they should receive frequent applications of manure. Being easily heated they are the earliest crops, and most liable to suffer from the droughts of summer.

What is said of peaty soils?

These soils abound in swamps and marshes, where vegetable matter exists in excess, in consequence of their being habitually saturated with water, which prevents its decomposition. On being thoroughly drained, some of these soils, in which the vegetable has been reduced to something like soft black powder, or where the earths constitute a considerable portion of the surface stratum, have become very productive. But where the vegetable matters greatly preponderate, or are coarse and woody, it has been found necessary, in order to render them valuable, after draining, to bring on a decomposition by paring and burning the surface, or by the application of lime, or barn-yard manure; and sometimes a good dressing of sand, or loam, has induced fertility. The cause of sterility is not the want of vegetable food, but the want of this food in a soluble or cooked state, prepared for the mouths and nourishment of plants.

What are alluvial soils?

They are, first, those soils which have been formed by the action of the sea, and are composed principally of sand, with but little organic matter except marine shells, such as the great level sandy districts lying along the border of the Atlantic; and secondly, those which have been formed from the deposits of rivers, as upon the Mississippi, the Ohio, and most of the secondary and minor streams of our country.

What is said of the temperature of soils?

The temperature of a soil depends much on its humidity. Damp, wet land is generally cold, whereas dry land is usually warm. And land which contains a considerable quantity of mold, or manure, which has been exhausted, or even other substances in a state of putrefaction, is much warmer than that which is not well supplied with them. And calcareous soils, or those partaking of chalk or lime, are always warmer than others.

What most impoverishes the soil?

There is nothing which exhausts either the plant or the soil in which it grows, so much as the ripening of its fruit and seeds. No animal labors with greater effort to support its offspring than the poor plant to bring its seed to maturity. It pumps up sap with all its powers of suction; yet, if it has much seed to ripen, after having accomplished its task, it frequently perishes through exhaustion from the intensity of its efforts.

HOW TO OBTAIN GOOD CROPS.

Practical agriculture is wholly indebted to science for a knowledge of the elements which nature must have to form each plant, seed and fruit grown on the farm or in the garden. We have known a kernel of corn in Georgia to produce one thousand kernels equal in weight to itself. The matter in the parent seed could form but one kernel, leaving that contained in the other nine hundred and ninety-nine, as well as that which exists in stems, leaves, roots, and cobs to be derived from the substance of the earth, air, and water.

Now, why should one kernel of seed corn give a harvest of one thousand kernels in one soil; and a harvest of only one hundred in another? The same degree of sunshine, dews, and rains; the same atmospheric gases and metereoric influences affect both corn plants alike. But if you analyze the soil critically, one will be found to abound in the elements consumed in organizing a large yield of this important grain; and the other will show a lack of some one or more of the things which God has appointed for the formation of corn. For years we have labored to convince our readers and hearers not only that good crops of the fruits of the earth cannot be formed out of *nothing;* but that each plant must have its appropriate constituent elements within reach of its living germ, in due quantity and in available form.

A wise farmer husbands all of these raw materials, out of which his grain, grass, roots, apples, and other fruit are literally made. He studies to accumulate in his soil the substances known to be indispensable to produce bread, meat, milk, wool, and cotton. To render land more and more fertile, is better than to deposit money in increasing sums in any bank in the world. A rich soil has an intrinsic value for hungry human beings that appertains neither to gold nor any other precious metal. The fertilizing of whole farms as a general practice will never obtain in the United States, till our children are taught a knowledge of those natural laws, by the operation of which poor soils may be transformed into fertile ones; and rich soils changed into sterile fields. The growth of plants is governed by laws as fixed and enduring as those which cause day and night, winter and summer, rain and snow.

Why then will not American farmers believe this simple truth, and permit their sons to study tillage, the formation of crops, and the improvement of cultivated earth as a science? If plants must be well fed to be fat, as well as animals, and their food must come from somewhere, why not learn how to feed the germs of corn, wheat, potatoes, oats, and apples, with the highest attainable skill and economy? Experience demonstrates that bone earth, or bones themselves, sulphur and lime, or gypsum, chlorine and soda or common salt, and other elements of crops, serve to augment the harvest. There are millions of tons of the elements of bones and flesh wasted in this country, before they are organized in any living plant as food for man or beast.

This waste accrues from a defective system of husbandry —one that permits no inconsiderable share of the dissolved minerals and organic matter in good soils to run with the water that holds them in solution, into creeks, rivers, lakes, and the ocean. This loss, more than the removal of crops, often exhausts ploughed and hoed land. The loosened soil is washed and leached till the food of corn and wheat, potatoes and turnips, in an available shape, becomes scarce indeed. The substances in the earth which form crops are lime, potash, iron, soda, magnesia, silica, sulphur, phosphorous, chlorine, carbon, nitrogen, oxygen, and hydrogen. Every substance that grows takes up from the soil through the pores in the roots most, if not all these elementary bodies, and fixes them in its organized tissues.

Cultivate a field and permit no plant whatever to grow therein, and both its vegetable mold and earthy salts of lime, potash and soda for instance, will be slowly dissolved and washed away. Partial, if not complete sterility can be induced without cropping at all. Nature renovates poor soils by constantly augmenting the annual yield of vegetation. This she does without the aid of tillage or manure of any kind applied from abroad. The farmer should study nature while at work drawing the food of plants from the subsoil and the atmosphere, to be organized and decay to enrich the surface soil.

This is one way to improve land. Another is to carry on, and spread over it just such things as are known to be indispensable in forming the crop. What are they? They are salts voided in the liquid and solid excretions of all animals, from man down to the bottom of the cast. These salts come from our daily food which took them from the soil. In cities and villages these elements of crops accumulate in stables and privies; because all tillers of the earth send something from the soil to market. Go then to the nearest city or village and take back on to your farm what will pay the debt you owe it. There are many soils in which an ounce of the voided elements of wheat, or corn, perfectly dry, will give a gain of a pound of dry corn at the harvest, if skilfully applied. By greatly extending the roots of a plant at a proper season, it is able to draw a much larger quantity of nourishment from any given soil, in addition to all that the fertilizer yields to it.

An exhausted horse may be within twenty miles of rich pasture, which one good feed of oats will enable him to reach. So there may be food for the hungry corn plant just beyond where its roots extend; and a little of the matter stored up in the kernels of corn to feed the germs when they begin to grow, if taken from the pig-sty or the privy and brought into contact with the roots, will cause them to grow into fresh pasture. It is the elements of the plant in this fresh pasture, not really the manure, that double the harvest.

The way that a pound of guano or rich, light soil operates to produce sixteen pounds of grain, is what we want all young farmers to study. When they do this experimentally they will see that all soils possess far more of the things necessary to feed and clothe mankind than is generally supposed. It is not necessary in the economy of Providence to give the earth a pound in order to get a like weight back again without detriment to the soil.—HON. ISAAC HILL.

THE POSITION OF THE FARMER.

The intelligent farmer who directs his energies with the zeal and spirit which begin to characterize his class—who looks at his profession with pride and pleasure, and considers agriculture an art to be associated with, and assisted by scientific inquiry, is as far superior to the silken dandy, who may think him a clod-hopper, as one class of beings can be to another. The one is the prop of the State—the other a trifling excrescence upon it. To the intelligent farmer nature unfolds her beauties as well as her bounties. His is the honest heart, the liberal soul, the ardent mind, the fresh imagination. He makes the best of parents and citizens, the most disinterested of patriots.—Between the well systematized labors of his life are intervals of leisure for general reading and improvement, enough to give him all the information necessary for individual culture, and social enjoyment. Though every farmer should look first to the general fertility of his farm, as the foundation on which all improvements are to be laid, he would be utterly wanting in the true spirit of his profession, if he did not design, in due time, to crown his whole work, by every domestic comfort and appropriate rural ornament. The business of agriculture is not one of merely practical utility. The farmer is not necessarily a dull swain. His pursuits are consistent with the keenest admiration of the beautiful in nature and art, with the most refined taste, and with all the graces of cultivated life. He owes it to himself as a rational being, gifted with all the capabilities of his race, to the obligations of domestic duty, and above all, to the devotion which we will acknowledge, to that gentle sex, whose smiles are the crowning bliss of life, to provide, for his own and his family's enjoyment, all the comforts and embellishments, which belong to a mature civilization. Among other high duties, is that of properly educating his children. And to such of them as are destined to pursue his own profession, he should give much more than that teaching, which stops at a knowledge of the mere routine of farm-practice.

A good agricultural education is both scientific and practical. The knowledge which is necessary to make a thoroughly intelligent farmer is to be drawn from a great variety of sources. Geology and mineralogy must instruct him as to the formation of the crust of the earth, the qualities and elements of the substances which compose it, and the character, nature, and properties of all the minerals imbedded in it. Chemistry will teach him to analyze the soil, to trace out every element of its fertility, and its just proportion—will develope to him the principles of its exhaustion and replenishment, and guide him in every effort to improve and ameliorate. Indeed without analytic chemistry he can never know the money value of the ameliorators which he purchases. It will teach him also the properties and value of the different kinds of food for stock—what kind supplies the bone, what the muscle, and what lays on the fat. Botany will inform him of the nature and structure of plants, from the forest tree to the herb; of their uses, value, medicinal properties, and adaptation to the climate and soil of his residence. Entomology will teach him the habits of insects injurious to vegetation, and how to prevent or remedy as far as possible their attacks. Natural philosophy will explain to him the principles of mechanics, and from these he will know how to estimate the value of every mechanical contrivance employed or proposed in rural art. Much important knowledge may be gathered from this source. These principles, for example, regulate the construction of wheat threshers and horse powers—the forms of good ploughs and their use, making them easy of draught and efficient in turning over the tough sward or crumbling the broken fallow. Even in the digging of a drain, the construction of a good axle tree, or the proper harnessing a horse, the principles of natural philosophy are involved.

In all these branches of science, the highest genius and the most persevering research have long been devoted to the ascertainment of truths simple only when demonstrated, and of the greatest practical value to those who dream not of the learning and toil necessary for their discovery: and all these should be taught the agricultural pupil.

It is gratifying to see the attention beginning to be paid to this kind of education in our country—to know that agricultural chemistry is becoming one of the branches of collegiate instruction, and that institutions are projected, and, indeed, in existence among us, where the best methods of rural art and every branch of farming work will be taught experimentally, practically and scientifically. In such institutions, the labors of the field, the barn and the workshop, will be followed by the lessons of the school room. Thus practice and science will be combined, and the agricultural pupil will become the finished farmer. Let me commend such institutions to your favor, gentlemen, for the benefit of your sons.

I trust that I may be permitted, also, to recommend every farmer to subscribe and pay for at least one agricultural paper or periodical. Knowledge is power in agriculture, as well as in every other department and business of life. No man can keep up with the improvements of the times who does not thus avail himself of the experience and knowledge of the many active and practical minds which are annually condensed into the columns of these useful journals. What science applied to practical uses has done for other arts it is now beginning to do for agriculture. While no one thinks, or pretends to think, that it has explained all those mysteries which perplex the farmer, no well informed man can doubt that it has already solved many difficulties and ascertained truths of great practical benefit. As little can he question that in its further progress it will remove doubts, expose errors, and develop facts and principles of the highest practical utility. The process by which these results will be arrived at may not and need not be understood by the great body of farmers. But the fruits of such scientific inquiries will be found in our agricultural journals, not buried among technical terms in learned treatises, but made clear to the comprehension of the ordinarily intelligent farmer.

The book farmer may be a thriftless theorist, but the practical man, while he avoids visionary speculations and hazardous experiments, knows how to turn to account the experience and the suggestions of others. He adopts what is consistent with reason and his own observation—tests what is doubtful, and rejects the inefficient or extravagant. Such a man will read without either credulity or disdain, and, therefore will be benefited by what he reads. Indeed, the practical character of most of the agricultural periodicals of the present day is such as to make them entirely acceptable to practical men.—HON. JAMES A. PEARCE, *Maryland*.

THE AMERICAN SCHOOLMASTER.

It has been to me a source of pleasure, though a melancholly one, that in rendering this public tribute to the worth of one departed friend, the respectable members of two bodies, one of them the most devoted and efficient for philanthropy and learning, have met to do honor to the memory of a schoolmaster. There are prouder themes for the eulogist than this. The praise of the statesman, the warrior, or the orator, furnish more splendid topics for ambitious eloquence; but no theme can be more rich in desert, or more fruitful in public advantage.

The enlightened liberality of many of our state governments, by extending the common school system over their whole population, has brought elementary education to the door of every family. In New York, it appears from the Annual Reports of the Secretary of the State, there are, besides the fifty incorporated academies and numerous private schools, about ten thousand school districts, in each of which instruction is regularly given. These contain at present half a million of children taught in that single state. To these may be added nine or ten thousand more youth in the higher seminaries of learning, exclusive of the colleges.

Of what incalculable influence, then, for good or for evil, upon the dearest interests of society, must be the estimate entertained for the character of this great body of teachers, and the consequent respectability of the individuals who compose it? At a recent general election in that state, the votes of above three hundred thousand persons were taken. In thirty years the great majority of these will have passed away; their rights will be exercised, and their duties assumed, by those very children, whose minds are now open to receive their earliest and most durable impressions from the ten thousand schoolmasters therein employed.

What else is there in the whole of our social system of such extensive and powerful operation on the national character? There is one other influence more powerful, and but one. It is that of the MOTHER. The forms of a free government, the provisions of wise legislation, the schemes of the statesman, the sacrifices of the patriot, are as nothing compared with these. If the future citizens of our republic are to be worthy of their rich inheritance, they must be made so principally through the virtue and intelligence of their mothers. It is in the school of maternal tenderness that the kind affections must be first roused and made habitual—the early sentiment of piety awakened and rightly directed—the sense of duty and moral responsibility unfolded and enlightened.

But next in rank and in efficacy to that pure and holy source of moral influence, is that of the schoolmaster. It is powerful already. What would it be if in every one of those school districts which we now count by annually increasing thousands, there were to be found one teacher well-informed without pedantry, religious without bigotry or fanaticism, proud and fond of his profession, and honored in the discharge of its duties! How wide would be the intellectual, the moral influence of such a body of men! Many such we already have amongst us—men humbly wise and obscurely useful, whom poverty cannot depress, nor neglect degrade. But to raise a body of such men, as numerous as the wants and dignity of the country demand, their labors must be fitly remunerated, and themselves and their calling cherished and honored.

The schoolmaster's occupation is laborious and ungrateful; its rewards are scanty and precarious. He may indeed be, and he ought to be animated by the consciousness of doing good, that least of all consolations, that noblest of all motives.

But that, too, must be often clouded by doubt and uncertainty. Obscure and inglorious as his daily occupation may appear to learned pride or worldly ambition, yet to be truly successful and happy, he must be animated by the spirit of the same great principles which inspired the most illustrious benefactors of mankind. If he bring to his task high talent and rich acquirements, he must be content to look into distant years for the proof that his labors have not been wasted—that the good seed which he daily scatters abroad does not fall on stony ground and wither away, or among thorns to be choked by the cares, the delusions, or the vices of the world.

Indeed, such a schoolmaster must solace his toils with the same prophetic faith that enabled the greatest of modern philosophers, amidst the neglect or contempt of his times, to regard himself as sowing the seeds of truth for posterity and the care of Heaven. He must arm himself against disappointment and mortification, with a portion of the same noble confidence which soothed the greatest of modern poets when weighed down by care and danger, by poverty, old age, and blindness, still

"—— in prophetic dream he saw
The youth unborn, with pious awe,
Imbibe each virtue from his sacred page."

He must know and he must love to teach his pupils, not the meagre elements of knowledge, but the secret and the use of their own intellectual strength, exciting and enabling them hereafter to raise for themselves the veil which covers the majestic form of truth. He must feel deeply the reverence due to the youthful mind fraught with mighty, though undeveloped energies and affections, and mysterious and eternal destinies. Thence he must have learnt to reverence himself and his profession, and to look upon its otherwise ill-requited toils as their own exceeding great reward.

If such are the difficulties and the discouragements—such the duties, the motives, and the consolations of teachers who are worthy of that name and trust, how imperious then the obligation upon every enlightened citizen who knows and feels the value of such men, to aid them, to cheer them, and to honor them! But let us not be content with barren honor to buried merit. Let us prove our gratitude to the dead by faithfully endeavoring to elevate the station, to enlarge the usefulness, and to raise the character of the schoolmaster amongst us. Thus shall we best testify our gratitude to the teachers and guides of our own youth, thus best serve our country, and thus, most effectually diffuse over our land light, and truth, and virtue.—VERPLANCK.

SEED TIME AND HARVEST.

As o'er his furrowed fields which lie
Beneath a coldly-dropping sky,
Yet chill with winter's melted snow,
The husbandman goes forth to sow;

Thus, Freedom, on the bitter blast
The ventures of thy seed we cast,
And trust to warmer sun and rain,
To swell the germ, and fill the grain.

Who calls thy glorious service hard?
Who deems it not his own reward?
Who for its trials, counts it less
A cause of praise and thankfulness?

It may not be our lot to wield
The sickle in the ripened field;
Nor ours to hear, on summer eves,
The reaper's song among the sheaves;

Yet where our duty's task is wrought
In unison with God's great thought,
The near and future blend in one,
And whatsoe'er is willed is done!

And ours the grateful service whence
Comes, day by day, the recompense;
The hope, the trust, the purpose stayed,
The fountain and the noonday shade.

And were this life the utmost span,
The only end and aim of man,
Better the toil of fields like these
Than waking dream and slothful ease

But life, though falling like our grain,
Like that revives and springs again;
And, early called, how blessed are they
Who wait in heaven their harvest-day!

WHITTIER.

VEGETABLE MANURES.

What may be affirmed generally of vegetable substances for manure?

There is no plant that grows, but what contains the elements for its own reproduction at least, and when reduced to the simple substances of which it is composed, will subserve that end. And, there is so much similarity in the composition of the vegetable kingdom generally, that all plants may answer a good purpose, when in a decayed state for manures, although some are much better than others, both on account of the ingredients of which constructed, and the abundance in which they can be procured. Organic vegetable matters in various conditions, constitute the largest part of manure in use.

What is said of hay, straw, and chaff, as manure?

When they are ploughed into the soil, they are slow in decomposing, and act more slowly than when previously fermented. The question of applying straw without previous decomposition, is, in practice, only a question of time. It is doubtless true that it furnishes about the same amount of manure in both cases; but in the one case it has more speedy and powerful, and in the other a more prolonged effect. The more common way, and probably on the whole, the preferable one, is to mix these substances with animal excrements in the barn-yard.

What does chemical analysis show us?

It shows us that all plants, and all the products of plants are resolvable into a small number of simple bodies in various states of combination; and, that these bodies when disintegrated may and will go into the composition of other plants, if mixed in the soil as a manure.

What is said of saw dust for manure?

Saw dust has been highly recommended, because it mixes readily with other substances, whether animal or vegetable.

Fine specimen of the DEVON COW, owned by AMBROSE STEVENS, ESQ., of New York.

It is a peculiarly good absorbent of the gases and of liquid manures; and, as serviceable in rendering stiff and clayey lands loose and mellow. The objection to it is, a long time is wanted for saw dust to ferment, and of course, is not felt by the soil as a fertilizing agent for a year or two.

In what respect is charcoal valuable for manure?

Charcoal is not so useful to the soil on account of any element it may furnish the soil, as by an intermediate agency of absorbing gases and neutralizing offensive odors. As an absorbent it is especially valuable for holding and preserving those volatile matters, which plants require, and which might otherwise make their escape and be lost. However, it evolves carbonic acid in its decomposition, and is in this way directly useful to vegetable growth. And being also a powerful antiseptic it keeps the soil free from putrefying substances, that might bring disease on the spongelets, and hence it keeps the plant in a healthy and vigorous condition. Pulverized charcoal is frequently of great service about the trunks of fruit trees.

What is said of wood ashes for manure?

It is apparent that the ashes of vegetables consist of such elements as are always required for the perfect growth of plants, and, must hence furnish one of the best saline manures which can be had. They contain all the inorganic substances that enter into vegetable composition, and usually in the right proportion, being much the same to the nourishment of the plant, that milk is to the nourishment of an animal. They should never be wasted, and if purchased at fair prices are a cheap manure.

What is said of leached ashes?

They are indeed less valuable than before having been subjected to the process of being leached, yet contain all the elements of the unleached, having been deprived of only a part of their potash and soda. They may be drilled into the soil with roots and grain, sown broadcast on meadows or pastures, or mixed with the muck-heap. They are good on all soils, varying from twenty to fifty bushels to the acre.

Are coal ashes of value for manure?

Some have esteemed them of no value; not worth the labor of preserving them; but it is now more generally admitted that the ashes of both anthracite and bituminous coal are of considerable value, although less so than those made from wood and the smaller vegetables. They should, therefore, be saved and applied to the soil. If they contain cinders from not having been thoroughly burned, they are more suitable to light, than to heavy soils.

What is said of the value of peat ashes?

Sir Humphrey Davy did not consider soils as coming under the denomination of peat, unless consisting of vegetable fibre to one half their bulk. It is evident, therefore, that ashes of peat must be valuable from this fact, as well as from the mineral substances in the earthy parts of it. If it does not become perfectly reduced to ashes, it will make a beneficial dressing to the soil. It is better to reduce peat to ashes, than to attempt mixing it with the soil without being burnt, especially if it be at a distance from the land that is to receive it.

If not reduced to ashes, how should peat be used for manure?

It may be decomposed by long exposure to the air, or by mixing it with quicklime. But it is better to carry it to the barn-yard, and cover it over with the dung; or to mix it with the dung in alternate layers. In this case it should be previously dried by the sun and wind, and then not be used in too large quantities. When the fermentation has arrived at blood heat, the mass should be turned over into another heap.

How is sea-weed prized for manure?

In places where it can be had it is esteemed of much importance, although transient in its effects. The most convenient method of using it, is to convey it directly to the land and apply it fresh, as a top dressing to the growing crops. If left in a heap by itself, its more soluble parts are exhaled, and a dry, fibrous matter alone remains. If, therefore, it is not applied in its recent state, it should be formed into a compost with dung, or with a mixture of dung and earth. Sea-weed may also be burned to ashes and applied to the soil like wood ashes. This is preferable if the weed is far distant from the land where it is to be used.

How are green crop manures described?

Sometimes a crop of clover is ploughed under and completely buried up, where fermentation soon succeeds, and the elements again mix with the soil. Buckwheat and other seeds are sometimes sown, and ploughed under in the same manner. The period at which the plants are ploughed down, is when they have come into flower, for then they contain the largest quantity of readily soluble matter, and have the least exhausted the nutrient substance of the soil.

What is said of green crop manure?

The practice of applying it has been long known and not a little used, and is pronounced of great service. Nevertheless, it is chiefly suited to the warmer countries where vegetation is very rapid. In colder countries where we are able to raise food of any kind, it is better that we apply it in the first place to the feeding of animals, for then it not only yields manure, but performs the no less important purpose of affording food. If persons have more clover than they need for hay, it is doubtless better to plough it under than to let it rot on the surface.

How are leaves valued as a manure?

The opinion of agriculturists is divided in regard to the utility of them. Some esteeming them scarcely worth the labor of collecting them, and others esteeming them of great value.

Which of these opinions is most worthy of attention?

They are evidently deserving more general regard than they receive. The prejudice by some entertained against them is based on the small amount of substance in a single leaf, or in a given volume without reference to the weight, not considering that a ton of leaves when reduced to their original elements, may contain as much fertilizing matter as a ton of other vegetable substances.

How does it conclusively appear that leaves are valuable?

From the fact that the soil is never impoverished by natural vegetation. What is there to enrich the soil in the dense forest where has been a successive growth of trees for thousands of years, except the leaves which annually fall and mix with it? In this way nature is constantly enriching the soil. The soil becomes impoverished only by cultivation and removing from it the plants which it had produced. And, it is known that nursery men prefer the mold from the forest, which is mainly the product of leaves, to any other soil. A farmer in the neighborhood of a forest never need fear a deficiency of manure, if he will apply himself to the collection of leaves.

What is the best mode of collecting and using leaves?

They should be collected in the morning when a heavy

dew is on them, or immediately after a fall of rain; otherwise it will be difficult to confine them in a cart. For the purpose of gathering them up, a common rake may be used. An iron one is better. They make excellent litter for hogs; and in one week's time a single hog will so trample and tear them to pieces and mix them with his own excrements, that it would not be easy to tell of what the mass had been composed; so that in a week or ten days a new supply will be required. The process is hastened by throwing in daily upon them a few handfuls of shelled corn. In this way a hog will make perhaps manure enough to pay for his feed. Leaves are also good for the litter of horses and cattle, absorbing the urine and mixing with the dung they soon ferment and decay. Or they may be at once mixed with the compost heap.

What is said of muck?

Muck, which is the black earthy substance obtained from swamps, and other low lands, is of incalculable richness for a fertilizing agent. It being composed of decayed leaves, roots, and branches of trees, and the saline substances washed upon it and mixed with it from the surrounding uplands, it is one of the best substances for manure a farmer can have. It is like a mine of gold or silver from which he may at pleasure extract the most unfailing elements of wealth. With such a mine within his reach and at his control, he need not be poor. His prosperity is in his own hands.

What is said of the mud of rivers and fords for compost manure?

It has much the same to recommend it that muck has; and in addition to what has been collecting for ages from a drain of the adjacent uplands, there are in it the organic remains of fish, shells, reptiles, and water fowls. A moment of reflection must satisfy any one that here may be found, especially at the bottom of ponds having no deep outlet for the escape of what is thus deposited, substances of the richest material for improving the soil. It is generally easy in dry seasons, so to drain mill-ponds especially, as to make these substances available.

What is said of sods or turf for manure?

It is clear that here may be found a large amount of the roots of grasses and other decayed vegetable substances; here, too, have been deposited animal excrements; so that the entire sod is capable of being converted, in a well prepared compost heap, into a manure of great energy. Oftentimes by the road side, or other waste places, particularly by fences which cannot be cultivated, large quantities of turf may be obtained. Instances might be named, where hundreds of loads are annually collected, and then mixed with lime, making in the following season one of the best fertilizers.

What is the design of compost manures?

Composts are an artificial mixture of vegetable or animal matters, with earthy or mineral substances, and may be profitably resorted to in two contingencies—*first*, to arrest and detain, for useful purposes, fertilizing matters which might otherwise be wasted and lost, as the urine of animals, or the gaseous matters which are evolved from animal or vegetable substances while undergoing fermentation; and, *secondly*—to render soluble, or available as the food of plants, matters which are not already so, as muck, woody fibre, and the like.

AGRICULTURE AND THE HOMESTEAD.

Let us say a word on the importance of the pursuits of the husbandman. What rank does agriculture hold, in the scale of usefulness among the pursuits of civilized communities? We shall arrive at a practical answer of the question by considering, that it is agriculture which spreads the great and bountiful table, at which the mighty family of civilized man receives his daily bread. Something is yielded by the chase, and much more by the fisheries; but the produce of the soil constitutes the great mass of the food of a civilized community, either directly in its native state, or through the medium of the animals fed by it, which become, in their turn, the food of man. In like manner, agriculture furnishes the material for our clothing. Wool, cotton, flax, silk, leather, are the materials, of which nearly all our clothing is composed; and these are furnished by agriculture. In producing the various articles of clothing, the manufacturing arts are largely concerned, and commerce, in the exchange of raw materials and fabrics. These, therefore, to a considerable degree, rest on agriculture, as their ultimate foundation; especially as it feeds all the other branches of industry.

Agriculture seems to be the first pursuit of civilized man. It enables him to escape from the life of the savage, and the wandering shepherd, into that of social man, gathered into fixed communities and surrounding himself with the comforts and blessings of neighborhood, country, and home. The savage lives by the chase,—a precarious and wretched independence. The Arab and the Tartar roam, with their flocks and herds, over a vast region, destitute of all those refinements which require for their growth the features of a permanent residence, and a community organized into the various professions, arts, and trades. They are found now, after a lapse of four thousand years, precisely in the same condition in which they existed in the days of Abraham. It is agriculture alone, that fixes men in stationary dwellings, in villages, towns, and cities, and enables the work of civilization, in all its branches to go on.

Agriculture was held in honorable estimation, by the most enlightened nations of antiquity. In the infancy of commerce and manufactures, its relative rank among the occupations of men was necessarily higher than now. The patriarchs of the ancient Scripture times cultivated the soil. Abraham was very rich in cattle. in gold, and in silver. Job farmed on a very large scale : he had seven thousand sheep, three thousand camels, five hundred yoke of oxen, and five hundred she-asses. In Greece, the various improvements in husbandry, the introduction of the nutritive grains, and the invention of convenient instruments for tilling the soil, were regarded as the immediate bounties of the gods. At a later period, land was almost the only article of property ; and those who cultivated it, if they were freemen, were deemed a more respectable class than manufacturers or mechanics, who were mostly slaves.

Among the Romans, agriculture was still more respected than among the Greeks. In the best and purest times of the republic, the most distinguished citizens the proudest patricians, lived on their farms, and labored with their own hands. Cato the censor was both a practical and a scientific farmer, and wrote a treatise on the art ; and, who has not heard of Cincinnatus ? When the Sabines had advanced with a superior army to the walls of the city, at the earnest request of the people he left his plough, and was made dictator. And, after having raised an army and defeated the enemy, in sixteen days, laid down the dictatorship which he was authorised to hold for six months, and on the seventeenth day got back to his farm.

On the destruction of the Roman empire, the feudal system arose in Europe, a singularly complicated plan of military despotism. In this system, the possession of the land was made the basis of military defence of the country. The king was the ultimate proprietor; and apportioned the territory among the great lords, his retainers. Those who cultivated the soil were slaves, the property of their lord, and were bought and sold with the cattle which they tended. Sir Walter Scott, in describing, with his graphic pen, one of this class of the former population of England, after depicting the other peculiarities of his costume, says, there was about his neck a brass ring, resembling a dog's collar, but without any opening, having been soldered fast; so loose as to form no impediment to his breathing; yet so tight as to be incapable of being removed, except by the use of the file. On the collar was engraved the name of the wearer, and a record denoting, that he was a *born thrall*, or slave.

The first and lowest in the scale of those, by whom the soil is now cultivated in Europe, are the *serfs* of Russia. In the different provinces of this vast empire, about thirty millions of souls, nearly the entire population employed in husbandry, are found almost exactly in the state which has already been described, under the feudal system. Some ameliorations have been introduced in some provinces, and not in others; and in the south-western portions of the empire, as Courland and Livonia, principally settled by Germans, the system of actual slavery has been abolished by law. But with these local exceptions, the Russian peasantry continue the property of the land owner, and may be sold by him with or without the land, as he pleases. He has power to give or sell them their freedom, and power to keep them in slavery; the power to chastise them, and to imprison them; and in all respects to dispose of them, with the exception of taking life, or preventing their being enlisted in the army. But when a draft is ordered by the government, the landlord directs who shall march.

The wealth of the great landholder is estimated by the number of his peasants; and individuals in the Russian empire are named, who possess a hundred thousand and even a hundred and twenty thousand slaves. Each individual peasant, of either sex, is bound from the age of fifteen to pay the *avrock*, or capitation tax, of about four dollars per annum. This is taken in lieu of performing three day's labor, in each week, to which the landlord is entitled by law. In addition to this, the serf has to account to his lord for a certain part of all his produce ; and besides all, he is subject to the government taxes. If the peasant chooses to make an offer to rise above his condition, he must apply to his lord for permission to leave the spot where he was born, and pursue some other trade. If this occupation be a more lucrative one, his annual tax is proportionably increased. Such is the condition of the entire civilized portion of the Russian empire ; and it is needless to state, that it places this portion far below the wild Tartars, who own a nominal subjection to the Russian sceptre, and pay a trifling tribute for the privilege of roaming their remote steppes, unmolested and free.

On the continent of Europe, when the servitude of the feudal system was broken up, the peasants became *tenants by the halves ;* and such are a considerable portion of the cultivators of the soil, at the present day. It was calculated that in France, before the revolution, seven eighths of the agricultural population were of this class. The revolution has greatly increased the number of small proprietors, in consequence of the sale of the estates of emigrants, and of the public domain ; but one half of the cultivators of the soil, it is supposed, are still tenants at the halves. Such a tenancy is not wholly unheard of in this country. The estates in Lombardy, and in some parts of Italy, are cultivated in this way. The terms of the contract between the landlord and the tenant are not uniform ; in some cases a third, and in others a half of the produce belongs to the landlord. In some cases, the tenant has a property in the lease, which descends to his children ; in others, he is a tenant at will ; in others, the leases are periodically renewed at short intervals.

But, however the details may vary, the system resolves itself, in the main, into a general system of tenancy at the halves. It is considered highly unfavorable to the improvement of agriculture. There is a constant struggle, on the side of each party. to get as much as possible out of the land, with the least possible outlay. The tenant has no interest in using the stock with care and prudence, as this is to be replenished by the landlord. In France, the effect of this system is acknowledged by the best writers in that country to be pernicious. A better account of it is given in Lombardy. There the tenant has the whole of the clover, and divides only the wheat, Indian corn, flax, wine, and silk. The landlord advances nothing but the taxes.

It has been a question much debated in England, whether a system of this kind, by which the land is principally held in large farms belonging to the aristocracy of the country, and cultivated by the tenants on lease, is more favorable to the improvement of agriculture than the multiplication of small farms. It has been urged that the great and expensive improvements in farming, cannot take place without great capital, which can only be furnished by large proprietors. It is these alone, who reclaim wastes—convert sandy plains into fertile fields—drain extensive fens—or shut out the sea from large tracts of meadow. All this is true; but where great improvements are made by the application of large amounts of capital; the return is not to the tenant, but to the capitalist.

A judicious operation upon a poor soil may turn it into a good one—the soil may produce twice as much as it did before; but the rent, in that case will be doubled. The landlord has doubled his capital; but it will depend on other circumstances, whether any beneficial change is produced in the condition of the tenant. The neighborhood, it is true, will be improved by the new creation of property—the population will increase—and, indirectly, every individual will be improved, by living in a larger community; but, directly, I cannot perceive that the tenant is benefited; inasmuch as it is plain, that precisely as the land is rendered more productive, the rent increases.

As the landed interest in England is the main interest of the country, and the accumulation of large estates in land is the most important element in their system, everything is made to favor this mode of cultivating the land, and the small proprietor labors under great disadvantages. Wherever he moves, he has a wealthy rival to contend with, able to overbid and undersell him; and as things now are in England, it is very possible that the condition of the tenant in that country is more desirable than the small farmer. But this, I conceive, proves nothing in the argument, whether the condition of the tenant or the proprietor of a small farm is to be preferred. It is, in fact, justly made a leading objection to the English system of tenancy, by a learned French writer, that it tends to the extermination of the small proprietors, and to reduce the cottagers, peasants, and all those by whom, under whatever name, the labor of cultivation is performed, to a state of abject and servile dependence.

In a country like our own, where every man's capacity, industry, and good fortune, are left free, to work their way without prejudice, as far as possible; there will be among the agricultural, as well as among the commercial population, fortunes of all sizes; from that of the man who owns his thousand acres—his droves of cattle—his flocks of sheep—his range of pastures—his broad fields of mowing and tillage—down to the poor cottager who can scarce keep his cow over the winter. There will always be, in a population like ours, opportunities enough for those who cannot own a farm, to hire one; and for those who cannot hire one, to labor in the employment of their neighbors, who need their services; and when we maintain that it is for the welfare of the society, that the land should be cultivated by an independent yeomanry, who own the soil they till, we mean only that this should be the general state and condition of things, and not that there should be no such thing as a wealthy proprietor, whose lands, in the whole or in part, are cultivated by a tenant; no such thing as a prudent husbandman taking a farm on a lease; or an industrious young man, without any capital but his hands, laboring in the employ of his neighbor.

There is no way in which a calm, orderly and intelligent exercise and control of political power can be assured to the people, but by a distribution among them, as equally as possible, of the property of the country; and I know no manner, in which distribution can be effected, legally, permanently, and peacefully, but by keeping the land in small farms, suitable to be cultivated by their owners. Under such a system, and under no other, the people will exercise their rights with independence. The assumption of a right to dictate, will be frowned at, if attempted; and even the small portion of the people who may be tenants, will possess the spirit and freedom of the proprietors. But when the great mass of the land is parcelled out into a few immense tracts, cultivated by a dependent tenantry, the unavoidable consequence is a sort of revival of the feudal ages, when the great barons took the field against each other at the head of their vassals.

But, I own, it is not even on political grounds, that I think our system of independent rural freeholders is most strongly entitled to the preference. Its moral aspect, its connexion with the character and feelings of the yeomanry give it, after all, its greatest value. The man who stands upon his own soil; who feels that the laws of the land in which he lives—by the law of civilized nations—he is the rightful and exclusive owner of the land which he tills, is, by the constitution of our nature, under a wholesome influence, not easily imbibed from any other source. He feels—other things being equal—more strongly than another, the character of man as the lord of the inanimate world. Of this great and wonderful sphere, which, fashioned by the hand of God, and upheld by his power, is rolling through the heavens, a portion is his:—his, from the centre to the sky. It is the space, on which the generation before him moved in its round of duties; and he feels himself connected by a visible link, with those who preceded him, as he is, also, to those who will follow him, and to whom he is to transmit a home.

Perhaps the farm of this man has come down to him from his fathers. They have gone to their last home; but he can trace their footsteps over the daily scene of his labors. The roof which shelters him, was reared by those to whom he owes his being. Some interesting domestic tradition is connected with every enclosure. The favorite fruit tree was planted by his father's hand. He sported, in his boyhood, by the side of the brook, which still winds through his meadow. Through the field, lies the path to the village school of his earliest days. He still hears from his window, the voice of the Sabbath bell, which called his father and his forefathers to the house of God; and near at hand is the spot where he laid his parents down to rest, and where he trusts, when his hour is come, he shall be dutifully laid by his children. These are the feelings of the owner of the soil. Words cannot paint them; gold cannot buy them; they flow out of the deepest feelings of the heart; they are the life spring of a fresh, healthy, generous national character.

The history and experience of the world illustrate this power. Who ever heard of an enlightened race of serfs, slaves, or vassals? How can we wonder at the forms of government which prevail in Europe, with such a system of monopoly in the land, as there exists? Nothing but this explains our own history; clears up the mystery of the Revolution; and makes us fully comprehend the secret of our own strength. Austria or France must fall, whenever Vienna or Paris is seized by a powerful army. But what was the loss of Boston or New York, in the Revolutionary war, to the people of New England? The moment the enemy set foot in the country, he was like the hunter going to the thicket, to rob the tigress of her young. The officers and soldiers of the Revolution were farmers and sons of farmers, who owned the soil for which they fought; and many of them, like the veteran Putnam, literally left their ploughs in the furrow, to hasten to the field. The attempt to conquer such a population, is as chimerical, as it would be to march an army down to the sea-shore, in the bay of Fundy, when the tide is rolling in seventy feet high, in order to beat back the waves with their bucklers.

There are other countries that surpass us in wealth and power; in military strength; in magnificence, and the display of the expensive arts; but none, which can justly lay claim to that glorious character—a free and happy commonwealth; none in which the image of a state, sketched by the fancy of the philosophic poet, is so beautifully realized—

"What constitutes a state?
Not high-raised battlement, and labored mound,
Thick wall on moated gate;
Not cities proud, with spires and turrets crowned;
Not bays and broad-armed ports,
Where, laughing at the storm, proud navies ride;
Not starred and spangled courts,
Where low-browed baseness wafts perfumes to pride—
No! men! high-minded men,
Men who their duties know,
But know their rights; and knowing, dare maintain;
Prevent the long-aimed blow,
And crush the tyrant while they rend the chain;—
These constitute a state,

And sovereign law, that state's collected will,
O'er thrones and globes elate,
Sits empress, crowning good, repressing ill."

Address, Massachusetts Agricultural Society, 1833, *by* Hon. EDWARD EVERETT, LL. D.

AGRICULTURE IN MARYLAND.

It has been very much the fashion of late, with a certain class of writers, whose views are readily adopted by the unthinking multitude, to decry the state of agriculture among us. Maryland and Virginia are usually selected, by way of illustration. A few half-observed facts are hastily collected, dignified with the name of statistics, and in defiance of the true principles of the Baconian philosophy, made the foundation of a comprehensive theory. A stranger, who has never visited our cheerful firesides, seen our well-tilled fields, or enjoyed the elegant hospitality of our refined and enlightened people, has no conception of our true condition. He has been taught to believe that poverty grass, broom straw and old field pines, constitute our chief productions. And because our population has not kept pace with that of the manufacturing States of the East, or the new and teeming West, we are supposed to have reached a premature decay, exhibiting a melancholy picture of homes abandoned, flocks dispersed, and lands desolate and uncultivated.

Moral and political philosophers, eager to build a system, taking this to be our true condition, immediately set about to account for it. Some, with ready zeal in the cause of a sentimental philanthropy, find this blighting curse in our peculiar institutions, and the species of labor with which our fields are cultivated; others, who deem our labor the best and most productive in the world, trace, with certainty, our supposed decline to the grinding influence of Northern monopoly, and would find for it an effectual remedy in unlimited free trade; whilst a third class, at the head of which stands the Nestor of the agricultural press, attributes it, with equal confidence, to the dispersion of our population, and the separation of "the plough, the loom, and the anvil," and thinks we can be relieved from our condition of degrading inferiority, only by an "efficient tariff of protection."

It is true, our population has not rapidly increased. But can this be a subject of astonishment to any well-informed and reflecting mind? Should it not rather be a matter of wonder, that notwithstanding the adverse circumstances we have had to encounter, both our wealth and population have been steadily progressive? Ours are almost exclusively an agricultural people, of Anglo-Saxon descent, inheriting that strong desire for the possession of land, which distinguished our ancestors. The policy of the federal government, by throwing open for settlement, almost without money and without price, our boundless public domain, has encouraged and gratified this desire. And those among us, without land or the means of purchasing it, have very naturally sought independence by settling on the public lands in the West; whilst many of our large farmers have been allured from their old homes and associations by promises, too often fallacious, of greater profits to be derived from the cultivation of the rich staples of the South.

Nor is this all: "Westward the star of empire takes its way," and our young men of talent and education, deeming the field at home too narrow for their efforts, and burning with the strong desire for political distinction, (which, unfortunately, is far too common in the South,) have rushed in countless numbers to the West. Many of them, after a brief and fitful struggle for distinction, fall under the pestilence, or that more terrible moral scourge, which annually slays its thousands; some, more fortunate, attain to eminence in their new homes; whilst others, after years of absence, return to the councils of the nation crowned with honor, and throw into the laps of their old mothers a garland of fame, which, although it may gratify a generous pride, repays none of the treasure expended in fitting them for success in the contests of life.

The industry of a people who have not only sustained, without ruin, all these drains upon their resources, but, notwithstanding them, have steadily advanced in wealth and general improvement, cannot be unproductive. If other evidence of this fact be required, survey for a moment the crowded streets of this great city;* cast your eyes on the noble structures, public and private, that adorn it. Almost

* Baltimore.

within the memory of man it was a poor collection of wretched huts of fishermen; now, it is a great emporium, rivalling in enterprise the larger cities of the North, and pushing its commerce into every portion of the globe. What created, what sustains this noble city?

It may be answered, its commerce, its manufactures, its arts. But what sustains them? Who are the customers of your merchants, manufacturers and artizans? If on any fine autumnal morning you will look from your heights on that beautiful expanse of water, as far as the eye can reach, you will see it studded with the white canvas of the moving messengers of trade, which, pressing onward, as they near their port, crowd upon each other, like trained coursers panting for their goal They come freighted with the products of agriculture, from the James, the York, the Rappahannock, the Potomac, and from the numerous rivers and tributaries of your own State, which with ours serve to swell the flood of this beautiful inland sea, and to pour into your port the materials of a rich and profitable commerce.—HON. WILLOUGHBY NEWTON, *of Virginia.*

A SOLILOQUY OF THE CLOUDS.

I bring fresh showers for the thirsting flowers,
From the seas and the streams;
I bear light shade for the leaves when laid
In their noon-day dreams.
From my wings are shaken the dews that waken
The sweet buds every one,
When rocked to rest on their mother's breast,
As she dances about the sun.
I wield the flail of the lashing hail,
And whiten the green plains under,
And then again I dissolve it in rain,
And laugh as I pass in thunder.

I sift the snow on the mountains below,
And their great pines groan aghast;
And all the night 'tis my pillow white,
While I sleep in the arms of the blast;
Sublime on the towers of my skyey bowers,
Lightning my pilot sits,

In a cavern under is fettered the thunder,
  It struggles and howls at fits;
Over earth and ocean with gentle motion,
  This pilot is guiding me,
Lured by the love of the genii that move
  In the depths of the purple sea;
Over the rills, and the crags, and the hills,
  Over the lakes and the plains,
Wherever he dream, under mountain or stream,
  The spirit he loves remains;
And I shall the while bask in heaven's blue smile,
  Whilst he is dissolving in rains.

The sanguine sunrise, with his meteor-eyes,
  And his burning plumes outspread,
Leaps on the back of my sailing rack,
  When the morning star shines dead.
As on the jag of a mountain crag,
  Which an earthquake rocks and swings,
An eagle alit one moment may sit
  In the light of its golden wings.
And when sunset may breathe, from the lit sea beneath,
  Its ardors of rest and of love,
And the crimson pall of eve may fall
  From the depth of heaven above,
With wings folded I rest, on mine airy nest,
  As still as a brooding dove.

That orbed maiden, with white fire laden,
  Whom mortals call the moon,
Glides glimmering o'er my fleece-like floor,
  By the midnight breezes strewn;
And wherever the beat of her unseen feet,
  Which only the angels hear,
May have broken the woof of my tent's thin roof,
  The stars peep behind her and peer;
And I laugh to see them whirl and flee,
  Like a swarm of golden bees,
When I widen the rent in my wind-built tent,
  Till the calm rivers, lakes, and seas,
Like strips of the sky fallen through me on high,
  Are each paved with the moon and these.

I bind the sun's throne with the burning zone,
And the moon's with a girdle of pearl;
The volcanoes are dim, and the stars reel and swim,
When the whirlwinds my banner unfurl.
From cape to cape, with a bridge-like shape,
Over a torrent sea,
Sunbeam proof, I hang like a roof,
The mountains its columns be.
The triumphal arch, through which I march,
With hurricane, fire, and snow,
When the powers of the air are chained to my chair,
Is the million-colored bow;
The sphere-fire above its soft colors wove,
Whilst the moist earth was laughing below.

I am the daughter of the earth and water,
And the nursling of the sky:
I pass through the pores of the ocean and shores,
I change, but I cannot die.
For after the rain when, with never a stain,
The pavilion of heaven is bare,
And the winds and sunbeams with their convex gleams,
Build up the blue dome of air,
I silently laugh at my own cenotaph,
And out of the caverns of rain,
Like a child from the womb, like a ghost from the tomb,
I arise and upbuild it again.

SHELLEY.

THE VINE AND THE OAK.

A vine that hung to an oak in its pride
And drank the nourishment drawn from its side,
Grew strong and broad in its coiling height,
But stronger still in its giddy sight,
Broke from the oak in an ill-starr'd hour,
And toss'd its head to display its power;
The storm-king gnashed his teeth at the sight,
And swept it off in retributive might;
For the thing that reaches too high and too wide,
Shall draw the lightning's stroke to its side.

It clung round each tree as it swept along,
But it passed unheeded by all the throng ;
None cared to look at a false one so vile
With bow, or nod, or with welcoming smile,
And the vine was thrown in its early prime
Amid nettles and weeds in filth and slime.

But the oak stood still in its lonely glade
With its furrowed sides that the vine had made,
Like the bird that had given its own life's blood
To cherish and feed its featherless brood,
The deep winding grooves, like the serpent's track,
Were pierced by the storm and the sap shrunk back
(The mark of guile that it touched in its rise,
Was the track of a fiend in Paradise ;)
And soon with a solemn and rustling sound
The leaves fell withered and dead to the ground :
The sun shone forth and the moistening rain
Was shed upon hill, and dale and plain ;
The trees put forth their foliage green,
Nature was dressed in her vernal sheen,
But the oak stood shorn of its dark green dress,
The victim cost of a faithless embrace !
A beacon to warn a confiding one
To trust in nought but a cold heart of stone.

Thus upon earth when the heart's fondest tie
Is severed by faithlessness, both must die ;
The union of hearts is the soul's deep well
Where Truth in her purity loves to dwell,
As clear and bright in the heart's faithful love
As the crystal fountain that's floating above.
When the well is broken the deep clear flood
Runs bubbling and purpled with streams of blood,
And Truth in agony shrinking flies
To her sisters bright, the stars in the skies,
(The glittering sentinels, night and day
That watched in the well where their sisters lay.)

The pledges of love we may never reclaim
Without perjury, treachery, sin and shame ;
The bolt that strikes such true friendship apart

SEED-SOWER AND CORN-PLANTER. The above is a correct representation of an implement manufactured by EMERY & Co., Albany, N. Y., for planting Indian Corn in hills or drills at any given distance; and also, for sowing turnips, carrots, beets, parsnips, and all other seeds, in the same way. The largest of these implements, with two horses and a man, will plant twenty acres of corn in a day; smaller ones, with one horse, about half as much. The latter are preferable for common farmers, because they afford all desirable speed. In using it, the operator takes the handles as with a wheelbarrow, and walks erect. The Machine making its own furrow, counting and measuring its own quantity of seed, deposits it in hills or drills at pleasure, covering the seed after it is dropped, and compressing it after it is covered, by means of a roller, and doing the whole at one and the same time. It would be difficult to imagine how it can be more complete, unless one can impart to it the locomotive power of the horse or steam-engine. It is believed, that every farmer planting no more than six or eight acres, will save labor enough in a single season to pay for it. The land, however, should be free from stones, and well prepared with the plough and harrow.

Comes back to the breast that directed the dart;
The strong one may pull down the temple's proud walls,
But its ruins shall cover them both when it falls.

JUDGE LEWIS, *Pennsylvania.*

ANIMAL MANURES.

What is said generally of animal manures?

Animal substances are better fertilizers than those of vegetable origin, on account of their chemical constitution and the facility with which they decompose; thus acting more promptly and rapidly; and they also furnish more manure in proportion to their bulk.

What are the most common animal substances used for manures?

The liquid and solid excrements of domestic animals; night soil, or human excrements usually manufactured into what is called poudrette; guano, or the excrements of sea fowls; the bones of all animals; all kinds of fish; and, also the flesh, blood, wool, hair, horns, and hoofs of animals.

What is said of the excrements of horned cattle?

They are generally esteemed more valuable than those of sheep and horses; fermenting more slowly on account of the smaller quantity of their nitrogen; but, on this account they retain their fertilizing energy longer, and produce more lasting effects on the soil. They also contain much less heat than the dung of the horse, which is in part owing to the greater amount of water in them.

What is said of the excrements of the horse?

They have more nitrogen than those of horned cattle, and consequently are liable to a rapid fermentation; and in a few weeks will lose half of their original weight. On this account they should as speedily as possible be mixed with charcoal, muck, or earth rich with vegetable matters. For the same reason, horse dung may to advantage be at once ploughed into the soil before fermentation occurs; and from its tendency to ferment and develope heat, it is admirably adapted to enter into all composts.

To what is the animal body compared?

To a furnace in which vegetable substances are burned, where every thing assumes a gaseous form, and escapes into the atmosphere, with the exception of the inorganic or mineral constituents of the plants ; these remain in the furnace in the form of ashes, whilst from the animal body they are evacuated in the fœces and urine. The ashes of the furnace, and the fœces and urine of animals, consist of exactly the same ingredients.

What is said of human excrements, as a manure?

They are very active, and differ essentially in composition from those of all domesticated animals. Their own quality probably varies according to the food from which they are produced. The excrements voided by human beings who live chiefly on animal food, are much more active and efficient as manure, than those which proceed from persons whose food is principally composed of vegetables.

What is said of the excrements of sheep?

They are particularly beneficial to soils containing much vegetable matter. The dung placed in the soil, before decomposition has taken place, or is voided over it, produces a speedy and energetic effect, but is soon exhausted. When used abundantly it often gives too much vigor to the first crop, and accelerates vegetation in too great a degree. It should, therefore, be used in smaller quantities, both as regards weight and volume, than any of the other kinds of manure. In most cases its action does not extend beyond the second crop.

What is said of the excrements of swine?

Although there may be some difference of opinion on the subject, it is generally conceded that they form a rich manure. Having, however, a strong and unpleasant odor, and often imparting a rank taste to crops upon which they are used, it is advisable that other manure be used in the culture of roots designed for family food. The excrements of swine are colder and less inclined to ferment than those even of the cow, and should therefore be combined with other manures, or made into composts. If hogs are kept supplied with vegetable substances with which to mix what falls from them, an amount of manure is formed far greater than would have been imagined—perhaps equal in value to their feed—at least, till the period of being fattened.

What circumstances operate to effect the value of dung from farm animals?

It is effected first, by the season of the year; second, by

the age of the animal; third, by the sex; fourth, by the mode of employment; and fifth, by the kind of food. Accordingly the dung voided in warm weather is better than that in winter; that from young animals is not as good as that from older ones; that from males is better than that from females; that from animals at work less valuable than that from those not laboring; and, that from those fed on grains and seeds better than those fed on straw or hay.

What is said of poudrette?

It is a preparation of night soil or human excrements, by being mixed with powdered charcoal, half burnt peat, or soil which is rich in vegetable matter. Quicklime has sometimes been used for the same purpose; but, although it destroys the odor, it dissipates at the same time a large portion of its ammonia. During the decomposition of night soil, an evolution of carbonic acid, ammonia, sulphuretted and phosphuretted hydrogen takes place. After the escape of these gases, the odor ceases, and the remainder, when dried, constitutes what is sold under the name of poudrette. The odor of recent night soil may be destroyed, and the volatile elements retained, by adding to it gypsum or diluted sulphuric acid.

What is said of the importance for manure of the human liquid excrement?

It has been calculated that the urine of one man will produce in a single year a sufficient supply of nitrogen for the formation of 800 pounds of wheat, or 900 pounds of barley; and, that if all this human excrement were applied to the purposes of agriculture, there would be no necessity for other animal manures.

Of what value is the liquid excrement of a cow?

This is said to be of even more value than that of the human species, because it contains more solid soluble matter, being not less than 900 pounds in a year, and worth at least ten dollars, when guano is sold for twenty dollars per ton, being sufficient to manure one acre and a quarter of land. The value of the urine of a cow is nearly double that of the dung.

What does Squarry say of human excrements?

That the liquid and solid excrement of man used together, forms, from its combination of ammonial salts with the phosphates of magnesia and soda, the most valuable compound that can be devised, and its extensive use, will confer a double benefit to the farmer and to the public, as well by the removal of matter which is now considered as a nuisance, as by increasing the produce of the soil. Macaire says that 100 parts of human urine are equal in their fertilizing power, to 1300 parts of fresh dung of the horse, or 600 of those of the cow.

Wherein generally are farmers neglectful of their interests?

In not having their stables, and cattle stalls, and barnyards so constructed as to save the liquid as well as the solid excrements of their stock. If they did this in connection with other available means for increasing the amount of fertilizing agents, rarely would they be under the necessity of purchasing manure. In this way they would soon double the amount of their products.

What is said of the excrements of birds?

They are so valuable as fertilizers as to deserve the attention of every agriculturist. The droppings in the poultry yard should never be permitted to be lost. They are worth far more than the labor of saving them. And the excrements of pigeons are particularly valuable. In some parts of Europe they are used to advantage for flax crops. The dung of birds owes its fertilizing power to the large amount of ammonia and phosphates which it contains. To be rendered in the highest degree efficacious, it should be mixed with other substances before undergoing fermentation.

What account is given of guano?

Guano consists principally of the dung of sea birds, and is found on islands of the Pacific and Atlantic oceans, in tropical latitudes. There are also mixed with it the remains of food and the carcasses of these birds, and seals; those islands having been their places of resort for rearing their young through unknown ages. Here there is but little rain, and of course there has been but little loss to the substances thus deposited; and, in some places it is found to the depth of fifty or sixty feet. The food of those animals being fish, their fœces are rich in nitrogen, and consequently of the first grade of fertilizing efficacy.

For what properties are bones valuable?

Both for the organic and mineral matters they contain. The bones of different animals may differ in their composition; but the phosphate of lime constitutes the greatest proportion of the matter of dry bones; the amount being from forty to sixty per cent. of their weight. Eight pounds of bone dust are equal in phosphates to 1000 pounds of hay or wheat straw. And the value of bones is additionally valuable on account of the gelatine and other organic matters in their composition.

How are bones prepared for use as manure?

They may be ground or otherwise reduced to powder, or small fragments, and then mixed with the soil. They are generally boiled before being ground, to extract from them their oily substances. This, it has been said, does not diminish their fertilizing power beyond the weight of what was removed. In the neighborhood of large cities, where bones may be collected in large quantities, there are mills for grinding them. Where mills do not exist, bones might be crushed, and the fragments put in compost manure heaps.

What is said of fish as a manure?

They become a powerful fertilizer. On the sea coast, and in some instances at the mouth of large rivers, where found in abundance, they are used for this purpose. The most common way is to spread the fish on the surface, and in a few days to plough them under. But the better mode is to cover them with quicklime, and subsequently to mix them with earth. In a short period they are decomposed. Or they may be strewed in layers, on compost beds, with peat, ashes, slacked lime, charcoal, and vegetable matters. Any kind of fish that can be had, and is of but little or no value for food, is taken for manure.

What other animal substances are good for manure?

The blood, and flesh that has become unfit for food; animals that have died from disease; the horns and hoofs; hair and wool; woolen rags, feathers, and old hats; old shoes and boots, and the fragments of leather from the shops of those who work it, are all good for manure, and should be carefully preserved and mixed in compost heaps. Also the refuse of the shambles, filth and all, should be used in the same way. The same may also be said of the animal offal of tan-yards. English agriculturists consider that five or six hundred pounds of the shavings of horn found about the shops of turners and comb-makers, and of the hoofs when chopped fine, are sufficient for an acre of land; yet more would doubtless be better.

What is said of the action upon the soil of different kinds of animal substances?

Animal substances containing much water, as with flesh and blood, decay rapidly, and operate immediately and powerfully; but those which are dry, as horn, wool, and hair, decompose and act slowly, and last perhaps several seasons; while bone, like horn, may act for several years, as they are very productive of earthy matter.

How may the value of some animal substances, very frequently lost by neglect, be estimated?

It is said that the dead body of a cow, ox, or horse, that has died from disease, if properly buried in a bed of peat, or other similar vegetable substance, will yield at least a dozen loads of rich manure. And butchers' offal, when thus preserved and used, will yield ten times its weight of more valuable manure than is found in the barn-yard.

What estimate has been placed on human excrements?

If every human being voids annually enough urine to manure an acre of ground, then a family of ten persons, if so minded, could save enough to enrich ten acres; and the inhabitants of the city of New York, provided there were 500,000, if means were provided to collect and convert their liquid evacuations into manure, would fertilize 500,000 acres of land; which if well cultivated would yield vegetable food double the amount for their own consumption; and enough also to rear and fat the farm animals required for their nourishment. The same estimate may be made of the inhabitants of all other localities.

How might the liquid excrements be best preserved?

In the country there should be for each barn, stable, and barn-yard, a large cistern or tank for the purpose, of dimensions to contain all that could be collected. To this there should be gutters from the stalls to convey whatever is voided from the horned cattle and horses. Into this cistern or tank might also be carried in slop pails whatever is taken from the chambers of the mansion. Here also, could be deposited the soap-suds formed in the various operations of the kitchen. The value of what would thus be saved in a single year would balance the cost of the fixtures.

In what other way might they be preserved?

In the barn-yard and the cattle stables there might be successive coats of vegetable mold placed to receive them; and when sufficiently saturated, to be removed to a compost heap, that a new coat may succeed it. And in a retired place at a convenient distance from the mansion, might be a heap of this mould, on which daily the chamber slops might be emptied; and, as frequent as needful, to receive on its upper surface, a fresh coat of this mold, so that by the end of the year it would contain many cords of the very best of manure.

How might the human liquid excrements of the city be saved?

If there were to each house two thirty gallon oil casks placed in the yard, to be alternately used for receiving them, the process would be simple. Let it be supposed that one cask would be filled in the first half of the month. At the middle of the month the teams from the neighboring farms would simultaneously appear, and in a single night would remove these casks from a whole city. From the middle of the month to the end of it, the alternate casks would be filled; when the same teams would return with the casks they before took, to be again filled, and at the same time taking away as before, the ones filled in the two preceding weeks.

How much profit would this probably yield to each family?

According to the above scheme, each family will have filled in the year twenty-four casks, which at five New-York shillings each, will be a saving of fifteen dollars to a family annually, and more than a million of dollars to the city of 500,000 inhabitants. At that low price, it is believed, there would be the greatest competition for the privilege of obtaining it, as it would be the cheapest manure to be had.

What other gain would be derived from such a plan?

In a few years, if it were fully carried into effect, the lands in the neighborhoods of large cities would become so fertile, there would at least double this sum be saved, in the reduction of the price of summer vegetables and milk, from being produced in so much greater quantities.

LIGHTS AND SHADOWS OF AGRICULTURE.

We live and move in a world of wonders. Every blade of grass, every leaf that flutters in the breeze, and every germ is an organized and living body. Every plant and vegetable is as capable as the human system of imbibing and digesting its appropriate food, and although it becomes me to speak with modesty on a subject in which I myself am but a learner, yet allow me to say that the application of science to agriculture has already settled many of the laws that regulate the growth of trees and plants, with a certainty approximating that which attends the calculations of the astronomer.

For instance, by an analysis of wheat, we ascertain the ingredients and the food it requires for growth and productiveness. We know that it needs phosphate of lime, and that it is useless to attempt its cultivation where the soil is wholly deficient in this element. Hence we are as competent to feed a crop of wheat, as a flock of sheep, or a brood of chickens; but without this knowledge, which science alone can furnish, we might apply a kind of manure which would be injurious, and perhaps destructive. But, suppose, however, such food be not administered, that the ground is prepared, and the grain sown; it may flourish for a season, because it may find its proper nutriment in the soil, but let it be sown year after year, and it will prove less and less productive, and ultimately fail.

It has been the practice of countries producing wine, to bury the prunings of the vine at its root; and chemical analysis has lately discovered that it contains a large proportion of potash, which is essential to its growth and productiveness. Again, it has long been known, that a tree planted in a soil in which one of the same species has previously grown, will flourish but poorly. Why is this? If a chemist analyses both the tree and the soil, the former will be found to contain, and to require for its growth and fruitfulness, elements of which the latter is deficient. Hence we learn with what kind of material that soil should be fertilized. We have seen instances also, in which barn-yard manure had been so abundantly applied, as to retard or prevent vegetation, and where sand, gravel, virgin loam or clay, was worth more to that soil than these manures; and we have seen other instances in which mineral manures, as lime, had been so profusely applied as to lose all efficacy. Why was it? Chemical analysis affords the reply, and discovers to us that the soil was surcharged with these elements, and makes known the materials and the proportion requisite to revive productive energy.

If by the application of science to agriculture, we can fathom the depths of nature, and bring up to the light, for the admiration and benefit of mankind, her previously hidden treasures, shall we hesitate to do it? Or, if others, fired with greater zeal, and endowed with more ample means, venture into the labyrinths of science, explore the springs of nature, learn how her curious machinery acts, and then returning, unfold and explain her various processes, and teach how to practise art more successfully, shall we refuse to avail ourselves of the benefits of their labor? What vast quantities of vegetable and mineral manures now lie buried in the earth, which might, by the application of these sciences, be appropriated to the fertilization of the soil!

By a natural law every tree, plant, and herb, from the cedar of Lebanon, to the flag on the Nile; from the loftiest oak of the forest, to the humblest daisy of the meadow; from the fantastic parasite luxuriating in solstitial air, to the little flower that peeps from Alpine snows; every thing endowed with vegetable life, requires its own peculiar aliment to sustain its vigor, and propagate its growth. However varied the sustenance may be, and whether derived from earth, air or water, if it be withheld, or mixed with uncongenial elements, deterioration and decay are inevitable.

One of the greatest embarrassments of the farmer is the want of proper education for his calling. In other arts and professions we employ only those who are properly trained for their business. The reason is evident. We do not expect others to succeed. But why do we not apply the same logic and practical sense to agriculture? We do not encourage an uneducated physician or a mechanic who is not master of his trade; why then do we expect men to succeed in farming who know no more of the nature of soils, nor of the adaptation of different species of manures to the various kinds of grain, grass, vegetables, and fruits, than they do of the rotation of day and night, or the seasons in one of the newly discovered planets?

Why have so many of our sons forsaken the farm, for the office, the counting-room, the warehouse, and the professions? Why such a rush by sea and land from the homes of their childhood, for the glittering dust of California? Why have they not retained

"That fond attachment to the well known place,
Where first they started into life's long race,
Which keeps its hold with such unfailing sway,
We feel it e'en in age at our last day ?"

Alas! what has driven them from the homestead overshadowed by the elms which their fathers planted, and under which in their boyhood, they wrought out so many youthful wonders? Nothing, save that lack of interest and skill in farming, which would have rendered it as lucrative and honorable as other pursuits, and which education alone can supply. Such examples which have fallen under our own observation, create a demand which I only reiterate, when I say that our farmers must be educated.

"Our fathers," it is said, "were not educated, yet they were successful farmers." True, but they possessed advantages which we cannot enjoy; then the soil was new, and of course more productive; now when its fertility has been diminished by successive crops, it must be restored and increased by artificial processes, to the success of which knowledge is indispensable. Besides, the progress of the other arts enables men to realize better profits than they then received, and corresponding improvements not having been made in agriculture, labor has here been less liberally rewarded.

Others insist that common sense alone is needful to be a good farmer. Common sense, indeed, such as they recommend, is a very good thing; yet, if it were possessed by all, why not rely upon it to make skillful mechanics, artists and teachers, as well as farmers? When common sense can manufacture a steam engine, construct a rail road, or teach mathematics, without education, we may expect it will successfully conduct the operations of the farm. Till then, let us not rely upon common sense for miracles, nor offer it as an apology for ignorance or idleness. Common sense is as valuable as it is rare, but let us remember that it never yet made a plough or planted an orchard, till it was properly instructed.

Our country boasts of men who have distinguished themselves in arts, letters, and morals; of men whose fame is our inheritance and glory. We are proud of the names of Rittenhouse in astronomy; of Franklin in philosophy; of West,

Allston and others in the fine arts, and in politics of John Hancock, of Patrick Henry, of Samuel Adams, of John Adams, and of his illustrious son John Quincy Adams, and not least of Fisher Ames, who gives to this town* an enviable distinction; whose hands planted many of the beautiful elms that adorn this village, and whose bounty distributed trees among his fellow-citizens, of the fruit of which its present inhabitants partake.

But where are the men whose names will go down to posterity honorably associated with these in the art or science of agriculture? True, we might speak of the farmer of Mount Vernon, who first called the attention of Congress to this subject—of Washington, whose name awakens the most grateful sensations in all our hearts; of the farmer of Monticello, whose genius first gave proper curvature to the mold board of the plough, and whose taste for rural life sought gratification in the perusal of his favorite classics in the bowers of his garden, with the earliest songsters of the morning; of Charles Cotesworth Pinckney, and Madison, and of other worthies, distinguished in this art, whose names are embalmed in the memory of their grateful countrymen.

At the present time we especially need young men who will devote exclusively to agriculture their talents, their fortunes, and their lives, and will rely on posterity to appreciate their improvements and discoveries, and to honor their memory. And, here the aspirant for fame has a fairer prospect of distinction, than he can find in any kindred art or science; first, because less progress has been made, and secondly, because the successful farmer of New England must be well educated for his profession, or he can never compete with the cultivator on the prairies and intervals of the West. There the soil is new and productive, and nature does at present do for him, what education must accomplish for us; and, it is capable of demonstration, that with the aid of science, the farmer of Massachusetts can compete successfully either with the southern planter or the western cultivator.

Nearly one thousand are added to the population of New England every week. And how are they to be fed? By the surplus products of the West? But how are the latter to be purchased? By the proceeds of the arts and manufac-

* Dedham, Mass.

tures? Highly as we prize these handmaids of this art; much as they have benefited the farmer in increasing the value of his land, and creating a ready home market for his productions; much as we think it the duty of our country to protect its own industry, and much as we believe that this has added to the independence, wealth, and importance of Massachusetts; yet shall the descendants of the Puritans, of Brewster, of Endicott, of Winthrop, spurn the chief inheritance of their fathers, and leave the natural resources of wealth and power for the uncertain results of trade and manufactures? *No!* A remote remonstrance breaks on my ear! It comes from a thousand cottages and happy firesides! It rings through our valleys and echoes among the hills. No!! Our descendants shall range the hills which their fathers cultivated; they shall eat the fruits of their gardens and orchards, and shall fling on the passing zephyrs a melody in praise of agriculture, sweeter than any songs which Grecian or Roman bards ever sung in honor of Ceres.

There is even in New England, much land to be possessed, but it consists not so much of forests to be converted into cultivated fields, as of deteriorated land, bogs, and meadows to be reclaimed, and of barren hills and plains to be fertilized, and covered with waving grass and grain. For such purposes, science alone is adequate and indispensable. Let, then, the immigrants who throng our shores, go and settle in the far West; but let the sons of New England, with their muscular frames, industrious habits, and generous hearts, divide among themselves the farms of their fathers; so that with less land and higher cultivation, they may be able to say with the poet,—

"Rough is her soil; yet blessed in fruitful stores;
Strong are her sons, though rocky are her shores;
And none, ah-! *none*, so lovely to my sight,
Of all the lands that heaven o'erspread with light."

What wonders has science wrought in other departments within the last half century? With a power and skill almost divine, man has seized on the very elements of nature, and made them subservient to his will. In obedience to established physical laws, he generates an agent which works for him in air, earth, or water; and from the tips of his fingers, with as much ease as one plays on an instrument, he sends forth the "winged lightning" to do his bidding.

But why should not these agents work for the farmer, as well as for the mechanic, the manufacturer, or the navigator? Why should not steam aid in the production of manure, as well as in the manufacturing of acids and alkalies? Doubtless it would ere this, if thought, enterprise, and capital had sought its application here, with equal zeal and perseverance as in other departments of labor; and we should to-day have been driving our ploughs as well as our cars, filling our barns as well as our warerooms and storehouses, and expediting the various processes of agriculture, as well as those of the other arts, by its magic power.

We may be deemed chimerical, but we have long ago ceased to wonder or be surprised at any discovery or invention. The improvement of to-day supersedes that of yesterday. No project, of whatever magnitude, whether the building of a rail road from the Atlantic to the Pacific; the tunnelling of the Rocky Mountains; the traversing of old ocean's bed with the mystic wires, or winding them round the globe, is too great for the enterprise of the nineteenth century.—*Address before the Norfolk Agricultural Society*, 1849, *by* HON. MARSHALL P. WILDER.

AGRICULTURE FAVORABLE TO MORALS

Agriculture, pursued as a mere branch of trade or commerce, or a mere instrument of wealth, will be found to have influences upon the mind, narrowing and restricting its operations and aspirations, corresponding with any other of the pursuits of mere avarice and acquisition, and which even those of the learned professions, when pursued wholly with such views, are sure to have. But when followed without exclusive views to mere gain or profit, it is far from being incompatible with a high state of intellectual cultivation. Many of the sciences are the handmaids of agriculture, and serve, as well as ennoble it. Its practical pursuit, though it occupies, yet it does not exhaust the mind; but, within certain limits, inspirits and invigorates all its faculties. A spiritual mind may spiritualize all its operations; a religious mind sees, in its wonderful and curious processes, and their marvellous results, many of the adorable miracles of a beneficent Providence.

It is believed that the agricultural profession is highly favorable to good morals; I shall not presume to say more so, than any other; but it will not be too much to say more so than many others. Perhaps it will be said, that the agricultural districts of England and other countries yield their full proportion of crime. I will not peremptorily deny what is often confidently asserted; but I am not ready to concede to it until other proof than I have yet received, is furnished. As far as my own personal observation and experience go, my conviction is the reverse of this. Two fruitful sources of crime are to be found in excited passions and in powerful temptations. Agricultural occupations, so far from exciting, tend to exhaust and allay the passions; and the retirement and seclusion of the country present fewer temptations than the tumultuous life, the opportunities for vicious association, the disorderly hours, and the infinite variety of attractions and engagements of city life.

Among, however, a degraded population, poor and half fed, without education, without any interest in the soil, without friends to take an interest in their welfare, without any sentiment of the value of character, without self-respect, accustomed to pass their unoccupied time in drinking-houses and in degrading pleasures, and treated and lodged without distinction of sex, and without any regard to the common decencies of life, it is not surprising to find a nursery and hot-bed of crime, where it shoots up in startling luxuriance. My acquaintance with many of the villages and rural districts of England and Scotland, satisfies me that the favorable moral influences which might be looked for from rural life and agricultural pursuits, are there found in full operation; and under a system of more general and improved education, and especially under institutions which would give those encouragements to labor which are the most powerful motives, as well as the proper rewards of industry and good conduct, these influences might be expected to be even more general.

Let me speak of a district of country with which I have been many years familiar—I mean the State of Vermont;—it is a purely agricultural district; it contains about three hundred thousand inhabitants; its climate is cold and severe; its soil, with some exceptions, of moderate fertility, and requiring the brave and strong hand of toil to make it productive. It has public and free schools in every town and parish, and several seminaries of learning of a high character, and where the branches of a useful and literary education are taught, at an expense so moderate, that it is placed within the reach of persons even of the most humble means. It has everywhere places of religious worship, of such a variety, that every man may follow the dictates of his own conscience, where religious services are always maintained with intelligence and decorum, sustained wholly by voluntary contributions; and sects of the most discordant opinions live in perfect harmony, recognizing in their mutual dependence the strongest grounds of mutual forbearance and kindness.

Taken as a community, the people of Vermont are the best informed I have known; and they have numerous and well chosen circulating libraries in almost every town. They have no connection with any large market; and the produce which they have for sale goes through intermediate hands to the great marts. They have few or no poor, and those only the immigrants who may stroll there from neighboring provinces. The sobriety of the people is remarkable; they are every where a well dressed people; their houses abound in all the substantial comforts and luxuries of life: and their hospitality is unbounded. They understand their rights and their duties, and have often distinguished themselves by an extraordinary bravery and manliness, in their vindication and defence.

No where is public order more maintained, or public peace better preserved, than in Vermont; large portions of the inhabitants never bolt a door, nor fasten a window, at night; and in a village of some thousand inhabitants, I have known a garden stored with delicious fruit, with no other fence than one which served as a protection against cattle, as entirely secure from intrusion and plunder, as if it had been surrounded even with a prison wall. In this State crimes are comparatively rare; courts of penal justice have little occupation; the prisons are often without a tenant, and there has scarcely been a public execution for half a century. From such an example of a community almost exclusively agricultural, I have a right to claim for agriculture and rural life, all the beneficial moral and social influences to which its enthusiastic admirers pretend.

The present excited state of the civilized world ought more than ever to call the attention of philanthropic individuals and of governments to the immense importance of agriculture. I have been in France during the exciting scenes of political revolution, in which I have seen very many thousands of workmen without the means of support from their labor, and large bodies of them actually dependent on public charity for their daily bread. It is not the danger to public liberty and order, growing out of such large unemployed and destitute multitudes, which so much disturbs me, as the actual suffering to which they are exposed, and the melancholy future that lies before them.

In London I have encountered, with an extreme depression of heart, thousands of squalled, ragged, miserable poor, without resource but from crime or charity. A distinguished manufacturer in one of the most industrious counties in Enland, states that there are at least five hundred thousand operatives without employment, (1848), and many on the borders of starvation. Tradesmen and professional men will tell you that every trade and profession is over stocked; and one is daily saluted with the melancholy, not to say presumptuous exclamation, that there are too many people. This reminds one of the sad shipwreck of the French frigate, the Alceste, when many of the wretched survivors, who were floating upon a raft composed of fragments of the ship, deemed it necessary to their own safety to drive by force a large portion of their suffering companions into the sea—a sad and horrible alternative.

Must we affirm, that there are too many people in the world? and that thousands and millions are born into it, for whom there is no place at the table of a beneficent Providence? Why, in France there are more than nineteen millions of untilled and unoccupied acres, and in England more than eight millions, all capable of yielding food and clothing to countless human beings; and here and in other lands there are millions of acres, for the want of labor which might be applied, that produce not a moiety of what they might be made to produce. In ancient Rome, seven acres were the ordinary size of farms, on which a family might be sustained. In Flanders, on a soil which was once sterile, but which human labor made productive, two and a half acres will give ample support for a man and wife, and three children, or what is considered equal to three grown up men and a half; and add to it three acres more, which this amount of labor is more than sufficient to cultivate, and you add a considerable surplus for other purposes.

The great cause, then, of the evils complained of, is, that the cultivation of the earth is deserted; and that such innumerable multitudes pour into cities and towns, and, filling every profession and every mechanical art and trade, destroy each other by a competition in articles of which the demand is necessarily limited. There may be too many physicians, too many lawyers, and too many ministers, for them all to get a sufficient and honest living; and too many hatters, and too many printers, and too many shop-keepers; for, besides that these persons furnish more of a particular article or service than the community require, their work is in general only formal; they only manufacture—they do not produce; they do not, like the grower of bread and clothing, create that which may be said to have a substantial and permanent value. For when was the time when there was too great an abundance of the materials—I mean particularly of those which can be kept from year to year—for food and clothing, for human subsistence and comfort? As long as this state of things continues, there must be misery in the community; as the population increases, this misery must increase.

In cities, money becomes the standard of prosperity. Wages are paid in money; money is the instrument of subsistence, of gain, and of pleasure. Avarice, under these circumstances, becomes stimulated to excess, and often leads to crime. Men's happiness becomes dependent upon that which has no intrinsic, but only an arbitrary value,—a value which is always capricious, and continually changing. If men could be induced to cultivate the earth, and, trained to simple habits of laborious and rural life, be satisfied with what that affords them; if they would measure their prosperity and wealth, not by so many shining pieces of gold or silver, which they have hoarded in their closets, but by the produce of their labor in bread and clothing, and the various and innumerable simple luxuries of life, with which a kind Providence so often blesses the labors even of the most humble, how changed would be their condition.

If men could be as well satisfied to breathe the fresh air of their native mountains and forests, as the corrupt and pestilential atmosphere of crowded streets and confined dwellings, from which both sun and light are shut out; as well content to enjoy the simple and healthful sports of the country, as the exciting and exhausting pleasures of city life; if their taste could be better satisfied to contemplate the verdant fields, waving with crops or enamelled with flowers, than carpeted and gilded halls; if they could be taught to prefer skies painted with clouds of brilliant hues, and studded with stars whose lustre never grows dim, to palaces blazing with artificial lustres and adorned with the far inferior magnificence of man's genius and taste; if, indeed, by any possible means, you could induce men and women, and, above all, the young, to love the country; if, in a word, you could keep them in the country by an attachment to its simple labors and recreations, and prevent their crowding cities to repletion, and thus destroying by competition the ordinary professions and trades which prevail there, where so many vigorous young men, and so many fair and blooming maidens rush in, like flies in a summer evening into a blazing taper, to find too often the grave of their health, hopes, happiness, and virtue,—what an immense gain would be achieved for morals and for humanity.

But while matters continue otherwise, while such millions of acres remain unoccupied, while such thousands upon thousands crowd into the learned professions, and into the mechanic arts and trades, and fill cities to excess, under the powerful stimulus of a vain ambition, an inordinate avarice, or a love of excitement, luxury, and pleasure as inordinate and unrestrained, we shall continue to complain of a superabundant population; and that superabundance, wherever the wave accumulates, will bring with it crime and misery. The decrees of Providence cannot be violated with impunity. Every inordinate and unrestrained passion will yield its bitter fruits. Every infraction of the laws of man's moral constitution, will be followed with its just and inevitable penalty.

Competition, which, when excessive, is so hurtful and serious in the mechanic arts and trades, is, in agriculture, always a good. Under proper management the earth cannot be made to produce too much. It is a generally received theory, that as yet there has been no surplus produce; and that one year's entire failure of the crops would cause the destruction of the human race. I shall not speculate upon this theory, which, possibly may be well founded, but which Heaven forbid that it should be put soon to experiment. In some years there may be a surplus of some products, and then there may be a dearth of others. But I have never known too much grown. I have never known the great mass of mankind enjoying too much bread, or too much clothing, or too many of the substantial comforts of life. If they get the comforts, or their substantial necessities are supplied, then certainly we should desire that they should have the luxuries of life in addition,—above all, those simple luxuries which are the produce of their own honest labor, and to which that circumstance alone will always give a peculiar zest.—REV. HENRY COLMAN.

THE GLADNESS OF NATURE.

Is this a time to be cloudy and sad,
  When our mother Nature laughs around;
When even the deep blue heavens look glad,
  And gladness breathes from the blossoming ground?

There are notes of joy from the hang-bird and wren,
  And the gossip of swallows through all the day;
The ground-squirrel gaily chirps by his den,
  And the wilding bee hums merrily by.

The clouds are at play in the azure space,
  And their shadows at play on the bright green vale,
And here they stretch to the frolic chase,
  And there they roll on the easy gale.

There's a dance of leaves in that aspen bower,
  There's a titter of winds in that beechen tree,
There's a smile on the fruit, and a smile on the flower,
  And a laugh from the brook that runs to the sea

And look at the broad-faced sun, how he smiles
  On the dewy earth that smiles in his ray,
On the leaping waters and gay young isles;
  Ah, look, and he'll smile thy gloom away.

BRYANT.

THE VERNAL SHOWER.

In a valley that I know—
    Happy scene!
There are meadows sloping low,
There the fairest flowers blow,
And the brightest waters flow
    All serene;
But the sweetest thing to see,
If you ask the dripping tree,
Or the harvest hoping swain,
    Is the rain!

Ah the dwellers of the town,
    How they sigh,
How ungratefully they frown
When the cloud king shakes his crown,
And the pearls come pouring down
    From the sky!
They descry no charm at all
Where the sparkling jewels fall,
And each moment of the shower,
    Seems an hour!

Yet there's something very sweet
    In the sight,
When the chrystal currents meet,
In the dry and dusty street,
And they wrestle with the heat,
    In their might!
While they seem to hold a talk,
With the stones along the walk,
And remind them of the rule,
    To "keep cool!"

But in that quiet dell
    Ever fair,
Still the Lord doth all things well,
When the clouds with blessings swell,
And they break a brimming shell
    On the air;
There the shower hath its charms,
Sweet and welcome to the farms,
As they listen to its voice
    And rejoice!

REV. RALPH HOYT.

CHEVIOT BREED OF SHEEP.

MINERAL MANURES.

What is said of mineral manures?

It is well known that various substances belonging to the mineral kingdom are capable of promoting the growth of plants. These substances have been termed *stimulating manure* in contradistinction to manures derived from the animal and vegetable kingdoms, which are called *nutritive manures.* This distinction, however, was applied before it was known that mineral substances are nutritive; and the present theory is, that they act upon the soil by improving its texture, or by rendering soluble the parts of it which are insoluble,—or by otherwise fitting it to promote the growth of plants; and, that they act immediately upon the plant itself, by being received into its substance.

Is the process of this action understood?

It may not be fully understood. Nevertheless, it is well ascertained, that certain earths, oxides, and alkalies, or earths, oxides, and alkalies, combined with acids, pass into the substance of the plant, absorbed it may be, in part, from the atmosphere, but chiefly, along with the aqueous portion of the sap, from the earth in which the roots are fixed.

What mineral substance is said to be most prominent as a manure?

Of all mineral substances known to us, lime is that which performs the most important part in improving the soil and promoting the growth of vegetables. It is found in nearly all soils that are capable of sustaining vegetation, and in combination with different acids, in nearly all vegetable substances.

What is said of lime in its natural state?

Limestone is the natural state of lime, or quicklime, and is by chemists called the carbonate of lime—that is quicklime and carbonic acid; 28lbs of the former and 22lbs ot the latter, making 50lbs. of limestone. In its natural state, limestone is too hard and compact to be diffused in the soil; and even quicklime would be too solid, were it not, that through its combination with water and carbonic acid from the atmosphere it splits and crumbles to powder.

How is limestone converted into quicklime?

By exposing it to a very strong heat. The process is by first reducing or breaking the stone into lumps of the size of a child's head, and then placing these in a kiln prepared for the purpose, so that a fire being made at the bottom, the heat will ascend through the whole, and cause the carbonic acid to disappear, or to be disengaged from it. What remains is called quicklime, because it is of a caustic, burning nature, having a strong affinity for water and carbonic acid, from which it has been thus forcibly separated.

Are there different kinds of limestone?

There are. The most common is hard, and of a grayish white, and is the one most used for mason work as well as for manure. Then there is a soft limestone, of which chalk is a sample. There is also a yellow limestone, called magnesian limestone, from which magnesia is obtained. And it is found in dark colors of numerous shades even to black, as may be seen in different kinds of marble.

In what states is lime most frequently known to exist?

First, as a carbonate, that is, in the compact state of limerock, combined with water and carbonic acid; *second*, as hydrate of lime, that is, immediately after being slaked and when it retains its caustic properties; *third*, as the sulphate of lime, which is the same as plaster of Paris or gypsum; and, *fourth*, marl, which is the same as lime stone, or the carbonate of lime, reduced to a powder and mixed with earthy matter.

By what names is limestone called after being burnt?

Most frequently quicklime, because of its caustic, burning quality; and, it is also called burned lime, caustic lime, and hot lime; but, all these names denote the same thing; that is, limestone, or the carbonate of lime, from which the carbon is expelled by heat.

What is the slaking of burnt lime?

Its literal signification is to extinguish thirst. Hence, when the term slake is applied to burnt lime, it is to pour water upon it. The quicklime as it were drinks in, or absorbs the water, and immediately becomes hot, swells up, and gradually falls to powder, The heat thus generated is sufficient to ignite or set fire to any combustible substances contigious to it. In this way vessels freighted with lime are sometimes consumed when water accidentally gets admission to it.

Why does quicklime become slaked from exposure to the air?

Because of its power to absorb from the atmosphere both water and carbonic acid. The process is less rapid than when water is poured upon it; but, is equally complete, the large lumps all crumbling to fine powder.

Where is lime found besides in limestone?

Sea shells subjected to heat in a furnace or otherwise become quicklime of equal value, and some think of more value than that obtained from limestone. Shells also without the agency of heat, when lying on the ground in masses or mixed with it, become broken and reduced to fine particles, and are then called shell sand. As such they make a valuable manure, either for top dressing or composts, essentially the same as the pulverized carbonate of lime, or slaked lime.

What is said of gypsum?

It is the same as sulphate of lime, that is a salt formed from the oxidation of lime. It is reduced to powder by being pounded or ground in a mill, and then spread on the ground, when the leaves of clover and other plants begin to cover the surface; and the operation should be performed, if possible, during slightly showery weather, it being beneficial that the leaves should be somewhat moistened, so as to retain a portion of the dust. On clover crops gypsum is of most use.

What is said of salt as a manure?

Salt, or the muriate of soda, as it is more correctly termed, is probably essential to the health of vegetables, as well as animals, and we may presume that a mineral thus widely diffused, performs important functions. It exists in all plants, is a constituent of almost every kind of animal and vegetable manure, and is found in most soils in sufficient quantity for the purposes of vegetation. Yet experience has shown that the application of salt as a manure, save in a moderate measure, is attended with no advantage, and usually with injury.

What is said of marl?

Marl consists, as before stated, of a mixture of clay and calcareous earth or lime, in various proportions. The great advantage of marl is, that it dilates, cracks, and is reduced to powder by exposure to moisture and air. Marl in masses would be totally useless on the ground; yet it is necessary to begin by laying it on the surface in heaps, for the more it is heaped the more it dilates, splits, and crumbles to dust; in which state it is fit to be spread upon the ground.

With what substances should lime be mixed?

The best earthy materials for mixing with lime are those which contain a certain proportion of decomposing organic matter; such as the scouring of ditches, the sediments of pools, mud deposited by rivers and tides, and similar substances. The lime may be applied at the rate of two bushels to the cubic yard, and fifty cubic yards of this mixture to the acre, will form an efficient manuring for almost any soil.

What would be the consequence if quicklime were applied to plants?

If quick lime were applied immediately to plants, it would be to them like poison; it would burn them up; but, when spread on the earth, it rapidly attracts water and carbonic acid from the atmosphere, and it is only when thus modified that it promotes vegetation.

How can the greatest benefit be had from lime?

To do this, it must be kept as near the surface as possible. The reason is this; its weight and minuteness give it a tendency to sink; and after a few years of cultivation, a large portion of it will be found to have gone beyond the depth of its most efficient action. Hence it is advisable to spread it on the ground after ploughing; then harrow it well in; and allow it to remain in grass as long as good crops can be had. When the lime is settled down below the reach of a common plough, the subsoil plough will prolong its effect, by enabling the atmosphere and the roots of plants to penetrate the subsoil likewise.

How should lime be proportioned to different soils?

The quantity of lime applied to soils is various, and is dependent upon the nature of the soils, the climate, and other circumstances. In warm countries, a smaller quantity need be used than in those which are cold and humid. The stiff clays for the most part, require a larger proportion of it than the lighter soils; and in the case of such soils as contain much undecomposed vegetable matter, as peat, a quantity should be applied sufficient to decompose effectually the inert matter.

How much lime is to be applied to the acre?

On common soils the first dressing is ordinarily in the neighborhood of an hundred bushels; and, then in four or five years, half as much more. On some heavy clays abounding in vegetable mold, there have been applied six hundred bushels to the acre with decided beneficial results to the land;

yet, it is not impossible nor improbable that half that quantity would have answered as well.

Why does lime require to be repeated?

Partly for the same reason that other manures are to be repeated. And the reasons may be stated, as follows: *first*, because the crops eat up and carry off a portion of the lime; *second*, because of its sinking into the subsoil; and, *thirdly*, because the rains are always washing a portion of it out of the land, and carrying it away to brooks and rivers, where it becomes mixed with the mud and decaying vegetables.

In the ashes of what plants is lime found?

Every plant that has been analysed, with one exception, contains a portion of lime in some form or other, which it must have derived from the soil in which it grew. Wheat in flower, when ripe, the straw, the bran, all yield lime when analysed; so likewise do barley, oats, vetches, and the leaves of various trees, the bark, and the timber; indeed this substance is so universally present in all portions of the vegetable structure, that it may fairly be assumed to be an integral part of all, varying however, according to the quantity existing in the soil in which plants are cultivated.

How is it proved that lime invigorates plants?

It is known that the roots of some plants actually penetrate rough limestone, and in a manner decompose it. This is particularly true of sainfoin, the tap root of which penetrates from ten to twenty feet deep in calcareous stones, and then puts forth new clusters of lateral roots which render the stone loose and friable all around. The deeper the roots of this plant penetrate, the more vigorous does it shoot, even on calcareous rocks or stony places which are only covered by a very thin layer of poor soil.

How have salt and soot been used for manure?

Mr. Sinclair mixed six and a half bushels of salt with the same quantity of soot, and used the mixture on lands sowed with carrots. The result was, that unmanured land gave twenty-three tons of roots to the acre, and the manured yielded forty tons per acre; and Mr. Cartwright found that where unmanured soil gave 157 bushels of potatoes per acre, thirty bushels of soot and six bushels of salt made it produce 240 bushels per acre.

In what manner are salt and lime mixed for manure?

Dr. Dana directs that the proportions shall be at the rate

8

of two bushels of lime to one of salt; that the salt shall be made into a brine, with which the lime is to be slaked, and then both well mixed together, and remain ten days. It is now to be mixed with peat or analagous vegetable substances, and shovelled well together, and after six weeks, is fit for use in the hills of corn, or otherwise. In such a compost, for every five bushels of salt and ten bushels of lime, there may be fifteen cords of peat.

THE OLD FARM HOUSE.

I love these gray and moss-grown walls,
  This ivied porch, this trellis'd vine,
The lattice with its narrow pane,
  A relic of the olden time;
The willow with its waving leaves,
  Through which the low winds murmuring glide,
The gurgling ripple of the stream,
  That whispers softly at its side.

The spring house in its shady nook,
  Like lady's bower is shadowed o'er
With clustering trees and creeping plants,
  That cling around the rustic door—
The rough hewn steps, that lent their aid,
  To reach the shady, cool recess,
Where humble duty spreads a scene
  That hourly comfort learns to bless.

Upland and meadow lie around,
  Fair smiling in the sun's last beam;
Beneath yon solitary tree
  The lazy cattle idly dream,
After the reaper's stroke descends,
  While faintly on the listening ear
The teamster's careless whistle floats,
  Or distant song or call I hear.

And leaning on a broken stile,
  With woods behind and fields before,
I watch the bee who homeward wends
  With laden wing—his labors o'er;

The happy birds are warbling round
  Or nestling in the rustling trees,
'Mid which the blue sky glimmers down,
  When parted by the passing breeze.

And slowly winding up the road,
  The wain has reached the old barn floor
Where plenty's hand has firmly heaped
  The golden grain in richest store.
This 'mid the dream land of my thought,
  With smiling lip I own it real,
Yet fancy's fairest visions blend
  With all I see and all I feel.

Then tell me not of worldly pride
  And wild ambition's hopes of fame,
Or brilliant halls of wealth and pride,
  Where genius sighs to win a name;
Give me this farm-house quaint and old,
  These fields of grain, the birds and flowers,
With calm contentment, peace and health,
  And memories of my earlier hours.

MARY A. LAWSON.

THE EVENING WIND.

Spirit that breathest through my lattice, thou
  That cool'st the twilight of the sultry day,
Gratefully flows thy freshness round my brow;
  Thou hast been out upon the deep at play,
Riding all day the wild blue waves till now,
  Roughening their crests, and scattering their spray
And swelling the white sail. I welcome thee
To the scorched land, thou wanderer of the sea.

Nor I alone—a thousand bosoms round
  Inhale thee in the fulness of delight;
And languid forms rise up, and pulses bound
  Livelier, at coming of the wind of night;
And, laughing to hear thy grateful sound,
  Lies the vast inland stretched beyond the sight.
Go forth into the gathering shade; go forth,
God's blessing breathed upon the fainting earth.

Go, rock the little wood-bird in his nest,
Curl the still waters, bright with stars, and rouse
The wide old wood from his majestic rest,
Summoning from the innumerable boughs
The strange, deep harmonies that haunt his breast;
Pleasant shall be thy way where meekly bows
The shutting flower, and darkling waters pass,
And where the o'ershadowing branches sweep the grass.

The faint old man shall lean his silver head
To feel thee; thou shalt kiss the child asleep,
And dry the moistened curls that overspread
His temples, while his breathing grows more deep;
And they who stand about the sick man's bed,
Shall joy to listen to thy distant sweep,
And softly part his curtains to allow
Thy visit, grateful to his burning brow.

Go—but the circle of eternal change,
Which is the life of nature, shall restore,
With sounds and scents from all thy mighty range
Thee to thy birthplace of the deep once more;
Sweet odors in the sea-air, sweet and strange,
Shall tell the homesick mariner of the shore;
And, listening to thy murmur, he shall deem
He hears the rustling leaf and running stream.

BRYANT.

A RILL FROM THE TOWN PUMP.

Noon, by the north clock! Noon, by the east! High noon, too, by these hot sunbeams, which fall, scarcely aslope, upon my head and almost make the water bubble and smoke in the trough under my nose. Truly we public characters have a tough time of it! And among all the town officers, chosen at March meeting, where is he that sustains, for a single year, the burden of such manifold duties as are imposed, in perpetuity, upon the Town Pump? To speak within bounds, I am the chief person of the municipality, and exhibit, moreover, an admirable pattern to my brother officers, by the cool, steady, upright, downright, and impartial discharge of my business, and the constancy with which I stand at my post.

Summer or winter, nobody seeks me in vain; for, all day long, I am seen at the business corner, just above the market, stretching out my arms to rich and poor alike; and at night, I hold a lantern over my head, both to show where I am, and to keep people out of the gutters.

At this sultry noontide I am cup-bearer to the parched populace, for whose benefit an iron goblet is chained to my waist. Like a dram-seller on the mall, at muster day, I cry aloud to all and sundry in my plainest accents, and at the very tiptop of my voice—Here it is gentlemen! Here is the good liquor! Walk up, walk up, gentlemen, walk up, walk up! Here is the superior stuff! Here is the unadulterated ale of father Adam—better than Cognac, Hollands, Jamaica, strong beer, or wine of any price, here it is by the hogshead or the single glass, and not a cent to pay! Walk up, gentlemen, walk up, and help yourselves.

It is a pity if all this outcry should draw no customers. Here they come. A hot day, gentlemen! Quaff, and away again, so as to keep yourselves in a nice cool sweat. You, my friend, will need another cupful, to wash the dust out of your throat, if it be as thick there as it is on your cow-hide shoes. I see that you have trudged half a score of miles to-day; and, like a wise man, have passed by the taverns, and stopped at the running brooks and well-curbs. Otherwise, betwixt heat without and fire within, you would have been burnt to a cinder, or melted down to nothing at all, in the fashion of a jelly-fish. Drink, and make room for that other fellow, who seeks my aid to quench the fiery fever of last night's potations, which he drained from no cup of mine. Welcome, most rubicund sir! You and I have been great strangers hitherto; nor, to express the truth, will my nose be anxious for a closer intimacy, till the fumes of your breath be a little less potent. Mercy on you, man! the water absolutely hisses down your red hot gullet, and is converted into steam, in the miniature tophet which you mistake for a stomach. Fill again, and tell me, on the word of an honest toper, did you ever, in cellar, tavern, or any kind of dram-shop, spend the price of your children's food for a swig half so delicious? Now for the first time these last ten years, you know the flavor of cold water. Good-by; and, whenever you are thirsty, remember that I keep a constant supply at the old stand.

Who next? Oh, my little friend, you are let loose from school, and come hither to scrub your blooming face, and drown certain taps of the ferule, and other school-boy troubles, in a draught from the Town Pump. Take it, pure as the current of your young life. Take it, and may your heart and tongue never be scorched with a fiercer thirst than now! There, my dear child, put down the cup, and yield your place to this elderly gentleman, who treads so tenderly over the stones, that I suspect he is afraid of breaking them. What! he limps by without so much as thanking me, as if my hospitable offers were meant only for people who have no wine cellars. Well, well, sir—no harm done, I hope! Go draw the cork, tip the decanter; but when your great toe shall set you a roaring, it will be no affair of mine. If gentlemen love the pleasant titillation of the gout, it is all one to the Town Pump. This thirsty dog, with his red tongue lolling out, does not scorn my hospitality, but stands on his hind legs, and laps eagerly out of the trough. See how lightly he capers away again! Jowler, did your worship ever have the gout?

Are you all satisfied? Then wipe your mouths, my good friends; and while my spout has a moment of leisure, I will delight the town with a few historical reminiscences. In far antiquity, beneath a darksome shadow of venerable boughs, a spring bubbled out of the leaf-strown earth, in the very spot where you now behold me on the sunny pavement. The water was as bright and clear, and deemed as precious, as liquid diamonds. The Indian Sagamores drank of it from time immemorial, till the fearful deluge of fire-water burst upon the red men, and swept their whole race away from the cold fountains. Endicott and his followers came next, and often knelt down to drink, dipping their long beards in the spring. The richest goblet then was of birch bark. Governor Winthrop, after a journey afoot from Boston, drank here out of the hollow of his hand. The elder Higginson here wet his palm, and laid it on the brow of the first town-born child.

For many years it was the watering place, and, as it were, the washbowl of the vicinity.—whither all decent folks resorted, to purify their visages and then gaze at them afterwards—at least the pretty maidens did—in the mirror which it made. On Sabbath days, whenever a babe was to be bap tized, the sexton filled his basin here, and placed it on the communion-table of the humble meeting-house, which partly covered the site of yonder stately brick one. Thus one generation after another was consecrated to heaven by its waters, and cast its waxing and waning shadows into its glassy bosom, and vanished from the earth as if mortal life were but a flitting image in a fountain. Finally the fountain vanished. Cellars were dug on all sides, and cart loads of gravel flung upon its source, whence oozed a turbid stream, forming a mud-puddle at the corner of two streets. In the hot months, when its refreshment was most needed, the dust flew in clouds over the forgotten birthplace of the waters, now their grave.

But, in the course of time, a town pump was sunk into the source of the ancient spring; and when the first decayed, another took its place—and then another, and still another—till here stand I, gentlemen and ladies, to serve you with my iron goblet. Drink and be refreshed! The water is pure and cold as that which slaked the thirst of the red Sagamore beneath the aged boughs, though now the gem of the wilderness is treasured under these hot stones, where no shadows fall but from the brick buildings. And be it the moral of my story, that, as the wasted and long lost fountain is known and prized again, so that the virtues of cold water, too little valued since your fathers' days, be recognized by all.

Your pardon, good people! I must interrupt my stream of eloquence, and spout forth a stream of water, to replenish the trough of this teamster and his two yoke of oxen, who have come from Topsfield, or somewhere along that way. No part of my business is pleasanter than that of watering cattle! Look! how rapidly they lower the water-mark on the sides of the trough, till their capacious stomachs are moistened with a gallon or two apiece, and they can afford to breathe it in, with signs of calm enjoyment. Now they roll their quiet eyes around the brim of their monstrous drinking vessel. An ox is your true toper.

But I perceive, my dear auditors, that you are impatient for the remainder of my discourse. Impute it, I beseech you, to no defect of modesty, if I insist a little longer on so fruitful a topic as my own multifarious merits. It is altogether for your good. The better you think of me, the better men and women you will be yourselves. I shall say nothing of my all important aid on washing days; though on that account alone, I might call myself the household god of a hundred families. Far be it from me also to hint, my respected friends, at the show of dirty faces which you would present, without my pains to keep you clean. Nor will I remind you how often, when the midnight bell makes you tremble for your combustible town, you have fled to the Town Pump, and found me always at my post, firm amid the confusion, and ready to drain my vital current in your behalf.

No; these are trifles compared with the merits which wise men concede to me—if not in my simple self, yet as the representative of a class—of being the grand reformer of the age. From my spout, and such spouts as mine, must flow the stream that shall cleanse our earth of the vast portion of its crime and anguish, which has gushed from the fiery fountains of the still. In this mighty enterprise the cow shall be my great confederate. Milk and water! The *Town Pump* and the *Cow!* Such is the glorious copartnership that shall tear down the distilleries and brew-houses, uproot the vineyards, shatter the cider presses, ruin the coffee and tea trade, and finally monopolise the whole business of quenching thirst. Blessed consummation! Then, poverty shall pass away from the land, finding no hovel so wretched, where her squalid form may shelter itself. Then disease, for lack of other victims, shall gnaw its own vitals, and die. Then sin, if she do not die, shall lose half her strength.

Until now the phrensy of hereditary fever has raged in the human blood, transmitted from sire to son, and rekindled in every generation, by fresh draughts of liquid flame. When that inward fire shall be extinguished, the heat of passion cannot but grow cool, and war—the drunkenness of nations—perhaps will cease. At least, there will be no war of households. The husband and wife, drinking deep of peaceful joy—a calm bliss of temperate affections—shall pass hand in hand through life, and lie down, not reluctantly, at its protracted close. To them, the past will be no turmoil of mad dreams, nor the future an eternity of such moments as follow the delirium of the drunkard. Their dead faces shall express what their spirits were, and are to be, by a lingering smile of memory and hope.

Ahem! Dry work, this speechifying; especially to an unpracticed orator. I never conceived till now, what toil the temperance lecturers undergo for my sake. Hereafter they shall have the business to themselves. And, my dear hearers, when the world shall have been regenerated through my instrumentality, you will collect your useless vats and liquor casks into one great pile, and make a bonfire in honor of the Town Pump. And when I shall have decayed, like my predecessors, then, if you revere my memory, let a marble fountain, richly sculptured, take my place upon the spot. Such monuments should be erected everywhere, and inscribed with the names of the distinguished champions of my cause.

One o'clock ! Nay, then, if the dinner-bell begins to speak, I may as well hold my peace. Here comes a pretty young girl of my acquaintance, with a large stone pitcher for me to fill. May she draw a husband, while drawing water, as Rachel did of old. Hold your vessel, my dear ! There it is, full to the brim ; so now run home, peeping at your sweet image in the pitcher as you go ; and forget not, in a glass of my own liquor, to drink—" Success to the Town Pump !"

NATHANIEL HAWTHORNE.

SUCCESSFUL RURAL ENTERPRISE.

When the farmer has obtained the requisite information for his appropriate duties, his next step will be to provide himself with the most approved implements, the most profitable flocks and herds, the most commodious buildings for utility and comfort. Without regard to these particulars, successful results will be greatly impaired, if not wholly thwarted.

In the manufacture of tools, the problem first to be solved, is to combine lightness with strength and efficiency. Every ounce, added to the weight of an implement, to be wielded by the hand or drawn by animals, beyond what is necessary to fulfil its appropriate functions, diminishes the productive industry of man and beast. The strength of the laborer is exhausted, by unnecessary weight in the instrument he uses ; and, no advantage is gained by way of compensation. Within fifty years, the toils of the husbandman have been greatly relieved by the invention of light and convenient tools. The clumsy and heavy hoes, that were forged by our country smiths, but a few years ago, required nearly double the expenditure of strength which the light and sharp instrument, now used, demands. The heavy oaken ploughs which our fathers used, with mold-boards sheathed with old ox shoes and bits of sheet iron, required, at least, a third more strength to draw them and guide them, than the slender, keen-cutting implement now generally employed by farmers.

It has been ascertained by experiment, that a rake that weighs two pounds, will do the service of one that weighs ten pounds; and, that the rapid execution of the horse-rake, is far preferable to the slow and toilsome labor of the hands. In every department of husbandry, the labors of the farmer have been lightened by the introduction of improved tools and machinery. It is also fast becoming the prevailing opinion, that cheap outbuildings, leaky and shattered barns, and open sheds, are not the most convenient or profitable shelters, either for crops or animals. It is demonstrable that a large part of the food which animals consume is directly employed in the production of animal heat, consequently, artificial warmth is equivalent to an additional supply of food. The greater the degree of cold to which the animal is subjected, the more nourishment he requires to supply internal heat. Hence some of the most scientific farmers, in England, not only warm their barns by artificial heat, but cook the esculent roots which their animals consume. They deem this the dictate of true economy.

Another method of bettering our condition is the improvement of our domestic animals. When a farmer has once learned the value of the best breeds of animals, he will never tolerate the small, ugly, stunted, and contemptible races, which still linger about the barns of many of our husbandmen. There is a well authenticated account of a dairy in New-York State, of forty-one cows, which yields annually, in butter, cheese, and milk, a product of sixty-two dollars for each cow. The average income of cows, in the United States, does not, probably, exceed twenty dollars. Now supposing that the millions of cows, throughout the Union, were so improved as to raise their average income from twenty dollars to thirty-one, only one half of the income of the best herds, what untold wealth it would add to our farmers' estates. New-York alone would derive an annual revenue of $12,100,000 from this process. The outlays for the accomplishment of this result would be small; for the expense of keeping a good cow is but little more than is required for a poor one.

There are thirty millions of sheep in the United States The best of them yield double the amount of wool produced by ordinary sheep. Now supposing the wool clip to be increased one half, by the substitution of the best blooded animals for common breeds, this improvement would increase the annual income of farms, in our own country, at least fifteen millions of dollars, which would be equal to the creation of new capital to the amount of two hundred and fifty millions of dollars, yielding an annual interest of six per cent. General Washington, in one of his letters to Sir John Sinclair, says in substance, "that at the time he entered the public service in the war of the Revolution, his flock, of about one thousand, yielded five pounds of wool per fleece. Several years later, when he returned to his estate, his flock had so degenerated that it gave an average of only two and a half pounds per head." The deterioration of his flock was evidently the result of bad management on the part of his agent. There are more than six millions of horses and mules in our country. Let the value of these animals be increased to an average of twenty dollars a head, and the available property of the owners would be increased one hundred and twenty millions of dollars.

Another method of increasing the value of our products is by introducing new varieties of seeds, roots, and fruits. In accordance with an admitted law of nature, cultivated fruits and plants, as well as domestic animals, are subject to constitutional deterioration, when reproduced for a long series of years, upon the same soil. Hence the importance of exchanging seeds and importing new breeds of animals. Where the true spirit of improvement has for some time been active, and agricultural fairs have yearly presented better animals and richer specimens of fruits and grains for premiums, agriculture has not only become more profitable to the farmer, but it has conferred upon him dignity and honor. The efforts of wise and good men, in this department of industry, have been crowned with the most flattering success. Information has been disseminated, soils and grains have been analysed, the stock of the farm has been improved in value and utility; better implements of husbandry have been invented and adopted; and, the rewards of labor have been materially advanced, while the labor itself has been greatly diminished.

Indeed, every department of industry has been laid under contribution to secure these inestimable results. The mechanic has applied all his skill and ingenuity to the perfecting of agricultural implements; the chemist and geologist have summoned the whole encyclopedia of the sciences to their aid, in ascertaining the composition of various soils, and in preparing manures to enrich them, the inquisitive traveller and commercial adventurer have brought to our doors the fruits and grains of foreign climes, while men of the highest professional eminence, have devoted their time and attention to the cultivation of grains, grasses, and roots, and have imported the best blooded cattle, sheep, and swine, to feed upon them. Our public halls have therefore been adorned with the portraits and statues of men who have been distinguished at the bar, in the senate, or on the battle-field; but we are fast changing all that. It is now becoming the common sentiment, that more true glory is won in providing honest labor with its appropriate implements, and increasing the fertility of the earth, the common mother of us all, than in improving the destructive engines of war, or in waging a profitless contest of words in the forum, or the halls of legislation.

"The plough and the sickle shall shine bright in glory,
When the sword and the sceptre shall crumble and rust;
And the farmer shall live both in song and in story,
When warriors and kings are forgotten in dust."

Another important method of bettering our condition, is by furnishing local markets, by bringing the plough, the loom, and anvil, into proximity with each other. All branches of productive industry are mutually related, and mutually dependent. Every laborer who is employed at the forge, in the mill, or factory, must be provided with food and raiment. Every little village which springs up around a factory, furnishes a local market for a little circle of farms in the vicinity. In such cases the cost of transportation to a distant market is saved.

The State of Connecticut illustrates the value of local markets as fully as any state in the Union. Her soil was once fertile and highly productive. By continual cropping, without properly enriching the land, it became comparatively barren. The people began to emigrate to the richer Western lands. Those who were left behind, sought to improve their condition at home. They resorted to manufactures for a livelihood. The town of Bristol, fifteen miles south-west of Hartford, forty years ago, was fast declining in population and wealth. Its soil was exhausted; its inhabitants were seeking new homes. There was then resident in that town, a single clock maker, who patiently wrought out his timepieces by hand, at an expense of sixty or seventy dollars each. About the year 1815, some individuals, stimulated by the industry of the solitary clockmaker, set up the business of making wooden clocks. This article went all over the United States, and brought rich returns to the inventors. When the market was supplied with these, cheap brass clocks were made by machinery, which could be afforded as low as two dollars each.

The business of making clocks by machinery has been set up in twenty other towns in Connecticut; and the whole world are now their customers. On the authority of Fraser's Magazine, it is asserted that "every hall and cottage in England is furnished with them." Every one of these towns has become a profitable market to the neighboring farmers, and has given a new stimulus and increased profits to agriculture. Other towns in the same state, are engaged in the manufacture of tin ware, cutlery, jewelry, hooks and eyes, pins and needles, and every kind of implement of brass and iron, used by our citizens. At Collinsville, fifteen miles west of Hartford, there is a manufactory of axes, with a capital of three hundred thousand dollars, where six hundred axes are made in a day. One hundred and seventy-five workmen are employed. These, with their families, must be fed from the products of the soil.

The farmers are, therefore, directly interested in the prosperity of such manufactories, and in the consequent increase of population. By the introduction of these kinds of manufactures, emigration has been arrested, and agriculture has become more profitable. The worn out lands have been recruited. The soil that was deemed worthless, by right culture has become almost priceless. In some instances, the value has risen from two or three dollars to fifty or sixty dollars per acre. The desert may be made a fruitful field, by patient industry; and, the unhealthy and unsightly swamp, may be converted into a rich garden. Such is the fact in some instances. The merest gravel beds, by skilful culture are changed into fruitful fields; and offensive bogs, which were made the receptacle of the loose stones from the higher lands, are now drained—the stones are taken out from their deposit, and the surface of the swamp converted into a rich meadow. The vicinity of the factory, the forge or the loom, produces such results.

Our state abounds in water power, and in the materials for many valuable manufactures. But neither water power nor steam are essential to the introduction of certain kinds of manufactures. The man who made the first pair of pegged shoes, ever seen, in this or any other country, is still living. The trade in this article in Massachusetts alone, amounts to eighteen millions of dollars, employing sixty thousand hands. How are these men fed ? Do not the farmers in the vicinity of Lynn and other towns, where this business is prosecuted, pocket the money which these sixty thousand laborers are annually earning—at least, a good portion of it ? Massachusetts has a comparatively barren soil, yet her agriculture is advancing. Within fifty years, her population has increased one hundred per cent., while that of one of the other states of the Union has increased but two per cent. What makes the difference ?

The mystery is of easy solution. The answer is at hand. The former employs every species of power ;—capital, machinery, labor, and intellect. She encourages manufactures and agriculture, by uniting them on her own soil. The latter, containining as much cultivated land as the whole of Great Britain, eschews those kinds of productive industry, which are employed in working up, at home, the heavy and bulky materials which the earth produces, pays tribute to England by importing her wares, and furnishes no stimulus to her farmers by the encouragment of local markets. And the little State of Rhode Island has one hundred and ninety-eight manufacturing establishments, engaged in working wool and cotton ; and one hundred and sixty of them has each a distinct village, and of course a local market. One little, tortuous river, of twelve miles in length, twenty-five feet wide and three feet deep, drives nearly forty thousand spindles, in twelve different factories, employing nearly one thousand operatives. These laborers are fed from the produce of the adjacent farms. Here agriculture lives and thrives near to the newly-created markets.—*Agricultural Address, Orford, N. H.*, 1850, *by* PROF. EDWIN D. SANBORN, A. M., *of Dartmouth College.*

A SOUTH WESTERN PRAIRIE.

The world of prairie which lies at a distance of more than three hundred miles west of this inhabited portion of the United States, and south of the river Arkansas and its branches, has been rarely, and parts of it never, trodden by the foot, or beheld by the eye, of an Anglo-American. Rivers rise there, in the broad level waste, of which, mighty though they become in their course, the source is unexplored. Deserts are there, too barren of grass to support even the hardy buffalo, and in which water, except in here and there a hole, is never found.

Imagine yourself standing in a plain to which your eye can see no bounds. Not a tree nor a bush, not a shrub nor a tall weed, lifts its head above the barren grandeur of the desert; not a stone is to be seen upon its hard beaten surface; no undulations, no abruptness, no break, to relieve the monotony—nothing, save here and there a deep narrow track, worn into the hard plain by the constant hoof of the buffalo. Imagine, then, countless herds of buffalo, showing their unwieldy, dark shapes, in every direction as far as the eye can reach, and approaching at times to within forty steps of you; or a herd of wild horses feeding in the distance, or hurrying away from the hateful smell of man, with their manes floating, and a tramping like thunder. Imagine here and there a solitary antelope, or perhaps a whole herd, fleeting off in the distance, like the scattering of white clouds.

Again, imagine bands of white, snow white wolves, prowling about, accompanied by the little gray collottes or prairie-wolves, who are as rapacious and as noisy as their bigger brethren. Imagine, also, here and there a lonely tiger-cat, lying crouched in some little hollow, or bounding off in triumph, bearing some luckless little prairie-dog, which it has caught straggling about at a distance from his hole If to this you add a band of Camanches, mounted on noble, swift horses, with their long lances, their quiver at the back, their bow, perhaps their gun, and their shield ornamented gaudily with feathers and red cloth, and round as Norval's, or as the full moon; and imagine them hovering about in different places, chasing the buffalo or attacking an enemy—you have an image of the prairie, such as no book ever described to me.

I have seen the prairie under all its diversities, and in all its appearances, from those which I have described, to the uneven, bushy prairies which lie south of the Red river, and to the illimitable Stake prairie, which lies from almost under the shadows of the mountains to the heads of the Brazos and of Red river, and in which neither buffaloes nor horses are to be found. I have seen the prairie and lived in it, in summer and in winter. I have seen it with the sun rising calmly from its breast, like a sudden fire kindled in the dim distance, and with the sunset flushing in the sky with quiet and sublime beauty. There is less of the gorgeous and grand character, however, belonging to it, than that which accompanies the rise and set of the sun upon the ocean, or upon the mountains; but here are beauty and sublimity enough to attract the attention and interest the mind.

I have also seen the *mirage*, painting lakes, and fires, and groves, on the grassy ridges near the bounds of Missouri, in the still autumn afternoon, and cheating the traveller by its splendid deceptions. I have seen the prairie, and stood long and wearied guard in it, by moonlight and starlight, and in storm. It strikes me as the most magnificent, stern, and terribly grand scene on earth.

A storm in the prairie is much like a storm at sea, except in one respect—and in that it seems to me to be superior—the stillness of the desert and illimitable plain, while the snow is raging over its surface, is always more fearful to me than the wild roll of the waves; and it seems unnatural—this dead quiet, while the upper elements are so fiercely disturbed! it seems as if there ought to be the roll and roar of the waves. The sea, the woods, the mountains, all suffer in comparison with the prairie—that is, on the whole; in particular circumstances either of them is superior.

We may speak of the incessant motion and tumult of the waves of the ocean; the unbounded greenness and dimness, and the lonely music of the forests; and the high magnificence, the precipitous grandeur, and the summer snow of the glittering cones of the mountains; but still the prairie has a stronger hold on the soul, and a more powerful, if not so vivid an impression upon the feelings. Its sublimity arises from its unbounded extent—its barren monotony and desolation—its still, unmoved, calm, stern, and most impressive grandeur—its strange power of deception—its want of echo—and, in fine, its power of throwing a man back on himself, and giving him a feeling of lone helplessness, strangely mingled at the same time with a feeling of liberty and freedom from restraint. It is particularly sublime as you draw nigh to the Rocky Mountains, and see them shoot up in the west, with their lofty tops looking like white clouds resting upon their summits. Nothing ever equalled the intense feeling of delight with which I at first saw them marking the western edge of the desert.—Albert Pike.

THE FIRST FALL OF SNOW.

I love to watch the first soft snow,
As it slowly saileth down,
Purer and whiter than the pearls
That grace a monarch's crown.
Though winter wears a freezing look,
And many a surly frown.

It lighteth like the feathery down
Upon the naked trees,
And on the pale and withered flowers
That swing in every breeze;
And they are clothed in such bright robes
As summer never sees.

It bringeth pleasant memories,
The falling, falling snow,
Of neighing steeds, and jingling bells,
In the happy long ago;
When hopes were bright, and health was good,
And the spirits were not low.

And it giveth many promises
Of quiet joys in store;
Of bliss around the blazing hearth,
When daylight is no more—
Such bliss as no where else hath lived
Since Eden-days were o'er.

God bless the eye that views with mine
The falling snow to-day;
May truth her pure white mission spread
Before its searching ray,
And lead, with dazzling garments, towards
"The strait and narrow way."

JULIA H. SCOTT.

NIGHT MUSINGS.

Hushed like the o'erweary child too late at play,
Who sobs to sleep upon its mother's breast,
Lies the vast city. The perturbed day,
With all its load of care and pain oppressed,
Sinks softly into slumber—down the West
Creeps the gray curtain of the peaceful night,
And hearts which break by day, in dreams are blessed—
While with a spring of fierce and far delight,
The soul unchained, resumes her heavenward flight.

Now, like a mother's blessing, the rich dews—
Tears of Heaven's pity—kiss the gasping flower,
Now every banished star its smile renews,
And winds new fragrance shed upon the hour—
Now the tired brain regains its wonted power,
And thoughts, like beauty bright, flash on the mind,
Which, upborne above earth's clouds that lower,
Seeks once more commune with her lofty kind,
And spurns the grovelling things of earth far, far behind.

Once more is nature beautiful. Once more
The bitter dregs of life have passed away
From my sick soul. Oh how Night's ministers pour
Their incense over me! How dim the day
Seems, with its dusty glare, to the soft ray
Of yon fair-smiling Goddess beaming light
And love and beauty o'er her starry way!
E'en *my* dark fortunes catch some hues of bright
And glowing radiance from the beauty-beaming night!

Now playful fancy from her world of dreams
Looks out and smiles; and a bright host of forms

With love and beauty sparkling, like the gleams
  Youth sees of glory, on my vision swarms,
  And makes my heart beat as when youth was warm,
And life one dream of rapture. Each bright hope
  Crushed to a fragrant ruin 'mid the storms
Of darksome life, feels its dry petals ope,
And once more boldly dares with life's sad ills to cope.

Now in this sweet and peaceful night, what fires
  On life's forgotten altars gleam anew!
What glowing visions and what high desires
  Like unsought spectres start upon my view!
  Young life's sweet garden, where my heart-flowers grew,
Blooms freshly round me—and the busy hum
  Of murmuring bees who not in vain pursue
Life's toil, distils its music round me. Come
Back to my weary heart, dreams of my childhood's home!

My father blesses me again—and she
  Whose tears like heart-gems glistened on my hair
Beneath her parting kisses, bends o'er me,
  As in that hour of hushed and solemn prayer!
  Mother! oh mother! since we gathered there
Around the hearth-stone I shall see no more,
  Darkness and gloom and anguish and despair
Have racked my soul to torture—but they bore
No power to tear *thee* from my heart's deep core.

Oh, bitter as the Dead Sea's hollow fruit,
  Which turns to ashes on the lips, has been
The cup of life to me—and cold and mute
  The music dreams which sang to me between
  The pauses of thy blessing. I have seen
The flowers of hope, leaf after leaf decay,
  And felt their canker-worm with poison keen,
Eat to my soul, where on its altar lay
My broken heart-strings, with their music passed away.

Yet mother, in this bright and hallowed hour,
  My soul is once more with thee; and again
The spells of love have o'er me their old power.
  Once more thine eyes beam into mine, as when
  Thy parting glance was on me; and again

Thine arms encircle me. Oh bless me now,
Ere yet my dream breaks up!—'Tis o'er! My pen
Blots as with tears the page which I endow
With the wild thoughts that swell beneath this aching brow.

G. G. Foster.

APPROVED MODES OF TILLAGE.

What is to be understood by the term tillage?

It is the operation, practice, or art of preparing the land for seed, and keeping the ground free from weeds which might impede the growth of crops. Tillage includes ploughing, manuring, harrowing, and rolling land, or whatever is done to bring it to a proper state to receive the seed; and, also, the operations of ploughing, harrowing, and hoeing the ground to destroy weeds and loosen the soil after it is planted.

Every boy in the country is familiar with the operation of ploughing, but it may be well that he be able to describe it; hence, it is asked—In what does the process consist?

It consists of cutting the ground into slices and turning them over. These slices usually vary in width from twelve to eighteen inches; and they are turned over in such a manner, that an entire new surface is presented to the atmosphere.

To what depth should these slices be cut?

The medium depth of good ploughing may be estimated at seven inches. When the circumstances, arising from the particular description of the crop and the nature of the soil, do not require deep ploughing, the depth may be less; but, in some cases where the upper soil has become enfeebled by long culture, or where the roots of the crop require more ample space, much deeper ploughing will be found expedient.

How is the ploughman enabled to regulate the width and the thickness of these slices?

Much depends on the particular construction of the plough. All ploughs are designed to effect this object as far as possible. Yet, it is perhaps impossible so to construct them, as to accomplish this work, unless aided by an experienced operator. As in the case of all implements for manual operations, good instruments are required, and skillful agents to apply them in the proposed work.

T. K. Van Zant

SHORT HORNED COW, Easterville. Bred by E. P. Prentice, Albany, N. Y.

In what consists the application of the ploughman's skill?

In this operation, the left hand or near-side horse, or ox, walks on the ground not yet ploughed, the right hand or off-side horse, or ox, walks in the furrow last made, and the workman follows, holding the handles. By means of these handles he guides the plough, and directs the animals of draught by the voice and the reins. If the plough cuts too deep, he is to press down the handles, so that the heel of the plough becomes a fulcrum, and the point of the share is raised upwards. If it does not cut sufficiently deep, this can be remedied by raising the handles and giving the point of the share a downward direction. And, if the plough inclines too much to the right or the left, in either case he is to give it a contrary pressure.

How can the ploughman discover defects in regard to the width or the thickness of the slices?

When the sods are considerably too wide in proportion to their depth, he will be admonished of it by their lying too flat, and too slightly overlapping each other. When their depth is considerably too great in proportion to their width, they will stand too upright, and be apt to fall back again into the furrow.

How much land can ordinarily be ploughed in a day by one team?

The common calculation, where good ploughing is practiced, is, that a pair of horses will plough an acre of grass land in nine hours. In very stiff soils less will be done; and in very light soils, more. When land is in a loose and pulverized state, from one-third to one-fourth part more may be done in the time mentioned. And it is very apparent that a pair of good horses well trained to the work, with an efficient man at the plough, will accomplish far more in the same time, than is performed by those of inferior quality. However, an acre a day, on average of different soils, different seasons in the year, and teams of varying quality, is a fair amount to be ploughed.

What should be a primary object in ploughing?

In the first place, the ground should be well prepared for the plough by the removal of large stones and other obstructions. In the second place, the farmer should use none but a first rate plough, and in prime order. And in the third place, he should employ a team well trained and well fed. With such appliances double the amount of labor will be performed, that could be effected by inferior instrumentalities; or where a miserable old plough is used, with a feeble or unmanageable team, and the whole brought to a dead stand from encountering a large stone in the ground, every ten rods.

For what purpose is ground ploughed?

To loosen or disintegrate the parts of which it is composed. This is necessary in order to enable the roots of the plants in the succeeding crop to extend themselves in every possible direction, downward and laterally, which they could not do, if the ground were hard and impenetrable. And, it is also necessary that the soil be rendered mellow and light, that the fertilizing substances designed for the growth of the plants, be thoroughly mixed with it, and especially that it be thus rendered capable of receiving the rain, which is to nourish the plants, instead of running off upon the surface.

Does any other benefit arise to the soil from ploughing?

There is. The lower strata are brought into a position to receive the influences of the sun and the atmosphere. If the soil is cold, it thus becomes warm. If too moist, it becomes drier; and, whether cold or warm—moist or dry, it is enabled to absorb from the atmosphere those gases which are so needful in the formation of vegetable substances.

What is sub-soil ploughing?

It is ploughing in the sub-soil, that is, in the stratum or layer of earth lying beneath the upper soil, or surface of the ground. The process is performed by what is called a sub-soil plough, which is drawn by four or six oxen, following in the furrow of a common plough, and penetrating this sub-stratum, or sub-soil, to the depth of ten, twelve, fifteen, or even twenty inches. If the sub-soil is very hard, six oxen may be none too much. If not, four will answer. It is not designed to raise the sub-soil above the surface soil; but only to break it up and pulverize it, and to mix a portion of it with the lower portions of the surface soil.

What are the supposed advantages of sub-soil ploughing?

The hard, sterile earth is thoroughly pulverized, thereby being exposed to the meliorating influences of the atmosphere, and furnishing increased supplies of food and moisture in dry seasons, for the roots of plants. These are the primary and general advantages of a deep wrought loam, and they are certainly such as to commend themselves to the attention of every reflecting farmer.

How does it appear that sub-soil ploughing is conducive to the growth of plants?

It has been said that it secures a better supply of heat and moisture, than where common ploughing only is performed. The fact is palpable, that in time of drought the vegetation of a garden will be much more vigorous than that in the adjacent field. This is mainly owing to the greater looseness of the soil.

What experiment has been made to prove this?

Mr. C. N. Bement, the distinguished agriculturist, some years since, sub-soiled several strips of a sandy knoll, which he planted with Indian corn. In the dry summer that followed, the corn of those strips was green and flourishing, while that on the other portions of the lot was almost burned up with the heat; and at the harvest, the difference in the yield was not less remarkable.

What is a more permanent benefit from sub-soiling?

The minute particles of the surface and sub-soils are gradually mixed together; the natural resources of the ground are wakened into life by the influence of the atmosphere; the thread-like web of roots with which it is filled decay when the plant dies, or is removed; and, in time, the sterile, unprofitable substratum becomes a valuable loam of great depth and fertility.

Is sub-soil ploughing always attended with profit?

In some instances it has been found of decided injury. It is where the sub-soil plough breaks through a more retentive stratum, and turns up a thirsty sand or gravel, which absorbs all the moisture and soluble manures of the surface soil. Land of this description ought to be kept in wood or permanent pasture, as under the most careful management it is always unprofitable for tillage.

How does the harrow contribute to the fertility of the soil?

The tendency of it, like the plough, the hoe, or the spade, is to reduce to fineness the large masses and hard lumps of earth that remain after ploughing. And, in preparing the soil for seed, it fills up caverns and brings down protuberances, thus better preparing it to receive the seed. By this means the seed is more uniformly scattered, and more likely to have a speedy and uniform germination. And, when the harrow is applied after the seed is sown, it is to mix it well with the soil, and to bury it at a moderate depth beneath the surface.

By the use of what other implements is the soil rendered fine and mellow?

The spade and the hoe have been most used; but of more recent origin the cultivator is applied to great advantage, especially as a time-saving implement. For deep action the spade answers admirably, both for the disintegration of the combined parts of the soil, and the removal of stones from it; but the process is extremely slow. Notwithstanding the invention of the sub-soil plough, in England, on some lease lands, the tenant is still required, in field culture, to upturn the soil with a spade.

What is said of the use of the hoe and cultivator?

The hoe has been so long used, and is so generally used for all purposes of the kind, that there is no occasion to speak of it. The cultivator is coming into general use for the extermination of weeds between the rows of Indian corn, and other crops in rows or drills; but, as it operates only on and near the surface, the plough, and subsequently the hoe, are more effectual for melioration. As a time-saving implement the cultivator is of great value; and every farmer may save the cost of it in a single year.

Of what ruinous neglect in regard to tillage are many farmers guilty?

Particularly in allowing weeds to grow, which exhaust the strength from the soil, and if allowed to mature, fill it with pernicious seeds, requiring years for their extermination. A farmer should no more allow weeds to remain on his lands, than burglars in his house, or wild animals to eat the grain wanted for his family, cattle, and swine.

What are the benefits from the use of the roller?

The first and most obvious benefit is, to break those clods or indurated masses of earth which have resisted the action of the harrow; or, at all events, to bury them in the ground, so that at the next harrowing, which, when thus buried, they cannot well escape—they must of necessity be somewhat diminished in size. It is for this reason that in countries where the soil is very tenacious, and tillage very carefully conducted, it is the custom, even after the preparatory ploughings, first to harrow, then to pass the roller over the ground, and then to harrow again.

What is a second benefit from the use of the roller?

This is to give a somewhat greater degree of compactness to a soil which is too light and friable, and to unite its component parts. The roller is not employed for this purpose to so great an extent as it might be with advantage. This application of the roller is particularly resorted to on the spongy soils of valleys. In such situations it cannot, indeed, be well dispensed with.

What is a third benefit arising from the use of the roller?

Again, the roller is used to press down and make firm the ground about newly sown seeds, and to cause the latter to adhere better to the soil. Sometimes, when very small seed is to be sown, it is found advantageous to pass the roller over the ground before the seed is sown, so as to level it thoroughly, and to facilitate a more equal distribution of the seed than could otherwise take place. Where the ground has been thus leveled, those seeds which happen to fall together, separate from each other; and it is seldom that two are found lying on the same spot. The harrow is then passed over the ground; and this operation is followed by repeated rollings, which obliterate the lines drawn by the harrow.

What is a fourth benefit from the use of the roller?

This is to cover with mold, or press against or into the ground, the roots of those plants sown in the preceding autumn, which have been detached by the frost. Soils rich in humus, such as those found in valleys, sometimes swell up in the spring to such a degree, that the roots of the plants contained in them are forced up. In such cases, if a fall of rain does not speedily occur, the roller is the only means of restoring them to their proper position. Hence, a grass field cannot be too often rolled; and, it is not going too far to assert, that the application of the roller in autumn to prepare the roots for resisting the winter frosts, and in spring to render them firm after the frosts, every year while the field remains in grass, will amply repay the expense.

THE AMERICAN FARMER.

What is the position of the American farmer, when compared with that of the merchant, the politician, the lawyer? Should he be content with his lot for himself and his children? Or should he leave his occupation and adopt some other? Like every other position, that of the farmer has its dark side as well as its bright one. And to decide on its comparative advantages, we must inquire what is the object of man's existence, and how he shall attain the end of his being?

To these questions, history and revelation, the world around us, and the spirit within us, answer, that the object of man's existence is happiness. Happiness here, and happiness forever. And the condition of that happiness is the diligent and proper exercise of his affections and faculties. If this be the case, 'does the situation of the American farmer offer a fair opportunity of insuring this happiness? To be happy is the object of life, and all that the world can give toward it, is health and competence. "Health of body is above all riches, and a strong body is above infinite wealth." And where is health to be found? There is no need of an audible answer. Look around on an assembly from the rural districts of the country, and we shall see bright eyes and blooming cheeks, as well as strong arms and untiring strength, giving testimony that the earth's first blessing is bestowed upon those who labor upon her bosom.

But health is often undervalued by its possessor, or only appreciated when lost. Wealth, the more obvious and immediate reward of labor, is the chief pursuit of the active. And here the farmer thinks he has a right to complain. The merchant will sometimes make more in a year than he can make in a lifetime; and it is not wonderful that he sometimes asks, would it not be better to leave small rewards, though regular and certain, for the chance of obtaining greater? To decide this question, we must ask, What is the price he pays? What is the reward he obtains? What is the price he pays? To say nothing of his moral exposures, in the great majority of cases, health of body and serenity of mind. Follow such a one into the crowded streets, or the close workshop. His strength for a time sustains him, but confinement and bad air soon deprive him of his healthful energy, and disease and premature decay become too often his portion. But supposing health can be preserved, where is his serenity of mind?

The risks attendant on rapid accumulation are always in proportion to the chances of success. The farmer sows his seed, and has no doubt but that the harvest will repay him. But he who embarks in speculations that promise sudden and great wealth, knows that he may be "sowing the wind to reap the whirlwind." And the constant fear of such a result imbitters his days and renders his nights restless. And if attained, success gives but little satisfaction. The higher the rise, the wider the horizon; the greater the accumulation, the more exorbitant the desire. And this is not the extent of the evil. A total want of independence is too often the result. Few men in our community have those resources that will enable them to carry on extensive operations on their own means. Almost all depend on borrowing, and the borrower is the servant of the lender.

But, even if success should be the portion of the aspirant for riches, when is he to attain it? Does it come forward to meet him? Years of anxiety may be repaid by wealth; but how seldom is this the case! More than ninety in every hundred, even in regular mercantile pursuits, fail. There are but few capital prizes in this lottery. The name of the fortunate holder may be seen at every corner, but where are the ninety and nine who draw blanks? And if attained, how uncertain its possession! Wealth "gotten by vanity," by which I suppose Solomon meant by speculation, "shall be diminished, but he that gathereth by labor shall increase," is a doctrine as true now as when first delivered, and is one which the experience of every age tends to corroborate.

And after all, what is the advantage of great wealth, or what is great wealth itself? It exists only in comparison. "A man is as well off," said the great capitalist of the United States, "who is only worth half a million of dollars, as he would be if he were rich." And one of the satirical papers of the day tells us that when Baron Rothschild, the Jewish banker, read that the income of Louis Philippe was only fifty dollars a minute, his eyes filled with tears, for he was not aware of the existance of such destitution. After the comforts of life are supplied, wealth becomes merely an imaginary advantage, and its possession does not confer any material for happiness, which an industrious and forehanded farmer does not possess. "We will conquer all Italy," said Pyrrhus, to his prime minister, "and then we will pass into Asia; we will overrun her kingdoms, and then we will wage war upon Africa, and when we have conquered all, we will sit down quietly and enjoy ourselves." "And why," replied the minister, "should we not sit down and enjoy ourselves without taking all this trouble?" And why may not you, it may be said to many an aspirant after wealth, enjoy in reality all you seek, in your present condition?

"Give me neither poverty nor riches," was the prayer of one of the sages of antiquity. And Lord Bacon, the wisest man of modern times, says, "Seek not proud riches, but rather such as thou mayest get justly, use soberly, distribute cheerfully, and leave contentedly." And can there be a truer description of a farmer's fortune? There is no greater independence than that possessed by the contented, forehanded farmer. "Tell your master," said a Roman general, to the ambassador of the King of Persia, who came to bribe him with great wealth, and found him washing the vegetables that were to constitute his dinner with his own hands, "tell your master that all the gold in Persia can never bribe the man who can contentedly live upon turnips." The answer was as true in philosophy, as it was elevated in patriotism. To be happy man must limit his desires. And when he has sufficient for his needs, he should remember that the temptations and perplexities incident to overgrown wealth more than counterbalance its seeming advantages. Health of body and competence of estate are all the requisites for organic happiness that the world can bestow. And to say that agricultural pursuits are eminently calculated to insure these, is only to reiterate the language of past ages, and to repeat the testimony of our own. If you leave such pursuits, the hazard increases as the profit augments. The amount of the premium is always proportioned to the greatness of the risk.

But health and the conveniences of life are not all that a man requires to make him happy. He desires to be useful, he wishes to be esteemed. And what profession can boast of a higher claim to utility than that of the farmer? The greater part of mankind must be agriculturists, and on their characters the well-being of every state must depend. Our free institutions are valued, but how shall they be preserved? By the virtue of the people. History gives no other answer. No truth is more clearly emblazoned on her pages than that if a nation would be free, they must be intelligently virtuous. And here the agricultural class becomes of the first importance to the state. The influence of a virtuous yeomanry on her character, like that of the air on the individual, is seen in the strength of those who are unconscious of its presence.

The agricultural life is one eminently calculated for human happiness and human virtue. But let no other calling or pursuit of honest industry be despised or envied. One cannot say unto another, "I have no need of thee;" and to every one there are compensations made that render all, in a great degree, satisfied with their lot. Envy not the wealth of the merchant; it has been won by anxieties that you never knew, and is held by so frail a tenure as to deprive its possessor of perfect security and perfect peace. While your slumbers have been sound, his have been disturbed by calculating chances, by fearful anticipations, by uncertainty of results. The reward of your labor is sure. He feels that an hour may strip him of his possessions, and turn him and his family on the world in debt and penury.

Envy not the learning of the student. The hue on his cheek testifies of the vigils by which it has been attained. He has grown pale over the midnight lamp. He has been shut up from the prospect of nature, while sound sleep and refreshing breezes have been your portion and your health. Envy not the successful statesman. His name may be in every one's mouth. His reputation may be the property of his country, but envy and detraction have marked him. His plans are thwarted, his principles attacked, his ends misrepresented. And if he attain to the highest station, it is to feel that his power only enables him to make one ungrateful, and hundreds his enemies, for every favor he can bestow.

Envy no one. The situation of an independent farmer stands among the first, for happiness and virtue. It is the one to which statesmen and warriors have retired, to find, in the contemplation of the works of nature, that serenity which more conspicuous stations could not impart. It is the situation in which God placed his peculiar people in the land of Judea, and to which all the laws and institutions of his great lawgiver had immediate reference. And when, in the fullness of time, the privileges of the chosen seed were to be extended to all his children, it was to shepherds, abiding in the field, that the glad tidings of great joy were first announced. Health of body, serenity of mind, and competence of estate, wait upon this honorable calling; and in giving these, it gives all that the present can bestow, while it opens, through its influence, the path to Heaven.—*Ad. before New-York State Ag. Society, by* HON. JOSIAH QUINCY, JR., *of Massachusetts.*

PERSEVERING LABOR AND WEALTH.

If there be any aristocracy of wealth in our country, it is a genuine, a native growth. It has been produced by the labor, enterprise, and persevering economy of the people themselves. Who, that has lived so long as I have lived, has not seen the progress of wealth in almost every department of acquisition? I can recollect a ship-boy, whose whole paternal wealth was a warm, and anxious parental blessing. What vocation can be more toilsome, what more perilous? Sleep, which in other callings seals the weary eye of labor on something like a bed of repose, weighs his eyelids down, and steeps his senses in forgetfulness, with no better pillow, it may be, than the head of the "high and giddy mast, when the wind takes the ruffian billows by the tops, curling their heads, and hanging them with deafening clamor in the slippery shrouds." Yet he, by perpetual toil, continued enterprise, and untiring economy, comes at last to be a wealthy and extensive ship-owner.

The vocation of the school-master is a life of toil, humble acquisition, and honest obscurity. His capital is altogether of the mind. There he is rich, in science, integrity, and habits of persevering labor and economy. How many of these men in our country, after toiling years, in this employment, gather up the earnings and saving of those years, and turning their attention to commerce or other industrial pursuits, become, by the practice of the same process of integrity, labor, and economy, men of wealth and eminence? And, how many, after a devotion in early life to these unpretending labors, have adopted one of the learned professions, and thence rising from one point of distinction to another, till ranked with the leading men of the nation?

I have seen the young mechanic, at the age of twenty-one, standing on the threshold of his father's humble dwelling. He was just about to step out into the world, and to begin life for himself; with no other earthly wealth than his own summer frock, trowsers, and straw hat. His whole capital was his hands, and his skill in the use of them. Notwithstanding all these discouragements, this same man, by perpetual toil and perpetual economy, became a wealthy and extensive manufacturer. Such instances are by no means rare. They are confined to no particular locality. They may be found in every part of our land. There are thousands, and tens of thousands of them.

The plough-boy belongs to another class of humble and toilsome employment; and who, that has ever shared in the toils or the sports of that vocation, can look back upon it, from any point in after life, without feelings of complacency and regret? The plough-boy drives his team across the field, when the first sunbeams of morning are spreading over the earth; when the world is bursting into life, and song, and action; and, buoyant with youth, and health, and hope, he, "as he turns over the furrowed land," joins the rude notes of his own voice to the jocund sounds of the merry morning. This laborious lad, by years of continual toil and continual economy, does at last become the owner of fields, and is himself rich in farms and plantations. He belongs, moreover, to that class of citizens, more numerous than all others, constituting the bulwark of the nation, and giving sustenance to the whole world.

The wealth of a nation is its labor, its skill, its machinery, its abundant control of all the great agents of nature employed in production. Wealth is power; and the defence of every nation depends on its wealth. Nevertheless, a large store of goods laid up for many years, was the wealth of a fool; but such a store is the poverty of a nation. A great annual consumption, alone, can ensure an augmented annual reproduction. The labor of a nation can no otherwise be sustained, than by the consumption of its products. The products of human labor in food and clothing, like the fruits of the earth, are annual; and God, in his wisdom, has adjusted human wants to the powers of production. Like the bread from heaven, that the Giver might not be forgotten, the dew of every night produced the crop, and the labors of every day gathered in the harvest.

What but a mighty phalanx of labor, an almost boundless power of consumption and reproduction, has defended and now sustains England in all the athletic vigor of the most glorious days of that extraordinary nation? Men who speculate on the duration of nations, seem to assign to them the several periods of human life; youth, manhood, and old age, and final dissolution. They draw their conclusions from the nations of antiquity, and apply them to those of modern times. They forget that those ancient nations were like beasts of prey, which find an enemy in every living thing; and must be, sooner or later, circumvented by stratagem, or overpowered and destroyed by force. Producing nothing by their own labor, and consuming all which, by violence, they could plunder from the labors of others, their whole existence was a burden to the human race, and they were finally destroyed as a common nuisance to mankind.

Not so with England; she is a glorious example of the self-subsisting, the all producing, and all defending powers of labor. With a valor purely Spartan, she builds no walls against the wars of the world. Her little island, accessible at a thousand points, and often within gunshot of the embattled fleets of her enemies, has not, for more than seven hundred years, been stept upon by a hostile foot. What has enabled her to do this? Her untiring labor; her unrivalled skill; her unequalled machinery; her exhaustless capital, and unbounded control over all the agents of production. Her goods, her wares, and merchandise, are in all the markets of the world; and wherever she wants a tongue to speak in her cause, or a sword to be drawn in her quarrel, if such things can be found in those markets, she can command them.

But, if there can be no wealth without labor, the former usually gives impulse and successful direction to the latter. Do we hear the sound of the axe, see the forests fall around, and the clearing extended over new regions. The idea of money in prospect, or in actual possession, gave energy to the first blow. Is the plough in motion on a thousand fields, and carrying culture to the very hill tops of our country? It is no hyperbole to say, that the idea of money sharpens the share, and feeds and invigorates the team. Do the sounds of our spindles and looms make music with the sound of our waterfalls; and are our fabrics sent into all parts of our country, and to foreign nations? The accumulated wealth of the people from previous labor was the great agent moving all this machinery. Are the saw and the hammer heard over the whole country, building work shops, warehouses, mansions, temples, villages, cities? What but a store of wealth, collected and laid up, by the labor and economy of the people, has called into activity the skill and the strength of mechanical labor, and thereby ornamented, as if by enchantment, the whole face of our country?

Nor is this all! What sea is left unvexed, by the oars or the keels of our fisheries or commerce? Let it be remembered, that not a line is drawn; not a harpoon thrown; not an oar-blade glitters in the sun; nor a sail whitens above the wave, without that invigorating current of vitality, the money of our country, which feeding and sustaining every department of labor, puts it all into animated and productive motion; and which, for that great purpose, has in former years, by so much toil, care, and economy, been earned, saved, and then made available in developing an increased and combined human industry throughout every member and limb, of the whole vast and gigantic body of our national labor. There is, therefore, a beautiful reciprocity between labor and wealth; if the latter is produced by the former, the former is invigorated and sustained by the latter. Thus it is with the husbandman in tilling the ground. His labor causes the fruits of the earth to spring up, and without it they would not be made to exist; yet without them for food the farmer could not live and toil.—TRISTAM BURGESS, LL. D.

GOD'S FIRST TEMPLES.

The groves were God's first temples. Ere man learned
To hew the shaft, and lay the architrave,
And spread the roof above them,—ere he framed
The lofty vault, to gather and roll back
The sound of anthems,—in the darkling wood,
Amidst the cool and silence, he knelt down
And offered to the Mightiest solemn thanks
And supplication. For his simple heart
Might not resist the sacred influences,
That, from the stilly twilight of the place,
And from the gray old trunks, that, high in heaven,
Mingled their mossy boughs, and from the sound
Of the invisible breath, that swayed at once
All their green tops, stole over him, and bowed

His spirit with the thought of boundless Power
And inaccessible Majesty. Ah! why
Should we, in the world's riper years, neglect
God's ancient sanctuaries, and adore
Only among the crowd, and under roofs
That our frail hands have raised? Let me, at least,
Here, in the shadow of this aged wood,
Offer one hymn; thrice happy, if it find
Acceptance in his ear.

Father, thy hand
Hath reared these venerable columns; thou
Didst weave this verdant roof. Thou didst look down
Upon the naked earth, and, forthwith, rose
All these fair ranks of trees. They in thy sun
Budded, and shook their green leaves in thy breeze,
And shot towards heaven. The century-living crow,
Whose birth was in their tops, grew old and died
Among their branches; till, at last, they stood,
As now they stand, massy, and tall, and dark,
Fit shrine for humble worshipper to hold
Communion with his Maker. Here are seen
No traces of man's pomp or pride; no silks
Rustle, no jewels shine, nor envious eyes
Encounter; no fantastic carvings show
The boast of our vain race to change the form
Of thy fair works. But thou art here; thou fill'st
The solitude. Thou art in the soft winds
That run along the summits of these trees
In music; thou art in the cooler breath,
That, from the inmost darkness of the place,
Comes, scarcely felt; the barky trunks, the ground,
The fresh, moist ground, are all instinct with thee.
Here is continual worship; nature, here,
In the tranquility that thou dost love,
Enjoys thy presence. Noiselessly, around,
From perch to perch, the solitary bird
Passes; and yon clear spring, that, midst its herbs,
Wells softly forth, and visits the strong roots
Of half the mighty forest, tells no tale
Of all the good it does. Thou hast not left
Thyself without a witness, in these shades,
Of thy perfections. Grandeur, strength and grace,

Are here to speak of thee. This mighty oak—
By whose immovable stem I stand, and seem
Almost annihilated—not a prince,
In all the proud old world beyond the deep,
E'er wore his crown as loftily as he
Wears the green coronal of leaves, with which
Thy hand has graced him. Nestled at his root
Is beauty, such as blooms not in the glare
Of the broad sun. That delicate forest flower,
With scented breath, and look so like a smile,
Seems, as it issues from the shapeless mold,
An emanation of the indwelling Life,
A visible token of the upholding Love,
That are the soul of this wide universe.

My heart is awed within me, when I think
Of the great miracle that still goes on,
In silence, round me—the perpetual work
Of thy creation, finished, yet renewed
Forever. Written on thy works, I read
The lesson of thy own eternity.
Lo! all grow old and die: but see, again,
How, on the faltering footsteps of decay,
Youth presses—ever gay and beautiful youth—
In all its beautiful forms. These lofty trees
Wave not less proudly that their ancestors
Molder beneath them. Oh! there is not lost
One of earth's charms: upon her bosom yet,
After the flight of untold centuries,
The freshness of her far beginning lies,
And yet shall lie. Life mocks the idle hate
Of his arch enemy Death; yea, seats himself
Upon the sepulchre, and blooms and smiles,
And of the triumphs of his ghastly foe
Makes his own nourishment. For he came forth
From thine own bosom, and shall have no end.

There have been holy men, who hid themselves
Deep in the woody wilderness, and gave
Their lives to thought and prayer, till they outlived
The generation born with them, nor seemed
Less aged than the hoary trees and rocks
Around them; and there have been holy men,
Who deemed it were not well to pass life thus.

But let me often to these solitudes
Retire, and, in thy presence, reassure
My feeble virtue. Here, its enemies,
The passions, at thy plainer footsteps, shrink,
And tremble, and are still.
O God! when thou
Dost scare the world with tempests, set on fire
The heavens with falling thunderbolts, or fill,
With all the waters of the firmament,
The swift, dark whirlwind, that uproots the woods,
And drowns the villages; when, at thy call,
Uprises the great deep, and throws himself
Upon the continent, and overwhelms
Its cities;—who forgets not, at the sight
Of these tremendous tokens of thy power,
His pride, and lays his strifes and follies by!
Oh! from the sterner aspects of thy face
Spare me and mine; nor let us need the wrath
Of the mad, unchained elements, to teach
Who rules them. Be it ours to meditate,
In these calm shades, thy milder majesty,
And to the beautiful order of thy works
Learn to conform the order of our lives.

BRYANT.

THE WINTER NIGHT.

'Tis the high festival of night!
The earth is radiant with delight;
And, fast as weary day retires,
The heaven unfolds its secret fires,
Bright,—as when first the firmament
Around the new made world was bent,
And infant seraphs pierced the blue,
Till rays of heaven came shining through.

And mark the heaven's reflected glow
On many an icy plain below;
And where the streams with tinkling clash
Against their frozen barriers dash,
Like fairy lances fleetly cast
The glittering ripples hurry past,

And floating sparkles glance afar
Like rivals of some upper star.

And see, beyond, how sweetly still
The snowy moonlight wraps the hill,
And many an aged pine receives
The steady brightness on its leaves,
Contrasting with those giant forms
Which, rifled by the winter storms,
With naked branches broad and high,
Are darkly painted on the sky.

From every mountain's towering head
A white and glistening robe is spread,
As if a melted silver tide
Were gushing down its lofty side;
The clear cold lustre of the moon
Is purer than the burning noon,
And day hath never known the charm
That dwells amid this evening calm.

The idler on his silken bed
May talk of nature cold and dead;
But we will gaze upon this scene,
Where some transcendent power hath been,
And made these streams of beauty flow
In gladness on the world below,
Till nature breathes from every part
The rapture of her mighty heart.

PEABODY.

AGRICULTURAL IMPLEMENTS.

What are the most important agricultural implements?

The plough, the harrow, the spade, the hoe, the roller, the fork, the scythe, the rake, the cultivator, the cradle, the seed-sower, the corn-sheller, the straw-cutter, the fanning-mill, the cart, the wagon, and many others of an unknown number.

What is said of the history of the plough?

It is the most valuable, and probably the most ancient of all agricultural implements. There are traces of it in the earliest written authorities; and Hesiod advised the Greek farmers to have a spare plough, so that in case of accident the work might not be interrupted. The ploughs of Rome were of the most simple construction, much resembling in form the anchor of a ship. Rivaling these in simplicity and rudeness of form, are the never-altered or improved ploughs of the Hindoos and Chinese, from whose implements it is probable the shape of those of Rome were borrowed. The ploughs of the early Saxons and Normans were furnished with wheels; and, it is supposed that the ploughs of the former were drawn by being fastened to the tails of their horses. The same practice also was known to exist in Ireland.

When and by whom did ploughs receive improvements indicating the present mode of construction?

Somewhere about the year 1730, the Dutch got up a plough which had some general resemblance to that now in use. It was constructed chiefly of wood; the draught-irons, share, and coulter, with the additional plating of iron to the mold-board and sole, being the only parts made of iron. Thirty years afterwards, in Scotland, ploughs were further improved by an iron mold-board and otherwise, so that in most respects, we now use much the same model, with less important variations.

What advantage have the ploughs now in use over those which preceded them?

The wood plough, as generally made, had many objectionable points and properties. It was very liable to get out of order from exposure to the soil and weather, and continually required repair. For want of a correct principle or guide, it would frequently happen that if two ploughs were made by the same maker, they would not work exactly alike; and it was a matter more of chance than certainty whether either would perform its work properly. Besides, it was so unwieldy and occasioned so much resistance, that double the draught power was required, now necessary; that is, with our present ploughs one yoke of oxen will perform about the same amount of work that two did with the old wood ploughs.

How are ploughs now made?

They are composed principally of cast iron; and this is not only much the cheapest, but for ordinary use it is much the best. The point being of cast iron, is easily replaced at a moment's notice, and with a trifling expense, when either

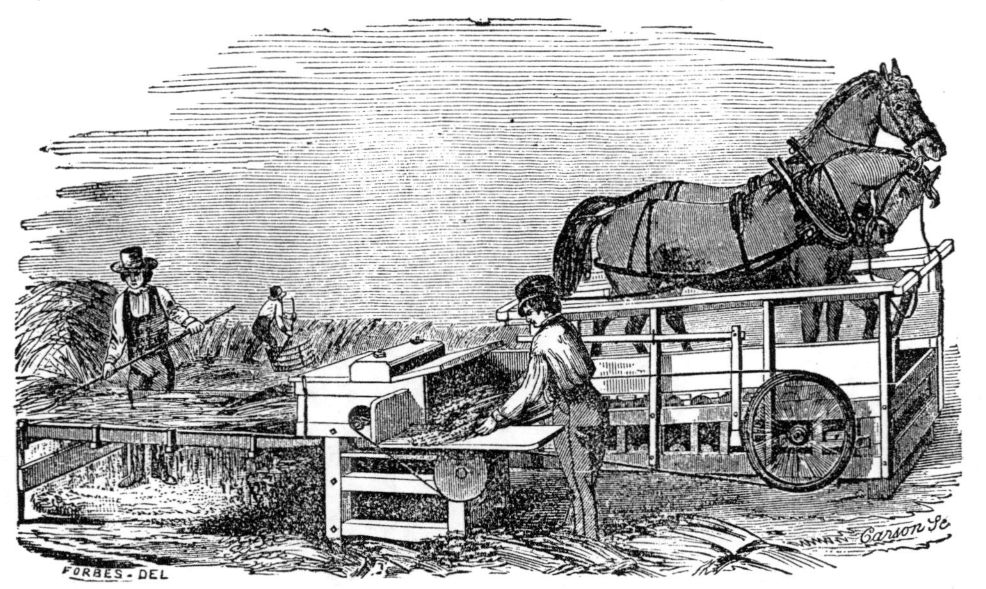

EMERY'S THRASHING MACHINE. By the use of this or any similar Machine, from fifty to one hundred bushels may be thrashed in a day; thus saving much in the cost of that operation. Besides, where it is left to be done in stormy days through the winter, ordinarily the rats and mice will destroy as much of it as would pay for it done by a Machine.

broken or worn out; and, in consequence of being made from a mixture of several kinds of the best metal, it has the tenacity, strength, and durability of steel.

How many ploughs are probably manufactured annually in the United States?

It is impossible to tell the exact number, or a near approximation to it. There are, however, about two millions of farmers, or persons who cultivate farms on their own account. Three ploughs are supposed to be the average number owned by each one; some have five or six, and no one can be expected to do with less than two, and rarely less than three. The wear on the whole cannot be less than one entire plough, so that each farmer, on an average, purchases a new one each year. This is a moderate estimate; and assuming it as a fact, two millions of new ploughs are demanded annually in the whole country. In the city of New-York there is one establishment which manufactures annually, ten thousand ploughs. There may be others still larger. If they average five dollars each, it is seen that the farmers in the United States, every year, pay ten millions of dollars for the single article of ploughs.

What is said of the construction of harrows?

They are of various sizes and shapes. The more common ones, triangular, square, or paralellogram, with unequal angles. The particular shape is not very material; but it is important that the teeth be so arranged that no two are on the same line as the harrow is carried forward; and that they form lines within two or three inches of each other, and at equal distances. Harrows that are made of two equal parts, and connected with hinges are best, because they work better on uneven ground, and when the teeth are required to be cleared of vines or any refuse vegetables with which they may become entangled, they may be cleared without lifting the whole weight of the harrow.

How are the best rollers constructed?

The most approved rollers are constructed wholly of iron, except the tongue, which is of wood. These rollers are usually made either eighteen or twenty-four inches in diameter, in separate sections, each one a foot long, turning on a wrought iron arbor, independent of each other, and without much friction. They can be of any weight, and used by hand, or so heavy as to require one or two horses. A roller of four sections only, thills or shafts being applied instead of a tongue, can be operated by one horse; but if it consist of six sections, two horses are necessary. It is a modern invention, but is coming into general use with the best farmers.

What is said of the cultivators?

They are designed to stir the earth and to destroy the weeds between the rows of corn, and other crops in rows or drills, and are among the most important labor-saving implements in modern husbandry, inasmuch as one man and horse, with one, will accomplish as much work as four men, at least. In form they much resemble the triangular harrow, with fixtures to expand and contract, according to the distance between the rows where they are to be used. They have teeth, but variously constructed; some to penetrate deep into the soil, and others resembling small ploughshares, so as the better to cut up weeds, and to change the position of the soil. No farmer, once acquainted with its use, would willingly do without one. They are also good for being passed over a field after the sowing of wheat, to mix the seed with the soil, instead of doing it with a plough, as often done.

How is the cradle described?

It is a broad scythe and sneath, with long bending teeth in a frame, to be used in cutting and laying in a swath, oats, wheat, and other grain, instead of doing it, as formerly done, by a sickle. It performs the labor with so much expedition, that the common sickle is almost wholly superseded. If the grain has not been blown down by storms or otherwise, a hand with the cradle will cut from two to four acres in a day.

What is the construction of reaping machines for horse power?

There are machines of this description which will reap, it is said, fifteen acres of wheat in a day, and will cut the grain as smooth and clean as it can be done with a sickle, or scythe. The machine has low wheels, and is drawn by a pair of horses, and cuts a swath five feet wide with twenty knives working horizontally, which require sharpening only once a day. A man sitting on the side of the platform with a rake, pushes off the grain as fast as it is cut. A field of oats or barley may be cut as neatly and expeditiously, as one of wheat or rye. They are of course designed for those extensively engaged in agriculture.

How are rakes constructed?

With the common hand farm rakes all are familiar; but

these will soon fall into disuse, except on small rough farms, or where the owners have not learned how much may be gained in raking hay by the use of horse-power. Horse-rakes are made of different patterns; and, what is called the revolving rake, is probably the best. With this, a man, a boy, and a horse, can rake over, from fifteen to twenty-five acres in a day—a saving of immense importance, at a season of the year when labor commands a high price, and cannot always be had. It is easy for any one to estimate in how short a time the cost of it would be saved. Besides, its importance is increased from the fact that sudden showers and storms are constantly liable to arise in the season of hay-making. It is of simple and durable construction, and is not expensive.

What is said of the modern inventions for sowing and planting different kinds of seed?

The machine for sowing garden seeds, and all small seeds in field culture, such as carrots, turnips, parsnips, and beets, is moved by hand. The operator pushes it before him, and by a single operation, the furrow or drill is opened, the seed is deposited, and the earth turned over it and compressed. The seeds are dropped with the greatest precision, at any desired distance from each other, whether in hills or drills. The horse-drill is for planting wheat, rye, Indian corn, oats, peas, beans, or any thing else, in large fields. A man with two horses and this machine, can put in from ten to twelve acres of wheat in a day; and with one horse he can plant twenty acres of Indian corn in a day. The drill is so arranged that it can be used on rough and hilly land; and, will deposit any desired quantity of seed to the acre.

What account is given of straw, hay, and corn-stalk cutters?

There are several different instruments in use coming under this denomination, more or less expensive, and varying in excellence for the purpose designed. Ordinarily, they have a crank, and are operated by hand; but can be made so firm as to be used by horse-power, cutting a ton in an hour and a quarter. There is a great saving in the cutting of corn-stalks, hay, and straw, in two ways. The animals do not waste it by drawing it out of the mangers, and trampling it under their feet, and time and labor are saved them in masticating. They obtain their supply of food readily, and then lie down to digest it. Fermentation also developes the nutritive matter, and requires less work for the stomach, and this, by saving muscular exertion, leaves more strength with the animal to be expended on ordinary work. Both hay and straw should be slightly wet; and if seasoned with a little meal and salt, is the better. The best machines are those with spiral knives and cutting against a roller of raw hide, being self-feeders, and without any complicated fixtures. Some feed their horses altogether with cut hay and straw, mixed with ground provender.

How are vegetable cutters made, and what is the benefit from them?

The cutting wheel of this implement is made of cast iron, faced on one side, through which are inserted three or more knives, like plane-irons. These cut the vegetables into thin slices with great rapidity, and then, by cross knives they are cut into slips of convenient form and size for cattle or sheep, without danger of choking. The pieces after cutting lie loosely, and can easily be taken up by the animal. One of these machines in good order, will cut fifty bushels of turnips or carrots in an hour.

How is the corn and cob crusher made?

In such a machine the cobs are first cut in short pieces by means of a strong spiral knife attached to the axle, and then pass between two grinding plates made of composition metal, that will last two or three years, when they are replaced by new ones. They are of sizes to be worked by horse-power, or with a wheel and crank by hand. This machine, where it is deemed an object to use the cobs for feed, and where other means of reducing them to small atoms are not within convenient distance, are of great value to the farmer. There are also grain mills of somewhat similar construction; one of which, with one horse-power, will grind four bushels of fine meal in an hour; if the meal is coarse, a greater quantity. In seasons of drought, when the common grist-mills are deficient in water, one of the above grain mills would be of great convenience.

What account is given of the fanning mill?

It is a contrivance employed for separating, by an artificial current of air, the chaff from the grain, after it has been threshed out of the straw. Various accounts are given of the introduction of this machine into England and Scotland, and of the many claimants for the credit of having been the first makers of this ingenious piece of mechanism. All, however, agree that the idea, model, or design was first furnished from Holland. Different patents have been taken out for those of different construction; but either, probably, of those now in use, will answer the purpose. The presumption is, that those are to be preferred which are the least complicated in their machinery.

Of the utility of corn shellers, what may be said?

The large quantities of corn raised in this country, render it, as it were, indispensable that there should be machines for shelling it from the cob in some expeditious manner. Formerly, to shell the corn by hand, growing on a farm, was a winter's occupation for one man at least. No wonder then that so many efforts have been made to supply this desideratum. Two men with one of the best, having a double hopper, can shell two hundred bushels of corn a day. So the venders of it affirm; but, if only half of that quantity were shelled, the inducements for possessing one would be ample. A sheller with a single hopper will be sufficient for all moderate corn growers. Of the different patents each purchaser can be guided by his own fancy and judgment, keeping in mind the amount of labor to be performed by it.

What is said of the relative value of wagons and carts for use on a farm?

It would be well for farmers to own both, if able, and their labors are extensive and of diversified character. For heavy burdens and for going long distances, especially on public roads, wagons are doubless preferable. The weight resting on two axles instead of one, the wheels press less heavily on the ground, and there is of course less friction on the axles, as the wheels turn round. The load also is kept, in ascending hills, from settling backward and thus tending to choke the animals upon draught; and in descending hills, settling forward and thus pressing too onerously upon them. But from the more convenient facility with which carts are turned round, much time will be saved in using them in doing most of the work on the farm, particularly the removal of manure and other objects for short distances only.

What is said of the importance of the grindstone?

It is surprising that any farmer should neglect to be furnished with a good grindstone, as it is found most necessary every few days. The individual that depends on his neighbor, if only a fourth of a mile distant, for this useful implement every time his edge tools require being sharpened, will be likely to lose the cost of two or three stones in time wasted in going back and forth. And, if mounted on friction rollers, which cost but a mere trifle, more than half of the toil in turning it will be saved; and, in addition to this, when putting an edge to small tools, and others requiring but little grinding, the labor of a person at the crank is saved altogether.

What may be said of the importance of being amply supplied with the implements generally needed in doing the different labors of the farm?

There is no economy or saving in not being furnished abundantly with them. It is better to have too many than not enough. The catalogue required would be too long for enumeration here. Suffice it to say, that in the saving of time they all pay for themselves many times over. The farmer who is destitute of the more constantly needful of them, would make a slow progress in his work, like that of the shoemaker who should attempt to manufacture shoes by using the tine of a table fork instead of an awl, or like that of the tailor who attempts to sew by fastening his thread to the head of a pin, instead of using a needle.

And, what is said of the expediency of using the more expensive kinds of agricultural implements?

It is very evident, that in all cases where they would lead to a saving of labor, varying from a hundred to a thousand per cent., the neglect to have them is an absurdity. This is especially true, where horse-power can be substituted for human labor. It is similar to a man, not owning a horse, spending a whole day in carrying to the grist-mill a bag of corn on his back, when he might hire a horse to do it, and pay for it in his own labor in one-fourth part of the time; or, like carrying to market in his own wagon a load of produce, when he might send it on the rail road, or some other public conveyance, for half the amount paid for tolls and other incidental expenses in the journey.

What is said of taking good care of farm implements?

The aggregate amount of loss to farmers who are inattentive to this subject, is incredibly great. A slight yielding of an implement is cheaply repaired, and then it answers the same as a new one; but, if neglected, it soon goes to ruin, and another is wanted. It is not unlike neglecting to close a small rent in our garments, which is almost sure to catch upon something that will further rend them, till completely worthless. All farm implements should be kept in perfect repair, every now and then as needed, receiving a coat of paint; and above all, when not in use, should be kept sheltered from the sun and weather, and well packed away. By such precautions a farmer well supplied with them will save annually one hundred dollars. This is a fair calculation.

ROMANCE IN REAL LIFE.

We have assembled here to-day to commune together, upon one of the great departments—the greatest indeed—of human employment. In the task assigned to me, I shall not enter into the practical details, which belong to this vast subject; interesting to all, however diversified their avocations in life. I feel my incompetence to perform the part of a teacher in the great art of agriculture; the art of directing and aiding nature in the performance of those functions, which were designed by Providence for the comfort and subsistence of man. A large portion of this intelligent assembly unite experience with observation, an intimate knowledge of practical operations, with a just theory of their application, and a full appreciation of the value of combining personal experience with those enlarged views, which are essential to progressive improvement. But omitting those practical details, which a familiar knowledge of the subject can alone furnish, there are still many general and important considerations, as well moral and statistical, as historical, which are neither inappropriate to the occasion, nor unworthy of your attention, and some of these I propose to present to you.

The more active portion of my life has been devoted to other and harsher duties; and much of it to ranging the forests in the pursuit of the red man in time of war, and to counsel with and restrain him in time of peace. And this is not the first time I have been here. I have been here, when he who now speaks, if he had spoken then, would have found no hearers. When the silence of the forest was unbroken by the cheerful hum of human industry, and its solitude uninterrupted, but by the wandering Indian, and the animals that ministered to his wants—when a world of primitive, gigantic vegetation extended its sway across our own beautiful Peninsula, and on to the very shores of the Pacific, where our fathers' flag and our own now waves in the breeze, that comes from the continent of Asia.

He alone, who has traversed these regions, day after day, in the freshness indeed, but in the silence and solitude of nature, almost appalled by a sense of loneliness and insignificance, amid these wonders of creative power, can justly appreciate the efforts of man in subduing and reclaiming the prairie and the forest, and preparing them for those scenes of improvement and cultivation, which cheer the eye and gladden the heart of the traveller; and, above all, of the traveller, who preceded the march of civilization, and now follows it in its glorous progress. Never has human industry achieved a prouder triumph, than in this conflict between nature and man. As in the exodus from Eden, he has been "sent forth to till the ground," and in the "sweat of his face" has he thus far fulfilled his mission. And a proud one it was; aye, and yet is; for, though it has done much, it has still much to do. It began at the beach of Jamestown, and the rock of Plymouth, where its first labors were broken by no sound but the surges of the Atlantic, and they will finish only, when the last echo of the woodman's axe shall mingle with the surges of the Pacific.

Do not these miracles of enterprise resemble the fictions of an Eastern imagination, rather than the sober realities of human experience? Do they not speak to us in trumpet tones of the value and dignity of labor, for by labor have they been wrought—persevering, unyielding, triumphant labor. There is no lesson more important to be taught to our young countrymen than that, which is taught by this great characteristic feature of American history; the immense conquest, which man has achieved, over the world of matter, that opposed his progress, and the scanty resources he brought to the work. His own exertions, and the axe and plough, have accomplished this mighty task; always indeed with toil and exposure, and sometimes under circumstances of privation and suffering, before which the stoutest resolution might give way. But if time brought its trials, it brought also its reward; it converted the wilderness into a garden, and spread over the face of the country, those beautiful habitations, and those fertile and productive fields, such as this region offers to view, and which are at once, the evidence and the recompense of that industry and enterprise, which quail not before toil or danger, but still go on preparing this goodly heritage, as well for ourselves, as those, who are to follow us, when our task is done.

And how would this great work of subduing nature and preparing the forest for the residence of man have been accomplished in the older regions of the globe, so long the theatre of human exertions? The answer to this pregnant question describes by a single trait the great marked difference between the condition of agricultural labor in the Eastern and in the Western hemispheres; between the laborer for others, and the laborer for himself. Across the water, which does not separate us more widely in space, than do the position and prospects of the people in their condition, great enterprises are never concerted, and conducted by the mass of the inhabitants. Counsel and capital are furnished by the few and fortunate; "the sweat of the face" by the many and the wretched. And the profitable results belong to the former; while the latter eke out a scanty and precarious subsistence, as poor and depressed at the termination of the most successful and gigantic undertaking, as at its commencement. Here it needs not that any one should tell you the difference. He, who runs, may read it in the history of our whole progress, individual and national.

The forest has fallen before those, who established their habitations in its dark recesses; dark till their toil made way for the light of Heaven to shine upon them. They labored themselves, and for themselves. No taskmaster directed their work, and no speculator garnered the profits. And thus exertion was stimulated by the most powerful motives, which can operate upon human nature; by the necessity of present subsistence, and the hope—the certainty, I should say, of future competence and comfort; and therefore it is, that upon the immense domain from Lake Erie, almost to the shadow of the Rocky Mountains, a vigorous, intelligent, and enterprising people have fixed their residence, and by their own labor, and for their own advantage, have prepared it for all the purposes of civilized life. And the time, within which this has been done, is not the least extraordinary feature in this great national migration—a migration going forth to invade the forest, and to fulfil the first command of the Creator, "to replenish the earth and subdue it," and not, as in the history of human conquest, to lay waste, and destroy, having before it fertile and flourishing regions, and behind it ruin and desolation. The man yet lives, who was living, when almost the first tree fell before the pioneer's stroke in this magnificent region, and the man is now living, who will live to see it contain one hundred millions of people. I have myself known it for half a century, and in that space, long indeed in the life of man, but brief in the life of communities, our own region of the Northwest, marked with its distinct boundaries upon the map of nature, by the Lakes, the Mississippi and the Ohio, has risen from infancy to manhood, from weakness to strength, from a population of a few thousands, to five millions of people; of freemen, owning the soil they occupy, and which they won by their industry, and will defend by their blood.

Where, in the long annals of the human race, can you find such an augmentation of the resources and numbers of a country, gained in so short a period, and under such circumstances of trial in its progress, and of prosperity in its issue? And may we not well say, that the mighty agent, which has built up this monument of productive power, deserves the gratitude and the fostering care of the American people? And that agent is labor, and our duty is to elevate it in the scale of employment. To show what it has done, and is doing, and is destined, I trust, yet to do. It has not founded a monarchy indeed, whose burthens are for the rejected, and its benefits for the chosen; whose splendor dazzles the eye, while its oppression sickens the heart. But it has laid the foundation of a Republic, broadly and deeply in the rights of man; whose equal protection covers all, as its equal honors are open to all; and whose career, if not checked by our own folly, or by the just judgment of God, promises a glorious and encouraging spectacle to the lovers of freedom through the world—aye, and an example too for long ages to come.

Human occupation should be measured by its useful consequences and by its moral tendencies, and by the principles and conduct of those, who are devoted to it, and whose character is formed by its pursuit. Tried by this standard, where shall we find an employment more worthy of honor and regard, than that which drew from Sir William Jones the eloquent panegyric, that "he who makes two spears of grass grow, where but one grew before, is a public benefactor, far in advance of the noblest chieftains, who, aided by armies and the enginery of war, sack cities, carry conquest onward, only to conquer, and subjugate and desolate kingdoms?" And yet so wayward is human nature, and so unjustly are its honors distributed, that the temple of Mars is thronged with the votaries of fame, while silent are the altars of Ceres, and those, who worship there must find their reward, not in public renown, but in the consciousness of a duty, self-imposed and faithfully performed.

But a better day has begun to dawn. Many old things are passing away, and with them is waning that military glory, which has so long led captive the best affections of our nature. The time is coming when the supporter of human life will find his station far higher in the world's estimation, than the destroyer. We are beginning to learn, that the splendor of victory is a fearful pageant, while conquest over the earth, and the multiplication of its products are acceptable sights in the eyes of God and man. He, who puts his hand to the plough, and does not look back upon more brilliant, but less useful employments, will not fail to find his reward in a happy and honorable life. What a perversion of terms, or rather what a perversion of moral sentiment does it exhibit to talk of the dignity of indolence, the dignity of doing nothing, and the unworthiness of useful, honest labor! Whatever of this feeling there is among us, and there is some, is exotic, not indigenous; imported whence many other notions, equally unreasonable and injurious, have come, to exercise a baneful influence upon our social system.

My own experience may not be without profit to some who hear me; certainly not, if it furnishes motives for encouragement or hopes to stimulate to exertion. It is fifty years and more, since I crossed the mountains on foot, a young adventurer, seeking that land of promise, which has been to me, as to so many others, a land of performance. I had many difficulties to contend with, many obstacles to encounter, and many privations, in peace and in war, to endure;* and I have probably undergone as large a share of fatigue and exposure, in the early part of my life, as often falls to the lot of our countrymen. But thanks to the nature of our institutions, to their glorious equality, and thanks above all to the favor of my countrymen, I have had a measure of political prosperity, far beyond what I merited, or even dared to anticipate. The youthful emigrant, now in the decline of life, communicating to his youthful hearers the result of his experience, that they may go and do likewise—in all but the errors he committed—has been borne onward by his generous fellow-citizens to the high posts of the nation, and has represented his country at the court of kings. And he has returned with the conviction, as abiding as his life, that the sun never shone upon as happy a region as this confederation embraces; nor one where human freedom meets less opposition, or the human intellect less restraint; nor where there are such powerful motives for exertion, or such distinctions for its reward.

Our history furnishes many striking examples of this progress from unprotected self-dependence to public confidence, and to the highest honors. The Father of his country commenced life as a land surveyor, and he died, leaving the brightest name that mere man has left in all the long annals that record the days and deeds of the human race. Greene, his friend, and undoubtedly his most confidential general, a distinction he well merited by his courage and conduct, was a blacksmith, and laid down the hammer, when he girded on the sword. Putnam, the very impersonation of hardihood and intrepidity, was driving his plough, when the musketry at Lexington aroused a continent. He left that plough in the furrow, and mounting his horse, repaired to Boston, and joined the throng of patriots who, with a devotion to freedom as true and holy, as ever animated the human breast, entered into a contest more unequal, perhaps, than any which oppression has ever waged against power.

And we learn from the Holy Scriptures, that the chosen champion of a higher warfare, the prophet Elisha, when summoned to his mission, was found by Elijah in his field, "ploughing with twelve yoke of oxen before him, and he with the twelfth." Morgan, the most enterprising partisan of our revolution, was a wagoner; Starke, whose memorable defeat of the British at Bennington, prepared the way for the capture of Burgoyne and his army, was, in early life, a field laborer. Roger Sherman, equally renowned for the power of his intellect, and his active and efficient exertions in the councils of the nation, was a shoemaker; and Franklin, whose name recalls his world-wide reputation, and the deeds in the Arts, in Science, and in the service of his country, which he did to acquire, and to deserve it, was a printer, laboring assiduously at his work, during many years of his life. But I need not add to these illustrious names. It would be easy to do so, were more examples of success required. But sufficient are these for all, who desire to profit by this characteristic and encouraging chapter of our history. Young men, whom I see around me, ponder over the lives of these great men of the past generation. Follow their course with probity and perseverance, and you will follow them also in their useful career. You cannot all, indeed, attain the highest stations, but you may all attain respectability and prosperity, and enough of both to satisfy the measure of a reasonable ambition.—*From Address of* LEWIS CASS, LL. D., *Michigan.*

SAGACITY OF THE SPIDER.

Animals in general are sagacious, in proportion as they cultivate society. The elephant and the beaver show the greatest signs of this, when united; but when man intrudes into their communities, they lose all their spirit of industry, and testify but a very small share of sagacity for which, when in a social state, they are so remarkable. Among insects, the labors of the bee and ant have employed the attention and admiration of the naturalist; but their whole sagacity is lost upon separation, and a single bee or ant seems destitute of every degree of industry, is the most stupid insect imaginable, languishes for a time in solitude, and soon dies.

Of all the solitary insects I have ever remarked, the spider is the most sagacious, and its actions, to me, who have attentively considered them, seem almost to exceed belief. This insect is formed by nature for a state of war, not only upon other insects, but upon each other. For this state, nature seems perfectly well to have formed it. Its head and breast are covered with a strong coat of mail, which is impenetrable to the attempts of every other insect, and its belly is enveloped in a soft, pliant skin, which eludes the sting even of a wasp. Its legs are terminated by strong claws, not unlike those of a lobster; and their vast length, like spears, serve to keep every assailant at a distance.

Not worse furnished for observation than for attack or defence, it has several eyes, large, transparent, and covered with a horny substance, which, however, does not impede its vision. Besides this, it is furnished with forceps above the mouth, which serves to kill or secure the prey already caught in its claws or its net.

Such are the implements of war with which the body is immediately furnished; but its net to entangle the enemy seems what it chiefly trusts to, and what it takes most pains to render as complete as possible. Nature has furnished the body of this little creature with a glutinous liquid, which, proceeding from the anus, it spins into thread, coarser or finer as it chooses to contract or dilate its sphincter. In order to fix its threads, when it begins to weave, it emits a small drop of its liquid against the wall, which, hardening by degrees, serves to hold the thread very firmly. Then receding from the first point, as it recedes, the thread lengthens; and when the spider has come to the place where the other end of the thread should be fixed, gathering up with its claws the thread, which would otherwise be too slack, it is stretched tightly and fixed in the same manner to the wall as before.

In this manner it spins and fixes several threads parallel to each other, which, so to speak, serve as the warp to the intended web. To form the woof, it spins in the same manner its thread, transversely fixing one end to the first thread that was spun, and which is always the strongest of the whole web, and the other to the wall. All these threads, being newly spun, are glutinous, and therefore stick to each other, wherever they happen to touch; and in those parts of the web most exposed to be torn, our natural artist strengthens them, by doubling the threads sometimes sixfold.

Thus far, our naturalists have gone in the description of this animal: what follows is the result of our own observation upon that species of the insect called the house-spider. I perceived, about four years ago, a large spider in one corner of my room, making its web, and though the maid frequently levelled her fatal broom against the labors of the little animal, I had the good fortune then to prevent its destruction, and, I may say, it more than paid me by the entertainment it afforded.

In three days the web was with incredible diligence completed; nor could I avoid thinking that the insect seemed to exult in its new abode. It frequently traversed its round, and examined the strength of every part of it, retired into its hole, and came out very frequently. The first enemy, however, it had to encounter, was another and much larger spider, which, having no web of its own, and having probably exhausted all its stock in former labors of this kind, came to invade the property of its neighbor. Soon, then, a terrible encounter ensued, in which the invader seemed to have the victory; and the laborious spider was obliged to take refuge in its hole. Upon this I perceived the victor using every art to draw the enemy from his stronghold. He seemed to go off, but quickly returned, and when he found all arts vain, began to demolish the new web without mercy. This brought on another battle, and, contrary to my expectations, the laborious spider became conqueror, and fairly killed his antagonist.

Now then, in peaceable possession of what was justly its own, it waited three days with the utmost impatience, repairing the breaches of its web, and taking no sustenance that I could perceive. At last, however, a large blue fly fell into the snare, and struggled hard to get loose. The spider gave it leave to entangle itself, as much as possible, but it seemed to be too strong for the cobweb. I must own I was greatly surprised when I saw the spider immediately sally out, and in less than a minute weave a new web around its captive, by which the motion of its wings was stopped, and when it was fairly hampered in this manner, it was seized and dragged into the hole.

In this manner it lived, in a precarious state, and nature seemed to have fitted it for such a life; for upon a single fly it subsisted for more than a week. I once put a wasp into the net, but when the spider came out in order to seize it as usual, upon perceiving what kind of an enemy it had to deal with, it instantly broke all the bands that held it fast, and contributed all that lay in its power to disengage so formidable an antagonist. When the wasp was at liberty, I expected the spider would have set about repairing the breaches that were made in its net; but those, it seems, were irreparable, wherefore the cobweb was now entirely forsaken, and a new one begun, which was completed in the usual time.

I had now a mind to try how many cobwebs a single spider could furnish; wherefore I destroyed this and the insect set about another. When I destroyed the other also, its whole stock seemed entirely exhausted, and it could spin no more. The arts it made use of to support itself, now deprived of its great means of subsistence, were indeed surprising. I have seen it roll up its legs like a ball, and lie motionless for hours together, but cautiously watching all the time; when a fly happened to approach sufficiently near, it would dart out at once, and often seize its prey.

Of this, however, it soon began to grow weary, and resolved to invade the possession of some other spider, since it could not make a web of its own. It formed an attack upon a neighboring fortification, with great vigor, and at first was as vigorously repulsed. Not daunted, however, with one defeat, it continued to lay siege to another's web for three days, and at length having killed the defendant, actually took possession. When smaller flies happened to fall into the snare, the spider does not sally out at once, but very patiently waits till it is sure of them; for upon his immediately approaching, the terror of his appearance might give the captive strength sufficient to get loose: the manner then is to wait patiently till, by ineffectual and impotent struggles, the captive has wasted all his strength, and then he becomes a certain and easy conquest.

The insect I am now describing lived three years; every year it changed its skin, and got a new set of legs. I have sometimes plucked off a leg, which grew again in two or three days. At first it dreaded my approach to its web; but at last it became so familiar as to take a fly out of my hand, and upon my touching any part of the web, would immediately leave its hole, prepared for defence or attack.

To complete this description, it may be observed, that the male spiders are much less than the female, and that the latter are oviparous. When they come to lay, they spread a part of the web under their eggs, and then roll them up carefully, as we roll up things in a cloth, and thus watch them in their hole. If disturbed in their holes, they never attempt to escape without carrying their young brood in their forceps away with them, and thus frequently are sacrificed to their paternal affections.

As soon as the young ones leave their artificial covering,

they begin to spin, and almost sensibly seem to grow bigger. If they have the good fortune, when even but a day old, to catch a fly, they fall too with a good appetite; but they live sometimes three or four days without any sort of sustenance, and still continue to grow larger, so as every day to double their former size. As they grow old, however, they do not still continue to increase, but their legs only continue to grow longer; and when a spider becomes entirely stiff with age and unable to seize its prey, it dies at length of hunger.

OLIVER GOLDSMITH.

PHYSICAL BENEFITS OF WINTER.

Winter has its physical benefits and its moral lessons. First in order we shall notice the *physical advantages* of winter. One of the most prominent of these is the repose which it secures to the earth after the productiveness of summer. It has been beautifully said, "The days of winter are the days of nature's rest. In the preceding months she has been exhausted with incessant labor for the good of man. How rich has the spring been in flowers; how the seeds have expanded and the foliage sprouted. What abundance of fruits the summer prepares for the autumn's maturing hand. Every month, every day, we receive some fresh gift from nature. As the tender mother provides for her young with anxious care, so nature is busied from morn to evening in supplying our wants, and in procuring us a succession of comforts and blessings to make life's fleeting moments smile with joy and with delight. Food, raiment, and the chief sources of our pleasures, are all derived from her fostering bosom. For us she makes the seeds to open and expand, the herbs to bud, the trees to look gay with foliage, beautiful with blossoms, and to pour forth their riches in fruit of every kind that can please the eye or gratify the taste. For us the golden grain waves over the fields, the vine offers her varied treasures, and the whole creation is clothed in verdure, and presents to the delighted observer an infinitely varied and beautiful field of attractions. Wearied by so many labors, nature, for a space, reposes, in order to acquire new force, that she may again be equally fruitful, and again be able to assume her wonted resplendency." The farmer well understands that land, to be made most highly and permanently productive, must have occasional rest. It must not be sown with grass or grain year after year, lest the strength and goodness of the soil should soon be exhausted, but it must be favored with a change of crops. The Jewish law required that every seventh year the land should rest and lie still; and farmers who have made the experiment testify that they find this to be good economy. On the same principle the temporary repose of the earth in winter, like rest in sleep, is favorable for augmenting the vigor and productiveness of the soil.

But winter is not to be regarded as the mere *repose* of nature; it is also a preparatory season—the seed-time of the coming year. Even when the ground is stiffened and covered with snow, secret processes are at work in the laboratory of nature, preserving and elaborating "the seeds, buds and roots of future plants and flowers," and unfolding the germs of the future harvest. The fall-sown grain, subjected to these processes, slowly developed till it just peers above the surface, and then covered with its warm blanket of snow to await the return of the sun, attains a fullness, vigor and maturity never reached by that which is sown in spring. One of the most beautiful sights in nature is that of a fresh green field of grain shooting up, when all around it is desolate, to indicate that nature's life is not extinct, and that seed-time and harvest shall not fail. Under the cold hard surface of winter lie hidden in the yet warm and nutritous earth, the spring, the summer, the autumn of the coming year. A new harvest is germinating there, to be quickened into life and fruitfulness by the warm summer's sun. Moreover there are plants and trees that preserve their verdure through the winter; flowers that spring up under the snow and bloom with a delicious fragrance; shrubs, herbs, and vegetables that come to maturity amid frosts and ice; showing that nature, though in comparative repose, is not dead—that winter is not all a waste. Some of the most kindly processes of nature in the preservation and development of the various species of grains and grasses, plants and vegetables, are carried forward in the dark recesses of winter.

But this is not the full extent of the physical operations of winter. The *diversity of climate* is one of the chief advantages of the present constitution of our earth, and of its relation to the solar system. Let us suppose that there was throughout the globe "an equal distribution of heat and cold, the same degree of fertility, and the same division of day and night." In the opinion of some, that would be to make the earth a Paradise. But the immediate effect of such a state of things would be the loss of that diversity in the appearance and the productions of the earth which is now so pleasing and so beneficial. A vast variety of the productions of the earth now so nicely adjusted to certain degrees of temperature, must perish by such a change, and men have everywhere the same limited and uniform means of subsistence. As a consequence of this, commerce must almost entirely cease; for neither the natural productions nor the artificial fabrics of one country would be needed in another,—each country affording a full supply of the same articles, and having nothing else to offer in exchange. There being no necessity for commercial intercourse, many of the arts and also of the physical sciences which have been developed or stimulated by commerce, would remain hidden in the secrets of nature, and the entire race would be far below its present position in cultivation and enjoyment. A uniformity of temperature would be a serious disadvantage. The withdrawal of winter from the earth, would render the heat intolerable—for even the heat of the torrid zone is now mitigated by the prevalence of cold in the polar regions. And though the temperature were moderate and endurable, but still uniform, we should be deprived of the benefit of winds,—those salutary disturbances in the equilibrium of the atmosphere which are owing to changes of temperature,—and should soon be transformed into a race of grovelling Cretins at the bottom of the sea of stagnation. One of the prime offices of winter with its storms is ventilation; the purifying of the atmosphere on a great scale, from the damps and noxious gases of the summer.

Nor should we overlook the effect of winter on the human system. This in general is invigorating. While some by reckless exposure, and others by excessive caution, make the season one of physical discomfort and ailing, most persons find the cold, clear, bracing air of winter a welcome tonic, after the lassitude and debility of the summer months. In general, winter is a season of health and physical vigor; when we breathe freer, and the blood, stimulated by our increased activity, courses more warmly and briskly through our veins.

Welcome then winter, its cold and storms; for amid its darkness and desolation, it keeps alive our vital energy, and rejuvenates the earth exhausted by the labors of spring and the fruits of autumn; it multiplies our comforts, by varying the products of the earth, and stimulates our commerce with the vast brotherhood of man; it sweeps away the pestilence, and nerves us for that hardihood and toil which produce the best physical development and the purest physical enjoyment. Thou, the beneficent Father of all, "hast made WINTER" for the good of thy creatures.—NEW-YORK INDEPENDENT.

THE TWILIGHT OF THE HEART.

There is an evening twilight of the heart,
  When its wild passion waves are lulled to rest,
And the eye sees life's fairy scenes depart,
  As fades the day-beam in the rosy west.
'Tis with a nameless feeling of regret
  We gaze upon them as they melt away,
And fondly would we bid them linger yet,
  But Hope is round us with her angel lay,
Hailing afar some happier moonlight hour;
Dear are her whispers still, though lost their early power.

In youth the cheek was crimsoned with her glow;
  Her smile was loveliest then; her matin song
Was heaven's own music, and the note of wo
  Was all unheard her sunny bowers among.
Life's little world of bliss was newly born;
  We knew not, cared not, it was born to die.
Flushed with the cool breeze and the dews of morn,
  With dancing heart we gazed on the pure sky,
And mocked the passing clouds that dimmed its blue,
Like our own sorrows then—as fleeting and as few.

And manhood felt her sway too,—on the eye,
  Half realized, her early dreams burst bright,
Her promised bower of happiness seemed nigh,
  Its days of joy, its vigils of delight;
And though at times might lower the thunder storm,
  And the red lightnings threaten, still the air

Was balmy with her breath, and her loved form,
 The rainbow of the heart, was hovering there.
'Tis in life's noontide she is nearest seen,
Her wreath the summer flower, her robe of summer green.

But though less dazzling in her twilight dress,
 There's more of heaven's pure beam about her now;
That angel-smile of tranquil loveliness,
 Which the heart worships, glowing on her brow;
That smile shall brighten the dim evening star
 That points our destined tomb, nor e'er depart
Till the faint light of life is fled afar,
 And hushed the last deep beating of the heart;
The meteor-bearer of our parting breath,
A moon-beam in the midnight cloud of death.

HALLECK.

AN EVENING SKETCH.

'Tis twilight now;
The sovereign sun behind his western hills
In glory hath declined. The mighty clouds,
Kissed by his warm effulgence, hang around
In all their congregated hues of pride,
Like pillars of some tabernacle grand,
Worthy his glowing presence; while the sky,
Illumined to its centre, glows intense,
Changing its sapphire majesty to gold.
How deep is the tranquility! the trees
Are slumbering through their multitude of boughs,
Even to the leaflet on the frailest twig!
A twilight gloom pervades the distant hills;
An azure softness mingling with the sky.
The fisherman drags to the yellow shore
His laden nets; and, in the sheltering cove,
Behind yon rocky point, his shallop moors,
To tempt again the perilous deep at dawn.
 The sea is waveless, as a lake ingulfed
'Mid sheltering hills—without a ripple spreads
Its bosom, silent, and immense—the hues
Of flickering day have from its surface died,

Leaving it garbed in sunless majesty.
With bosoming branches round, yon village hangs
Its row of lofty elm trees; silently
Towering in spiral wreaths to the soft sky,
The smoke from many a cheerful hearth ascends,
Melting in ether.
As I gaze, behold
The evening star illumines the blue south,
Twinkling in loveliness. O! holy star,
Thou bright dispenser of the twilight dews,
Thou herald of Night's glowing galaxy,
And harbinger of social bliss! how oft,
Amid the twilights of departed years,
Resting beside the river's mirror clear,
On trunks of massy oak, with eyes upturned
To thee in admiration, have I sat
Dreaming sweet dreams till earth-born turbulence
Was all forgot; and thinking that in thee,
Far from the rudeness of this jarring world,
There might be realms of quiet happiness!

BLACKWOOD'S MAGAZINE.

ROTATION OF CROPS.

What is to be understood by rotation of crops?

The word rotation signifies a turning round, as a wheel or any other body revolves on a real or imaginary axis, till a complete revolution is made; that is, till each part is brought to the point or place occupied by it before the revolution commenced. And, such revolutions may be continued to any fixed or indefinite length of time. When the term is applied to agriculture, it signifies a succession of different crops, instead of a succession of the same crops; no two years in the period assigned for a rotation or cycle, to have the same crop.

How may this be explained?

In garden culture, should early potatoes be planted on any given tract one year, sweet corn the second year, cabbage the third year, carrots the fourth year, peas the fifth year, beans the sixth year, and melons the seventh year, this would be called a rotation. And when completed, the same order of change might be observed, if judged best, for a second ro-

PAULER MERINO SHEEP, owned by Messrs. CAPEHEART, of Merry Hill, North Carolina, and which obtained a premium at the Fair of the American Institute, 1849.

tation; or for another period of seven years. This is the principle of the rotation of crops in agriculture, as well as in horticulture.

What is the reason for a rotation of crops?

It is known that the proportion of elementary substances that enters into the composition of plants, is not the same in all. Probably it is not precisely the same in any two plants. The soil containing the substances for the growth of plants imparts them as needed till nothing remains, when the plants will cease to grow. Supposing a particular ingredient for a particular plant were lime, it is evident that when the lime is all exhausted, or drained from the soil, that plant can no longer be produced on it. So also of all other plants and all other substances which compose them.

How did this become manifest?

The rotation of crops grew out of experience. The practical farmer observed that, in most cases, when the same plants were grown for two, three, or more years consecutively upon the same soil, it did not yield the same abundant harvest; whilst, when another crop was tried upon that soil, the production was satisfactory. Observation and experience subsequently and gradually established for different parts a different alternation of crops. In this, at first, science had no agency. The reason for it was wholly unknown.

In what way was the reason of it unfolded?

Whilst the practical farmer was content to rest simply on the facts supplied by experience, and remained satisfied with believing that some plants exhaust the soil, while others do not, the theorist endeavored to discover the key to this remarkable phenomenon, as it then appeared. Different theories were suggested, but it was a long time before one was adopted that seemed exempt from objection.

What is this theory?

The same as has been intimated; that the utility of the rotation of crops depends exclusively upon the circumstance that cultivated plants withdraw from the soil unequal amounts of certain ingredients for their nutrition. Assuming this as the hypothesis, all the known facts relating to it are satisfactorily explained. Thus science comes to the aid of experience, demonstrating what was before a mere matter of fact, without a knowledge of the reasons for it.

But if any one crop is sought successively every year will there be an entire failure?

There may not be an entire failure the second, third, or even the fourth year; but each succeeding year, all other things being equal, there will be a diminished crop. But other things may not always be equal. Drought or cold may destroy or greatly injure a crop of Indian corn one year, and the next year, being no drought and abundance of heat, the crop of corn may be far better than the preceding year. The soil too, may be so amply furnished with a particular elementary substance for vegetable growth, that several crops of the same plant may be raised in succession, before material diminution will be perceived; but this makes no exception to the principles for a general rotation. Sooner or later this substance will be exhausted, and there would then be a complete failure.

How may the theory for rotation be further illustrated?

If we take a field, the soil of which contains the mineral and saline materials required to produce wheat, and yet only in a quantity exactly sufficient to produce a single crop, it follows, of course, that a second crop of wheat cannot be reared on the same field. The soil is completely exhausted for the moment, and will remain so forever, if it does not contain substances which may, by disintegration and decomposition furnish a new supply of the ingredients necessary to the growth of plants, or if these essential matters are not artificially supplied.

Is such a complete exhaustion of soil common?

It is not. The case supposed is for illustration, and is not likely ever to happen in fact. But what really happens, and common enough, is, that although all the salts are not exhausted, yet being present in the soil in relative proportions very different to the amounts required by various plants, a single crop of wheat may deprive the soil so completely of one of its mineral constituents, that another crop of wheat would not grow upon it, and yet this soil may still contain abundant mineral constituents for the production of a good crop of clover or turnips.

What period is generally assigned by agriculturists for a complete rotation?

There is no fixed period followed by all. It depends upon the particular crops that constitute the rotation. Different individuals vary it according to fancy or to the results of their past experience, or the productions of which they have most need. Five, six, or seven years, is the usual time, unless it be for lands that may advantageously remain a long period in grass. In that case, as long as a good grass crop is yielded, they are permitted to remain.

In what case is the necessity for rotation prevented?

By keeping up an annual supply, by artificial means, of the fertilizing agents of the soil equal to what is taken away by the plants. Thus gardens are usually kept so highly manured as to require no rotation; and, it might not be necessary on the farm, if it were as highly enriched in the same way.

Has rotation of crops always been practiced?

It has not. Farms were formerly divided into meadow, plough or tillage land, and pasture; and each section was permanently used for these specific purposes, till the meadows were covered with moss, and the tillage ground was so impoverished as to yield inferior crops.

Under this system how were farms restored?

Meadows might have been, and frequently were restored by what is called a top-dressing, which is a scattering upon the surface a coat of fine manure of some kind or other. The tillage land was restored by what was called a fallow, which meant a suspension of cropping, letting the soil remain inactive or uncultivated, till nature should restore the equilibrium.

What is the objection to this mode of farming?

The objection is a serious one. For during the continuance of it the use of the land is lost, the same as the use of money is lost if permitted to lie idle without drawing interest. The very idea of a fallow is, the ground is let alone, to produce just what springs up spontaneously, or nothing at all; and whatever does thus spring up, is permitted to remain and decay where it grew.

Who first resorted to the rotation system for keeping the soil in a proper state?

The Flemings are the first known to have made it a fixed part of their system of agriculture. They insisted that where it was practiced, the land did not need rest; and, it was this system which gave their husbandry a pre-eminence over that of every other country at that period. They relied so much upon it, that in some instances they were able to obtain two crops in the same year.

Where else has it been found signally beneficial?

In Scotland it has been scrupulously pursued with the very

best results. The improvements in their agriculture were incredible. It was also introduced into England, and is become general there; and, it is now constantly gaining advocates in this country.

From what species of alternation in crops is the greatest benefit received?

It is that which is made between *culmiferous* and *leguminous* crops. The former include wheat, oats, barley, rye, Indian corn, tobacco, and most of the grasses. The latter include peas, beans, other pulse, potatoes, turnips, carrots, beets, cabbage, and clover. Accordingly it has by some been adopted, that good husbandry requires that these two classes should follow each other uninterruptedly, unless where grass is made to intervene; the farmer, however, selecting whatever particular ones from each of these classes he may think best.

What is the basis of this classification of plants?

Culmiferous plants are termed robbers or exhausters of the soil. They are particularly so during the process of maturing their seeds. Hence, if cut green, or when in blossom, they are far less exhausting. Leguminous plants, as a class, are less exhausting; in the first place, because only a few of them mature their seeds—and, in the second place, all of them having broad leaves, draw more moisture from the atmosphere than the narrow-leaved plants which compose the culmiferous class.

What is the difference of the roots in the two classes?

The roots of culmiferous plants are generally more fibrous and more divided, spreading themselves near the surface, and draw their nourishment principally from the upper stratum of the soil. Leguminous roots are generally spindle formed, having what is called a tap-root, with few radicals, and consequently draw most of their nourishment from the lower stratum of the soil, and through the lower extremities of their roots.

How does this difference in the roots effect the theory?

An eminent chemist says that plants exhaust only that portion of the soil which comes in contact with their roots; and a spindle root may be able to draw an abundance of nourishment from land, the surface of which has been exhausted by short, or creeping roots. The same writer remarks, that the roots of plants of the same or analagous species, always take a like direction, if situated in a soil which allows them a free development; and thus they pass through, and are supported by, the same layers of earth.

What fact may be given in confirmation of this theory?

It is proverbial that trees of the same species will not flourish in succession in the same place. Hence, if a worn out peach orchard is to be removed, and young trees of the same species are to occupy the same ground, instead of being planted in the holes from which the old ones were taken, they must be arranged in rows intermediate to the old ones. So likewise in regard to all fruit trees, unless a suitable period has been allowed for producing the decomposition of the roots of the removed trees, and thus supplying the earth with fresh manure.

What argument in favor of alternations in crops can be drawn from the natural course of vegetable growth?

In forest lands the new growth seldom resembles altogether that which has been felled. Hard wood frequently succeeds the pine and hemlock, while the pine and cedar, in innumerable instances, succeed the primitive growth of hard wood. In agreement with this tendency, we may see the strawberry and raspberry, and some other plants, sending out their roots or stollens to establish a new progeny in a soil they had not previously exhausted; thus by their own instincts changing their locality.

What may be said of the natural adaptation of particular soils to particular crops?

As a general thing it may be considered that calcareous and stony loams are better adapted to wheat than silicious gravels and sands; while the latter are better fitted for Indian corn, turnips, clover, and other tap roots, than clayey soils.

What are the plants raised in this country on a large scale?

The cereal grasses, including wheat, barley, oats, and partially rye, cultivated chiefly for the farina of their seeds; certain leguminous plants, as the bean and the pea; the turnip, the cabbage, and the potato, cultivated for their leaves, roots, and tubers; hemp and flax, cultivated for their fibres; and the plants cultivated mainly for their forage, of whatever name or description.

WEALTH AND LABOR.

There are instances in this country of enormous individual wealth—frequent instances of independent individual fortunes. But who are they that possess, and whence did they derive them? From some old ancestors, who won broad lands and proud titles in the field of battle—or in the senate—at the bar—or the counting house? If you look for such inherited fortunes as these, you will discover that they were long since dismembered—that with every revolution of the seasons, they are diminishing—and in a very few instances can one of their descendants call the roof-tree of his father's house his own.

No! These are the fruits of individual industry, skill, or enterprise. And you can seldom trace their history farther back than to find them commanding a trading sloop to the West Indies, purchasing fur in small quantities on the frontier, or selling excellent groceries at a first-rate stand for business. They are self-made men—the architects of their own fortunes; and I yield a thousand fold more respect to such as they, than I can ever feel for one who owes his wealth and his standing in the world to the mere accident of birth; and when their names are uttered in the marts of commerce, and the country rings from side to side with the story of their success, I feel that this, of all countries, is the best for human labor and enterprise.

A very important and striking feature in our political and social system, which indeed is the inevitable result of our institutions and laws, is, that there is no aristocracy amongst us—not even an aristocracy of wealth. An aristocracy cannot exist without peculiar and exclusive privileges and rights, recognized, sanctioned, and upheld by law. There cannot be, in this country, even a confederacy or combination among the rich men to acquire peculiar privileges. They have none to defend. There is no clanship, no esprit du corps among them. They are not like the hereditary nobles of Europe, whose names are enrolled in a heraldic college, set apart from the rest of mankind, designated by titles, marked by badges of honor, bound together by intermarriages, by a community of interests and of feelings, a distinct order in the state; nothing of all this, and they are as mutable besides as the motes that float in the summer air.

Death is ever busily at work in dismembering all overgrown fortunes. Misfortunes too—and, alas! they occasionally rain thick and fast, and do their part in the ceaseless work of distribution. The rich man of to-day is the poor man of to-morrow.

And, while from these causes, multitudes are passing out, thousands are, in the land, passing into this charmed circle; for, those who commenced life with no inheritance but poverty, are usually the individuals that rise to affluence. If a line could be drawn between the two classes, at any given moment, and then five years pass away, I doubt whether the smaller portion could be recognized as the same. Hundreds on hundreds would be found to have changed places. And to speak of a clan of men thus constituted as an aristocracy, is as sound and sensible philosophy as to point to the insects of summer as the emblems of eternity.

The condition of the laboring classes in the United States is universally admitted to be better than in any other country in the world. They are already in that position which the laborers of other countries are struggling to attain. The rate of wages is incomparably higher than in any other country—the means of comfort, not to say wealth, more easily accessible. Owing to their vast numbers, and to the possession of all political rights, their influence in the government is controlling and resistless, and all legislation is shaped in promotion to their interests rather than to those of any other class. Without having examined the laws of all the states, which would be a Herculean task, I dare to affirm, that not a statute can be found in force, in any one of the states, which establishes or recognizes any inequality of right or privileges between them and other persons; or if such a statute can be found, it is their fault that it remains on the statute book a single year. They have but to speak the word and it is done—to command, and it is repealed.

Nay, the universal sentiment among American statesmen is that the legislation and policy of the government should be such as to lend aid and encouragement to the poorer classes, and leave the rich to take care of themselves. They have accordingly been extremely liberal in granting acts of incorporation, by which men of small means may combine and compete with the richest capitalists in any branch of industry. With the laws of this state I profess to have some acquaintance, and in their general bearing and character I suppose them to be similar tô those of other states. And I challenge any man to put his finger upon a statute there, that gives a man of a million one jot or tittle more of right or privilege than to the laborer that ploughs his field, or the needy knife grinder, that spins his wheel at his door. What magic words were those which have been for years upon the lips of statesmen, to which the people have responded, as deep calleth unto deep? Not the protection of American wealth, but the "protection of American industry."

And what are all the Societies and Institutes, that are established in almost every state, and sustained at great expense, but the voluntary efforts of the people, who can afford it, to stimulate American industry? This great and splendid Institution, which I have the honor to address, is of itself a noble practical illustration of American policy. Here are the "merchant princes," the capitalists, nay, the very "aristocrats" of New-York, giving freely of their time, of their influence, of their wealth, not to obtain special privileges for themselves, but to stimulate and encourage art and industry, and to spread through the length and breadth of the Union, broadcast, those improvements in agriculture and the arts, which skill, thus stimulated, has made. There is not a laboring man, in the most distant and sequestered nook of this far spreading country, who is not, or may not be benefited by its patriotic efforts.

Yes, ye laborers, there is no land like yours. It is yours to possess, to enjoy. Here is a fair field for all to labor, in whatever vocation they please, and the rewards of diligence are ample and secure. There is not an avenue to wealth or distinction which is closed—not a post unattainable. There is no ground for any hostility or unkindness of feeling between the rich and the laboring classes, but the strongest reason, on the contrary, for mutual friendship and the most cordial union. It may well be questioned, whether they should ever be spoken of as classes, since the term presupposes a line of demarcation, which cannot be drawn. Both are striving with the same eagerness for the same object—some portion of wealth—and both are interested in the protection of property. If instead of spending time in mutual jealousies and recriminations, they would join heart and hand in all great and good undertakings, the one contributing the means, the other the skill and labor, they would accomplish more for themselves and their country in one year than by fifty years of dissension.

We should not forget that there are those who grace and gladden our festivities by their presence; who do not mingle with us, indeed, in the walks of business, but exert a more potent influence upon the affairs of men than we are always willing to acknowledge; whose empire is absolute over the world of fashion; whose appearance in the midst of dissensions is like the radiant bow that spans the storm. If their smiles do sometimes kindle dissension, they oftener allay it, and I would invoke their gentle influence in the work of reforming the national manners. If they would bestow more of their kind regards upon those athletic and manly forms that make our hill sides and valleys laugh and ring with the wealth of golden harvests, and less upon those whiskered and bedizened apes that infest the drawing room, we should love them better, and our country would regard them as her jewels.

What honest vocation can be named that does not contribute, in a greater or less degree, to the enjoyment of man? It may be humble, indeed, but it goes to swell the mighty aggregate; it may be the rill that trickles from the mountain side, but it diffuses fertility through the valley, and mingles its drops at last with the ocean. The true American motto is and must be—marked upon our foreheads, written upon our door-posts—channelled in the earth, and wafted upon the waves—INDUSTRY—LABOR IS HONORABLE, and idleness is dishonorable,—and I care not if it be labor, whether it be of the head or the hands. Whitney, whose cotton gin doubled the value of every acre of land in the South, raised more cotton with his head than any twenty men ever raised with their hands. Let me exhort those of you who are devoted to intellectual pursuits, to cherish, on your part, an exalted and just idea of the dignity and value of manual labor, and to make that opinion known in your works and seen in the earnest of your actions. The laboring men of this country are vast in number and respectable in character. We owe to them, under Providence, the most gladsome spectacle the sun beholds in its course—a land of cultivated and fertile fields, an ocean white with canvass.

We owe to them the annual spectacle of golden harvests, which carries plenty and happiness alike to the palace and the cottage. We owe to them the fortresses that guard our coasts—the ships that have borne our flag to every clime, and carried the thunder of our cannon triumphant over the waters of the deep.

Sir Walter Scott, a mere writer of poetry and romance, has given employment to ten thousand paper makers, type founders, printers, tanners, book-binders; and beyond that, has awakened the love of elegant literature in millions of minds. Sir Isaac Newton spent his days partly in sleep, and his nights in watching the stars in the midnight sky: and yet his discoveries have enabled the mariner to pursue his foaming pathway in the deep, as safely as on the land, and thus poured the products of every clime into the lap of labor. The benefactions of these men were indeed great and illustrious; but there are men in our midst engaged in similar pursuits every day of their lives, bestowing the same kind of benefits upon mankind. The merchant's life is a life of excitement and care, of risk and uncertainty, but of the first importance to every community; as indispensable to the laborer as the laborer is to him.

The village school master, who devotes the years of his youth or his manhood to the exhausting drudgery of instruction; who moulds the character and fixes the principles of an advancing generation—is as eminently useful, though he sink at last into the grave unhonored and unsung, as the demagogue whose presence is greeted in caucuses, or whose voice is heard in the halls of legislation, discussing the constitutional power of congress to buy a penknife.—*Address, Anniversary Am. Inst.*, 1842, *by* HON. H. G. O. COLBY *of New Bedford, Massachusetts.*

EXHAUSTION OF THE SOIL.

When will our statesmen awake to the necessity of viewing agriculture as a fundamental source of our national prosperity? So long as we have more land in the far West to cultivate, the wearing out of that in the older States seems to be looked upon as a matter of little consequence. The older States, with all their best land in cultivation, do not at this time raise half the quantity of wheat they raised a few years ago; and the consumers in the Atlantic States are paying nearly as much for transportation on a large proportion of their breadstuffs, as the farmers who grow it receive for their grain. The wheat crops of New-York are less than half per acre what they were thirty years ago, and still no effort is made to disseminate the necessary information for arresting the evil. Stern necessity has rendered such action as we shall soon require in this country, imperative in Europe. Ohio no longer surprises the seaboard farmers by large crops, and the same course of cropping and modes of tillage which have impoverished the lands of the older States, are daily producing similar results in the far West. Many farmers are still living and carting manures upon their poor farms, who in the Mohawk and Genesee Valleys threw their manures into the river when younger—this removal of manures from the vicinity of their stables and throwing it into the river, was called a *Bee*, and the winter time was chosen for this *suicidal frolic.* Whole neighborhoods would get together with their teams and sleds for this purpose, and the same practices are now followed in the Wabash Valley: the tributaries of the Mississippi and other rivers, are suffered to drain the very essence of our future prosperity, and to convey it to the ocean.

Nature's laws tell us that the decay of the crops of one year furnishes the raw materials for the creation of those of the next year; but we must retain the results of this decay, and not suffer them to part from us—present individual enterprise being answered, must not cause us to forget the debt we owe to posterity—and that we have no *moral right* to permit the ultimate constituents of plants, *the agricultural capital of our country,* to find its way to the ocean, or into the ocean of atmosphere, by sheer ignorance and negligence. It may be answered that Europe, when necessity demanded, found the means of restoring her worn out soils to fertility—this is true; but how will the same means continue through all time? If it had not been for the importation from our Continent of breadstuffs, raw materials of all sorts, guano, saltpetre, cubical nitre of Peru, and other materials, the results of which, by decay and by direct application to their soils, recovered their lost ultimate constituents of plants, they would long ere this have suffered from famine. Every bushel of corn, bale of cotton, barrel of resin or other commodity we now send to Europe, and which are consumed there, places just so much of the ultimate constituents of plants in their soil for continued and repeated reproduction, and removes it from ours.

If the importations were at all equal to the exportations, or if we took the same care as they do of what we have, or import, our exhaustion would be slower; but as it is we are rapidly parting with our capital, never to return. The ultimate constituents of vegetables pay at least one hundred per cent. profit when re-used for producing new growths, and this may be repeated almost yearly; but part with them to the ocean through our rivers, and they are lost forever. All this may appear very farcical to the casual observer, but nevertheless it is true; and the falling off of our crops is only unobserved from our great area of territory and consequent continuance of supplies. Should we be contented to render it necessary each year to bring our supplies further from the seaboard? Or should we adopt the proper means to produce an excess by keeping our Eastern lands in their present or in an improved condition? and thus, by cheaper products, be able to compete with Europe as manufacturers.

Individual farmers may continue to move West as they wear out their lands; but as a nation, what effect must this have on our general prosperity? It may be answered, that as our enterprising farmers look for better lands, the tide of emigration, composed of European agriculturists, will take their places, and thus restore the old lands by European styles of farming. If the best farmers of Europe were among those who come, this might be true; but those who understand their business seldom migrate; it is the laborers only who come to us; the more intelligent and better educated remain at home.

We have but one remedy, and that is entirely within our reach. Let our legislators spend part of the nine-tenths of the whole national income which is now paid by farmers in placing proper instructions within the reach of those who till the land—and that, too, in a way to be immediately effective. We must not wait to remedy the difficulty by educating the rising generation; we must inform the many what is doing by the few who are successful as agriculturists. Some farmers raise a hundred bushels of shelled corn to the acre, and some raise fifty bushels of wheat to the acre, but these are the one in ten thousand. Send competent persons among the ten thousand to tell them how the one manages his crops; let any well educated practical man be called from his plough, and employed solely in collecting and disseminating information, and instead of raising large crops himself he can cause *a thousand others* to do so. Every farmer should hear such a lecture at least once in each year; and should have an opportunity of propounding questions for his examination—such teachers would soon know what the farmers required, and could obtain the information for them from other and more successful practitioners.

It need not be urged that farmers will not listen to accredited teachers; we have lectured in many counties in New Jersey for three years, and in those where we first lectured most evident improvement has ensued. Farmers cannot leave home, and hence do not learn of the improvements of the day, unless they occur in their own immediate neighborhoods—nor will they have confidence in the recipes of mere book-makers; they must see those who would teach them, and have an opportunity by listening, and questioning, to form their own estimate of their capacity to teach; and, if they approve of the teacher, no set of men are more ready to be instructed. The improvements in agriculture in Europe are greater than at any former time—the free trade system, by lowering the prices of farm products, has rendered it imperative on Governments to enable the farmers to produce proportionate increased quantities to compete with foreign prices; and did they not pursue this course, revolutions would be inevitable, or their farm products must be protected by high duties; and while our ratio of crops have been yearly decreasing, those of England have as steadily increased, until the opponents of their new school of politics are daily becoming converts to the new system. Every county in England now receives per annum more benefit in the form of agricultural information, disseminated at the public expense, than the total amount paid for similar purposes since the formation of our Government.

Our politicians at Washington say that the powers of the general Government do not reach the case, and that it should be done by the States—if so, the States should not be inactive. New-York does much by publishing large editions of the Transactions of her State Society and of the American Institute, but not half so much as she should and could do by the appointment of a few lecturers to visit each county, collecting and disseminating information. Maryland has appointed a State Agriculturist, and already the good results have rendered both the office and the incumbent popular. No other State has acted as yet in any way to improve their greatest source of wealth. Two men might exchange their hats once per hour for a year, and neither of them would be improved in fortune at the end of the time, but if each of them could produce new merchandise, as does the farmer, not only themselves, but the body politic of which they form a part, would be benefited; and as one per cent. increase of crops would be more than equal in value to the whole of the present receipts of the Government, it is at least proper that less than a one-thousandth part of that receipt should be spent to produce a probable gain of many times one per cent.

PROF. MAPES.

A SIBERIAN WINTER.

The traveller in Siberia, during the winter, is so enveloped in furs that he can scarcely move, and under the thick fur hood, which is fastened to the bear skin collar and covers the whole face, one can only draw in, as it were by stealth, a little of the external air, which is so keen that it causes a very peculiar feeling to the throat and lungs. The distance from one halting place to another takes about ten hours, during which time the traveller must always continue on horseback, as the cumbrous dress makes it insupportable to wade through the snow. The poor horses suffer at least as much as their riders, for besides the general effect of the cold, they are tormented by ice forming in their nostrils, and stopping their breathing. When they intimate this by a distressed snort and convulsive shake of the head, the drivers relieve them by taking out the pieces of ice, to save them from being suffocated.

When the icy ground is not covered by snow, their hoofs often burst from the effects of the cold. The caravan is always surrounded by a thick cloud of vapor; it is not only living bodies which produce this effect, but even the snow smokes. These evaporations are instantly changed into millions of needles of ice, which fill the air, and cause a constant slight noise, resembling the sound of torn satin or thick silk. Even the reindeer seeks the forest to protect himself from the intensity of the cold. In the tundras where there is no shelter to be found, the whole herd crowd together as close as possible to gain a little warmth from each other, and may be seen standing in this way quite motionless. Only the dark bird of winter, the raven, still cleaves the icy air with slow and heavy wing, leaving behind him a long line of thin vapor, marking the track of his solitary flight.

The influence of the cold extends even to inanimate natures. The thickest trunks of trees are rent asunder with a loud sound, which, in these deserts, fall on the ear like a signal shot at sea; large masses of rock are torn from their ancient sites; the ground in the tundras and in the rocky valleys crack, forming wide yawning fissures from which the waters which were beneath the surface rise, giving off a cloud of vapor, and become immediately changed into ice. The effect of this degree of cold extends even beyond the earth. The beauty of the deep polar star, so often and so justly praised, disappears in the dense atmosphere which the intensity of cold produces. The stars still glisten in the firmament, but their brilliancy is dimmed.—*Travels in the North.*

THE LEAFLESS TREE.

Poor cheerless tree! how desolate!
  Bereft of all thy beauty now,
Which decked thy stately form so late,
  And crowned thy lofty, sunny brow.

Each leaf breathed forth a kindly air,
  In native strains most sweetly sung,
When gentle zephyrs wantoned there,
  With harps so delicately strung.

The birds too sang melodiously,
  Among those boughs once wreathed so gay,

With softest notes of minstrelsy,
Oft ushered in the rising day.

When first I saw thy verdure fade,
I turned my tearful eye away,
Lest sorrow should my mind pervade,
To see thy loveliness decay.

But ah! the rude winds did not spare,
Thy verdant robe—but reckless tore,
That lovely mantle—wrought with care,
By vernal beauty, seen no more.

Thy leaves were scattered by the blast,
And borne upon the eddying stream,
To mingle with the earth at last,
The grave for all which ere hath been

The freshest, fairest, coronal,
Which nature in its pride could boast,
Is shrouded in a funeral pall,—
To humble beauty in the dust.

The leafless tree, and faded bower,
In silent eloquence declare,
That man too has a dying hour,
And must the parting anguish share.

While I now their requiem sing,
I see my own approaching doom,
Sealed fast on time's expanded wing,
To bear me onward to the tomb.

The sullen wail which strikes my ear—
Sweeps o'er the lyre's responsive string,
As if the mournful bier was near,
Where beauty lies all withering.

But that is all that Death can claim—
The fragile form which must decay,
The nobler powers of mind remain,
To shine in virtue's sunny ray.

Then cease my troubling heart—be still,
And mourn no more o'er autumn's gloom,

A clime there is which has no chill,
 Or blight of death—beyond the tomb.

There friendship shall her wreath entwine,
 Composed of never fading flowers,
And bind it round love's sacred shrine,
 In Heaven's own bright perennial bowers.

MISS BREWER.

THE SEASONS OF THE YEAR.

 These, as they change, ALMIGHTY FATHER, these
Are but the varied God. The rolling year
Is full of thee. Forth in the pleasant Spring
Thy beauty walks, thy tenderness and love.
Wide flush the fields—the softening air is balm—
Echo the mountains round—the forests smile;
And every sense, and every heart is joy.
Then comes thy glory in the Summer months,
With light and heat refulgent. Then thy sun
Shoots full perfection through the swelling year;
And oft thy voice in dreadful thunder speaks;
And oft, at dawn, deep noon, or falling eve,
By brooks and groves and hollow whispering gales,
Thy bounty shines in Autumn unconfined,
And spreads a common feast for all that live.
In Winter, awful thou! with clouds and storms
Around thee thrown—tempest o'er tempest rolled:
Majestic darkness! on the whirlwind's wing
Riding sublime, thou bidst the world adore,
And humblest nature with thy northern blast.
 Mysterious round! what skill, what force divine,
Deep felt, in these appear! a simple train—
Yet so delightful mixed, with such kind art,
Such beauty and beneficence combined—
Shade, unperceived, so softening into shade—
And all so forming a harmonious whole—
That, as they still succeed, they ravish still.
But, wandering oft with brute unconscious gaze,
Man marks not thee, marks not the mighty hand,
That, ever busy, wheels the silent spheres—

Works in the secret deep—shoots, streaming, thence
The fair profusion that o'erspreads the spring—
Flings from the sun direct the flaming day:
Feeds every creature—hurls the tempest forth:
And, as on earth this grateful change revolves,
With transport touches all the springs of life.
  Nature, attend! join every living soul,
Beneath the spacious temple of the sky,
In adoration join—and, ardent, raise
One general song! To him, ye vocal gales,
Breathe soft, whose Spirit in your freshness breathes:
O talk of him in solitary glooms!
Where, o'er the rock, the scarcely waving pine
Fills the brown shade with a religious awe.
And ye, whose bolder note is heard afar,
Who shake the astonished world, lift high to heaven
The impetuous song, and say from whom you rage.
His praise, ye brooks, attune, ye trembling rills—
And let me catch it as I muse along.
Ye headlong torrents, rapid and profound—
Ye softer floods, that lead the humid maze
Along the vale—and thou majestic main,
A secret world of wonders in thyself—
Sound his stupendous praise, whose greater voice
Or bids you roar, or bids your roarings fall.
  Soft roll your incense, herbs, and fruits, and flowers,
In mingled clouds to him, whose sun exalts,
Whose breath perfumes you, and whose pencil paints.
Ye forests bend, ye harvests wave to him:
Breathe your still song into the reaper's heart,
As home he goes beneath the joyous moon.
Ye that keep watch in heaven, as earth asleep
Unconscious lies, effuse your mildest beams,
Ye constellations, while your angels strike,
Amid the spangled sky, the silver lyre.
Great source of day! best image here below,
Of thy Creator, ever pouring wide,
From world to world, the vital ocean round,
On nature write with every beam his praise.
Ye thunders roll; be hushed the prostrate world,
While cloud to cloud returns the solemn hymn.
Bleat out afresh, ye hills; ye mossy rocks

"The Downer Cherry."

Retain the sound : the broad responsive low,
Ye valleys raise ; for the great Shepherd reigns,
And his unsuffering kingdom yet will come.
Ye woodlands all, awake : a boundless song
Burst from the groves : and when the restless day,
Expiring, lays the warbling world asleep,
Sweetest of birds ! sweet Philomela, charm
The list'ning shades, and teach the night his praise.
  Yet chief, for whom the whole creation smiles ;
At once the head, the heart, the tongue of all :
Crown the great hymn ! In swarming cities vast,
Assembled men, to the deep organ join
The long resounding voice, oft breaking clear,
At solemn pauses, through the swelling base ;
And as each mingling flame increases each,
In one united ardor rise to heaven.
Or if you rather choose the rural shade,
And find a fane in every sacred grove—
There let the shepherd's flute, the virgin's lay,
The prompting seraph, and the poet's lyre,
Still sing the God of Seasons as they roll.
For me, when I forget the darling theme,
Whether the blossom blows, the Summer ray
Russets the plain, inspiring Autumn gleams,
Or Winter rises in the blackening east ;
Be my tongue mute, my fancy paint no more,
And, dead to joy, forget my heart to beat !
  Should fate command me to the farthest verge
Of the green earth, to distant barb'rous climes,
Rivers unknown to song ; where first the sun
Gilds Indian mountains, or his setting beam
Flames on the Atlantic isles ; 'tis nought to me ;
Since God is ever present, ever felt,
In the void waste as in the city full ;
And where He vital spreads, there must be joy.
When even at last the solemn hour shall come,
And wing my mystic flight to future worlds,
I cheerful will obey ; there, with new powers,
Will rising wonders sing—I cannot go,
Where Universal Love smiles not around.
Sustaining all yon orbs, and all their suns :
From *seeming evil* still adducing *good*,

And *better* thence again, and *better* still,
In infinite progression——but I lose
Myself in HIM, in LIGHT INEFFABLE!
Come then, expressive Silence, muse His praise.

THOMSON.

THE DISTRIBUTION OF PLANTS.

In what proportion are plants distributed in different latitudes?

Botanists compute that, at Spitzbergen, which lies near the eightieth degree of north latitude, there are only about thirty species of plants; in Lapland, which lies under the seventieth degree, about five hundred and thirty; in Iceland, under the sixty-fifth parallel, about five hundred and fifty; in Sweden, from the southern parts of Lapland to the fifty-fifth degree, about thirteen hundred; in Brandenburg, between the fifty-second and the fifty-fourth parallel, two thousand; in Piedmont, between the forty-third and forty-sixth, two thousand eight hundred; nearly four thousand in the Island of Jamaica, which is between the seventeenth and nineteenth degrees; and in Madagascar, situated under the tropic of Capricorn, between the thirteenth and fourteenth degrees, more than five thousand.

What may be said of the same plant growing in different latitudes?

Plants of the frigid zones, are also found in the torrid; but they occur only in situations where they find a temperature as low as that of the colder zones, viz., upon high mountains. Thus the plants of Greenland and Lapland are found not only on the Alps and Pyrenees, but even on the Cordilleras. Edwards says, that while no tropical fruits grow upon the mountains of Jamaica, many European fruits thrive admirably; European Alpine plants occur on the cold mountains of Terra del Fuego; and the pine occupies the extreme limit of arborescent plants in the mountains of America, of Switzerland, and Lapland. The same physical climate, therefore, favors the growth of the same plant.

Why are exceptions found to this principle?

If under the same climate we do not observe the same plants produced, we must attribute the difference to local peculiarities, such as the quality of soil, the degree of shade, the atmosphere, and other circumstances.

What constitutional difference in plants in regard to their power of growing in different climates, may be observed?

Some plants are found universally distributed, and are consequently adapted to every climate; while others are confined to very limited districts, beyond which they cannot be cultivated. Many plants, particularly the most useful ones, may be successfully naturalised by judicious management, in countries far distant from their original habitation, and under climates very different from that in which they were originally found. Most of our fruits, our corns, and edible vegetables are of foreign extraction.

By what means are plants transferred to new localities?

The migration of plants has been assisted by the wind, and graniverous birds and quadrupeds, as well as by the hand of man himself. Thus the seeds are carried from their native soils, over rivers and mountains to distant countries, by the former means, as well as over oceans by the latter agency.

Under what circumstances have men done it?

Many of our choicest fruits and vegetables were brought into Italy by the ancient Greeks and Romans, from territories they had subdued; were thence scattered over Europe; and finally transported to this continent by the first settlers. Others were brought from the East by the Crusaders on their return from the Holy Land. This is one of the means by which war has promoted civilization, and encouraged the arts of peace. And at the present day, it is our boast that the same blessings follow in the track of our navigators and missionaries, who, instead of inciting men to strife and bloodshed, teach the holy precepts of love and universal brotherhood.

Has there been any change in the character of these plants since their first cultivation?

The improvement in their appearance and quality has been surprisingly great. Those which were of little value, now rank with our most important crops. This marked change has been the result of skill and care through a great number of years, and it is by no means certain that these plants have yet reached their highest point of perfection.

Is the origin of these plants well ascertained?

In some cases it may be so considered, but in general, any

attempt to fix upon their native countries is a mere matter of speculation, or at best of strong probability.

What plants, or classes of plants, are most widely diffused?

The anti-scorbutic and edible plants seem to be most widely disseminated throughout the earth. Such are the different varieties of cresses, celery, parsley, and scurvey grass, which are found on every coast which has yet been visited by navigators. Many plants bearing edible berries are also of very general distribution, and form an important article of food for man. These graminea also, which are of most valuable service to man and to inferior animals, are very widely spread, although different species of them appear to thrive best in certain climates. The mosses and lichens, however, are of widest distribution. They are found in every part of the world, and in every situation.

What is known of the early history of Indian corn?

America is commonly called the native country of this grain, where it was cultivated by the natives at the time of the discovery. It is certain that it was unknown to those ancient writers whose works have come down to us, and that it did not attract attention in the eastern hemisphere until after the voyage of Columbus. Yet, as it has not been found growing wild on this continent, its origin has been variously attributed to India, Persia, and the western coast of Africa.

What is the early history of cotton?

Cotton appears to have been known in the earliest ages of antiquity. The Egyptians were familiar with its use, and Herodotus, in speaking of the Indians, says, "They possess a kind of plant which, instead of fruit, produces a wool of a finer and better quality than that of sheep." It was found in this country when discovered. Within the past few years its culture has increased to an astonishing degree; and, although grown in the Indies and Egypt, as well as North and South America, two-thirds of the whole annual crop is the product of our own soil.

What is known about the early history of tobacco?

The discoverers of America noticed the free use of tobacco by the aborigines, who claimed for it several important medicinal properties. It was first known in Europe about the year 1560, when some of the seeds were carried from Portugal by the French Ambassador to that country. The word *tobacco* in the Haytian language signifies the pipe used by the natives in smoking, and it was very naturally transferred by the Spaniards to the plant itself. In 1586, it was introduced into England by the colonists who had returned from Virginia. Although the practice of smoking was everywhere met in Europe by ridicule and persecution, yet it spread itself rapidly through England, as it had previously done through Portugal, Spain, and France.

Are we acquainted with the origin of wheat?

This valuable cereal is supposed to have been brought from the central table land of Thibet, where it is said that its original still exists as a wild grass. But, like the other grains, it has been so long and so extensively cultivated, that it is impossible to decide upon its native country. Travellers have found it growing spontaneously in many different regions, and it is doubtful whether the plant may not have been carried there at very distant periods, and have remained as an evidence of ancient civilization. It is now generally diffused, and as an article of food, maintains a high and undisputed rank.

Do we know the native country of rye and buckwheat?

Many have supposed that they originated in Northern Asia, because buckwheat was thence brought to Europe, and rye is said to be now growing there in its wild state. This is the most we know on the subject.

What is known about the natural locality of barley?

Barley has been cultivated from time immemorial, and its origin is altogether uncertain. It exists wild in the mountainous regions of central Asia, and is evidently a native of a warm climate. We have the best authority for its having been raised in Syria more than three thousand years ago.

Is any thing certainly known of the origin of the oat?

There is not, except the single fact of its being the natural inhabitant of cold latitudes, where it is extensively cultivated, being used not only for the food of horses and other stock, but also in some countries for the food of man, and in others is distilled for beer and ardent spirits. The oat is said to grow wild in Northern Africa.

What is the history of rice?

That it was carried from Africa to India, whence it was

taken to Egypt and Greece. It is more probable, however, that it originated in Eastern Asia. It has formed the chief support of the people of China and India for centuries back, and to its successful cultivation has been frequently attributed their early civilization. It has been said to have altered the face of the globe, and to have changed the destiny of nations. It was introduced into this country about the year 1697, and is now raised in large quantities in all warm climates. In the east, a population so vast that it almost surpasses credibility, is dependent on the rice crops, and when they fail, thousands of human beings perish of hunger.

What is the native country of the potato?

We are indebted for this invaluable root to South America It has been found in the highlands of Chili and Peru. In the natural state the tubers are small and of very inferior quality; the excellence of the varieties now in use is wholly owing to careful cultivation. It was first carried to Europe by John Hawkins, a slave trader, but did not attract much attention until its merits were proclaimed by Sir Walter Raleigh. He first raised it on his estate near Cork, in Ireland, from the seed brought from the colony of Virginia. Its progress in the favor of the people on the continent was slow, and not until the middle of the eighteenth century did it get into general use, nor until the beginning of the nineteenth century were the strong prejudices of the French overcome.

Where did the sugar cane originate?

There are several varieties of the sugar cane, but all of them appear to be natives of the eastern part of Asia. If not unknown to the ancient Egyptians, Greeks, or Romans, its value was only imperfectly understood by them. By the Chinese, however, it has been cultivated from remote antiquity, and their sugar occasionally found its way into Europe, in small quantities, by means of the Arabian merchants. Our term sugar is probably derived from the Bengalee *shukur*, the name by which it is still known in India. The cane was carried by the Saracens into Northern Africa, and thence into Southern Europe. It was taken by the Portuguese to the West India islands and the Brazils. Until the middle of the fifteenth century sugar was considered in Europe chiefly valuable as a medicine; but it has come into such general use that it now ranks next to wheat and rice among all the vegetable products of the world.

What is known of the history of flax, or linseed?

It has been cultivated from remote antiquity in Europe, Asia, and northern Africa; but its original locality is not known for a certainty. Olivier says it grows wild in Persia. The ancient Scandinavians and other barbarous nations were clothed with linen. The mummies of Egypt are inveloped with it, and immense quantities are still raised in that country, and especially about the mouths of the Nile.

Of what countries is hemp a native?

Of India and Persia, whence it was transported to Europe and more recently to our own country. It is probably more cultivated in Russia than anywhere else.

What is the original locality of our most common fruits?

The apple and pear trees are mostly from Europe. The currant and gooseberry came also from the south of Europe. The cherry, plum, olive and almond came from Asia Minor. The mulberry tree, walnut and peach from Persia. The citron from Media; the quince from the Island of Crete; and the chesnut from Italy. The whortleberry is a native of Asia, Europe and America; and the cranberry is a native of the two latter countries.

Whence are received our most valuable garden esculents?

The turnip and mangold wurtzel came from the shores of the Mediterranean; but the white turnip is a native of Germany. Celery is a native of the same country. Jerusalem artichoke is a Brazilian product. The carrot is by some supposed to have been brought from Asia, and by others from Germany. Spinach is attributed to Arabia; okra, to Tartary; the cucumber, to the East Indies; the melon, to Kalmuck; parsley, to Sardinia; the onion, to Egypt; the radish, to China; and the horseradish, to the south of Europe.

To what countries are to be referred our principal vegetables of the pulse species?

The origin of peas is unknown. The lentil grows wild on the shores of the Mediterranean. Vetches are natives of Germany. The chick pea was brought from the south of Europe; the garden bean, from the East Indies; the horse bean, from the Caspian Sea; and the lupin, from the Levant.

Where were obtained our principal plants for condiments, or for medicinal or narcotic purposes?

Hops come to perfection as a wild plant in Germany. The same may be said of mustard and carra-way seed. Dill is an Eastern plant. Anise was brought from Egypt and the Grecian Archipelago. Koriander grows wild near the Mediterranean. Saffron came from the Levant. The poppy was brought from the East; the sun-flower from Peru.

Where is the natural locality of some other well known farm plants not yet named?

The gourd is probably an Eastern plant; spurry is an European one; chickory grows wild in Germany; rape and cabbage wild in Sicily and Naples; one species of millet is a native of India, and another species is from Egypt and Abyssinia; white melilot is from Greece; most grasses are native plants; and so are clovers, except lucerne, which is a native of Sicily.

WONDERS OF VEGETATION.

The time, required to attain the magnitude that trees sometimes reach, is a subject of controversy among scientific botanists; but though actual precision cannot be expected, still we know, that the rate of increase diminishes as the size augments, and that many centuries are required to produce these wonderful monuments of the vegetable creation. And this power of longevity affords Isaiah an expressive illustration of the duration of God's chosen people. "As the days of a tree," says the prophet, repeating the promise of the Lord, "As the days of a tree shall be the days of my people."

In Oregon, pines are found upwards of 300 feet high, and in other regions trees are produced not less remarkable for their prodigious circumference; like the plane tree in the valley of Bouyouderck near Constantinople, which measures 150 feet in girth. Some have attained historical celebrity, and have been associated with remarkable events in the progress of nations. In the Province of Oaxaca is a cypress 122 feet round, said to have been mentioned by Cortez in his despatches, and to have afforded shelter from the sun, to the whole of his Mexican army. To this class belong the Parliament Oak, in Clipstone Park, in England, under which a parliament was held in 1290, during the reign of Edward the First; the Ankernyke Yew, at Staines, which witnessed the conference between King John and the Barons, and in sight of which Magna Charta was signed; and the Sycamore Maple in the Grisons, beneath whose branches their Grey League, the foundation of their freedom, was ratified in 1424.

At Morat, in Switzerland, was fought the great battle where Helvetian patriotism triumphed over the oppressor, and secured that liberty, which makes its dwelling place in the fastnesses of the Alps. A Linden tree, contemporary of this desperate conflict still marks the site where it occurred, and is approached with reverence by the Swiss patriot, when he performs his pilgrimage to this high place of his country, to recall the deeds and the dead, which gave her a station among the nations of the earth, and gave her also those equal rights, and a determination to defend them, in weal and in woe, which have made the name of Switzerland a name of honor through the world.

But the most interesting relic of the ancient vegetable creation is to be found upon one of the ridges of Lebanon, not far from the renowned temple of Baalbec. It consists of twelve gigantic cedars, the remains of the primitive forest, which once covered that great mountain chain of Syria, and which yet rear their heads, prodigies of vegetation, and each surmounted with a dome of foliage overshadowing the spectator, as in the time of Biblical story. One of them is 45 feet in circumference, and all, both in size and height, tell of the long ages that have swept over them, leaving them the the most striking natural monuments that the eye can rest upon. What interesting associations cluster round them! They have been consecrated by history, religion and poetry. Their beauty has been recorded by Ezekiel, and their excellence and perfume by Solomon, who placed them at the head of the vegetable creation, when he discoursed of trees "from the cedar, which are in Lebanon, even to the hyssop that springeth out of the wall." Could these mute memorials of bygone times tell of the scenes that have passed in the shadow of their foliage, what lessons of power and of instability might they not teach, in the long interval that has elapsed, since these hills resounded with the noise of the workman, preparing the timber for the Temple of Jerusalem, to the solitude, which establishes its dwelling place wherever the Moslem plants his standard.

I have worshipped in many of the high places of the old world; in the Cathedral of Christendom, the Basalic of St. Peter, when the Sovereign Pontiff, the head of the Catholic Church, ministered at the altar; and though educated, as I have been, in the simplicity of the Presbyterian faith, yet I could not look upon the imposing solemnities without feeling a reverential awe pass over me, as though I were in the presence of Him, whose visible glory descended upon the Temple of Mount Moriah. And yet a naked Greek mass, for it happened to be an annual fete when I was there, celebrated under the patriarch cedar, before a rude altar of unwrought stones, by a poor priest, surrounded by a little band of worshippers, with the cliffs of Lebanon around them, and the canopy of heaven over them, this act of primitive devotion, in a temple not made with hands, has left traces upon my mind and memory more powerful than the most gorgeous ceremonies, and which no subsequent event can eradicate.

And this power of association, which seems to make us almost contemporary with the earliest and the latest incidents of history, with Socrates and with Washington, is not confined to the giants of the forest, throwing out their broad branches for the sun to vivify, but it is connected with seeds, deposited in the dark recesses of the catacombs, companions of the bodies of the mummies of Egypt, which poor human nature attempted to rescue from destruction, and which for more than thirty centuries have found there a resting place. Seeds have been taken from these receptacles of the dead, with the power of germination yet existing, and have borne plants, the immediate descendants perhaps of those growing in the days of Abraham. This very season, it is said, a small plat of ground in Caithness, in Scotland, planted with wheat, thus resuscitated from its cemetry, produced a crop honorable to the character of Egyptian agriculture in the days of the Pharaohs, yielding, such is the report, a thousand fold; two seeds only having been planted upon every three feet square—a degree of fecundity heretofore unknown in modern husbandry.

The human imagination loves to revel in facts like this. It is interesting, as illustrating one of the most wonderful laws of nature; and it calls into action that beautiful and beneficent faculty of association, which enables us to withdraw ourselves from the present, by connecting it with the past, and seems to make us spectators of events in the earliest periods of recorded time. This Scottish wheat may be the offspring of grain, taken perhaps from the granaries of Joseph, *where was gathered corn, as the sand of the sea, against the seven years of famine.* But this triumph of the intellect, by which time and space are annihilated, is rebuked at the very moment it is achieved. The efforts of man to preserve the body, when death had done its work, and thus to reverse the decree of Providence, have signally failed. The misshapen matter is but earth, more revolting to the feelings, than if left to its natural decomposition. It will awaken into being but at the sound of the last trumpet, which will gather together the scattered members of every human body, however separated and however changed, as easily as it will rouse into life the mortal remains of Cheops, so long reposing in solitary magnificence, and guarded by the pyramids erected for that purpose. But the little seed obeys its law. Time passes harmlessly over it, and it is ready to start into life, whenever placed in a favorable position. The lesson is a pregnant one—may it prove a profitable one, teaching us to confine our operations to aiding the laws of nature, and not to endeavor to reverse them.—Lewis Cass, LL. D., *Michigan*

VISIT TO AN ENGLISH DAIRY.

Let the reader accompany us half a dozen miles out of town. We pass through Camberwell, through Peckham, and Peckham Rye, and we presently find ourselves in a district that looks uncommonly like "the country," considering how short a time it is since we left the "old smoke" behind us. We alight and walk onward, and certainly, if the sight of green fields, and cows, and hedges, and farmyards, denote the country, we are undoubtedly in some region of the kind. We pass down a winding road, between high hedges of bush and trees, then climb over a gate into a field; cross it, and then over another gate into a field, from which we commence a gradual ascent, field after field, till finally the green slope leads us to a considerable height. We are on the top of Friern Hill.

It is a bright sunny morning in September, and we behold to perfection the most complete panorama that can be found in the suburban vicinities of London. Step down with us to yonder hedge, a little below the spot where we have been standing. We approach the hedge—we get over a gate, and we suddenly find ourselves on the upper part of an enormous green, sloping pasturage, covered all over with cows. The red cow, the white cow, the brown cow, the brindled cow, the colley cow, the dappled cow, the streaked cow, the spotted cow, the liver-and-white cow, the strawberry cow, the mulberry cow, the chestnut cow, the gray speckled cow, the clouded cow, the black cow—the short-horned cow, the long-horned cow, the up-curling horn, the down-curling horn, the straight-horned cow, and the cow with the crumpled horn—all are here—between two and three hundred—spread all over the broad, downward-sloping pasture, feeding, ruminating, standing, lying, gazing with mild earnestness, reclining with characteristic thoughtfulness, sleeping, or wandering hither and thither. A soft gleam of golden sunshine spreads over the pasture, and falls upon many of the cows with a lovely, picturesque effect.

And what cows they are, as we approach and pass among them! Studies for a Morland, a Gainsborough, a Constable. We had never before thought there were any such cows out of their pictures. That they were highly useful, amiable, estimable creatures, who continually, at the best, appeared to be mumbling grass in a recumbent position, and composing a sonnet, we never doubted; but that they were ever likely to be admired for their beauty, especially when beheld, as many of these were, from a disadvantageous point of view, as to their position, we never for a moment suspected. Such, however, is the case. We have lived to see beauty in the form of a cow—a natural, modern, milch cow, and no descendant from any Ovidian metamorphosis.

We will now descend this broad and populous slope, and pay a visit to Friern Manor Dairy Farm, to which all these acres—some two hundred and fifty—belong, together with all these "horned beauties." We find them all very docile, and undisturbed by our presence, though their looks evidently denote that they recognize a stranger. But those who are reclining do not rise, and none of them decline to be caressed by the hand, or seem indifferent to the compliments addressed to them. In passing through the cows we were specially presented to the cow queen, or "master cow," as she is called. This lady has been recognized during twelve years as the sovereign ruler over all the rest. No one, however large, disputes her supremacy. She is a short-horned, short-legged cow, looking at first sight rather small, but on closer examination you will find that she is sturdily and solidly built, though graceful withal. "She is very sweet-tempered," observed the head keeper, "but when a new-comer doubts about who is the master, her eye becomes dreadful. Don't signify how big the other cow is—she must give in to the master cow. It's not her size, nor strength, bless you, it's her spirit. As soon as the question is once settled, she's as mild as a lamb again. Gives us eighteen quarts of milk a day."

We were surprised to hear of so great a quantity, but this was something abated by a consideration of the rich, varied, and abundant supply of food afforded to these cows, besides the air, attendance, and other favorable circumstances. For their food they have mangold-wurzel, both the long red and the orange globe sorts, parsnips and turnips. But, besides these articles of food, there is the unlimited eating of grass in the pastures, so that the yield of a large quantity of milk seems only a matter of course, though we were not prepared to hear of its averaging from twelve to eighteen and twenty quarts of milk a day, from each of these two or three hundred cows. Four and twenty quarts a day is not an unusual occurrence from some of the cows; and one of them, we were assured by several of the keepers, once yielded the enormous quantity of twenty-eight quarts a day during six or seven weeks. The poor cow, however, suffered for this munificence, for she was taken very ill with a fever, and her life was given over by the doctor. Mr. Wright, the proprietor, told us that he set up two nights with her himself, he had such a respect for the cow; and in the morning of the second night after she was given over, when the butcher came for her, he couldn't find it in his heart to let him have her. "No, butcher," said he, "she's been a good friend to me, and I'll let her die a quiet, natural death." She hung her head, and her horns felt very cold, and so she lay for some time longer; but he nursed her, and was rewarded, for she recovered; and there she stands—the strawberry Durham short-horn—and yields him again from sixteen to eighteen quarts of milk a day.

Instead of proceeding directly down the sloping fields toward the Dairy Farm, we made a detour of about half a mile, and passed through a field well inclosed, in which were about a dozen cows, attended by one man, who sat beneath a tree. This was the Quarantine ground. All newly-purchased cows, however healthy they may appear, are first placed in this field during four or five weeks, and the man who milks or attends upon them is not permitted to touch, nor, indeed, to come near, any of the cows in the great pasture. Such is the susceptibility of a cow to the least contamination, that if one who had any slight disease were admitted among the herd, in a very short time the whole of them would be affected. When the proprietor has been to purchase fresh stock, and been much among strange cows, especially at Smithfield, he invariably changes all his clothes, and generally takes a bath before he ventures among his own herd.

From what has already been seen, the reader will not be astonished on his arrival with us at the Dairy Farm, to find every arrangement in accordance with the fine condition of the cows, and the enviable (to all other cows) circumstances in which they live. The cow-sheds are divided into fifty stalls, each; and the appearance presented reminded one of the neatness and order of cavalry stations. Each stall is marked with a number; a corresponding number is marked on one horn of the cow to whom it belongs; and, in winter time, or any inclement season (for they all sleep out in fine weather) each cow deliberately finds out, and walks into her own stall. No. 173 once got into the stall of No. 15; but, in a few minutes, No. 15 arrived, and "showed her the difference." In winter, when the cows are kept very much in-doors, they are all regularly groomed with currycombs. By the side of one of these sheds there is a cottage where the keepers live—milkers and attendants—each with little iron bedsteads, and in orderly soldier fashion, the foreman's wife acting as the housekeeper.

These men lead a comfortable life, but they work hard. The first "milking" begins at eleven o'clock at night; and the second, at half-past one in the morning. It takes a long time, for each cow insists upon being milked in her own pail, *i. e.*, a pail to herself, containing no milk from any other cow—or, if she sees it, she is very likely to kick it over. She will not allow of any mixture. In this there would seem a strange instinct, accordant with her extreme susceptibility to contamination.

The milk is all passed through several strainers, and then placed in great tin cans, barred across the top, and sealed. They are deposited in a van, which starts from the Farm about three in the morning, and arrives at the dairy, in Farringdon street, between three and four. The seals are then carefully examined, and taken off by a clerk. In come the carriers, commonly called "milkmen," all wearing the badge of Friern Farm Dairy; their tin pails are filled, fastened at the top, and sealed as before, and away they go on their early rounds, to be in time for the early-breakfast people. The late breakfasts are provided by a second set of men. Such are the facts we have ascertained with regard to one of the largest of the great dairy farms near London.

DICKENS' *Household Words*

THE GALWAY WOMEN.

A very large fair is held at Galway, Ireland, in the county of Galway, called the Fair of Rose Mount, at which I was present. This was chiefly for the sale of ponies, or horses of a small breed, with some few cattle. On this occasion, the collection of people was surprisingly great; and I could then well understand what was intended by the public meetings in Ireland, called "monster meetings," in respect to which, until I saw this collection of people, I had always supposed the account of the numbers assembled had been much exaggerated. There were here, on this occasion, some cattle and sheep; but there were, also, four thousand ponies, the catching of which, for examination or sale, as they had, in general, neither bridle nor halter, was sufficiently amusing, and I was about to add, sufficiently Irish. The fair was held on the sea-shore, where the receding tide left a large bed of mud. The ponies, when required to be caught, were surrounded and driven into this mud; and here, in a very ignoble way, they were secured, though it was not always without some difficulty they were extracted after being caught.

There was another circumstance, perfectly unique in its character, to which I shall be pardoned for alluding. There was another species of live stock exhibited at the fair, which I cannot say is never seen at such places, but which does not always present itself under the same frank circumstances. The kind nobleman who accompanied me, and who, like many others, noble and simple, whom it has been my good fortune to meet with on this side of the water, left no effort unessayed for my gratification, after looking at the various objects of the fair, asked me, at last, "if I would like to see the girls?" I confess my natural diffidence at once took the alarm; and my imagination cast a few furtive glances over the sea at some precious objects I had left behind. However, upon a voyage of discovery, why should I not see what was to be seen? I gave him an affirmative reply.

Upon inquiring of one of the trustees, or masters of the fair, "if the girls had come," we were informed they would be there at twelve o'clock. At twelve o'clock we went, as directed, to a part of the ground higher than the rest of the field, where we found from sixty to an hundred young women, well dressed, with good looks and good manners, and presenting a spectacle quite worth any civil man's looking at, and in which, I can assure my readers, there was nothing to offend any civil or modest man's feeling. These were the marriageable girls of the country, who had come to show themselves, on the occasion, to the young men and others of the fair. I am free to say that I saw in the custom no very great impropriety. It certainly did not imply that, though they were ready to be had, anybody could have them. It was not a Circassian slave market, where the richest purchaser could make his selection. They were in no sense of the term on sale; nor did they abandon their own right of choice; but that which is done constantly in more refined society, under various covers and pretences,—at theatres, balls, and public exhibitions; I will say nothing about churches,—was done by these humble and unpretending people in this straightforward way.

Between the noble duchess, who presents a long train of daughters, rustling in silk, and glittering in diamonds, at the queen's drawing-room, or the ladies of rank and fashion, who appear at public places with all the beauty and splendor of dress and ornament which wealth, and taste, and art, and skill, can supply, meaning nothing else but "Admire me!" and these honest Galway nymphs, with their fair complexions and their bright eyes, with their white frilled caps and their red cloaks and petticoats,—for this is the picturesque costume of that part of the country—all willing to endow some good man with the richest of the gifts of Heaven, a good and faithful wife, I can see no essential difference.

"Let not ambition mock their useful toil,
Their homely joys, and destiny obscure."

I hope I shall be excused, if I say something more of these Galway women. I never saw a more handsome race of people. I have always been a great admirer of beauty—natural beauty, personal beauty, mental beauty, moral beauty. For what did the Creator make things so beautiful as they are made, but to be admired? For what has he endowed man with an exquisite sense of beauty, but that he may cultivate it, and find in it a source of pleasure and delight? As I have grown older, this sense of beauty—and I deem it a great blessing from Heaven—has become more acute; and every day of my life, the world and nature, nature and art, the animal, the vegetable, and the mineral creation, the heavens and the earth, the fields and flowers, men, women, and children, wit, genius, and learning, moral purity and moral loveliness, deeds of humanity, fortitude, patience, heroism, disinterestedness, have seemed to me continually more and more beautiful, as, at the setting sun, man looks out upon a world made richer and more glorious by his lingering radiance, and skies lit up with an unwonted gorgeousness and splendor.

But the human countenance seems in many cases to concentrate all of physical, of intellectual, and of moral beauty, which can be combined in one bright point. Why should it not, therefore, be admired? In the commingled beams of kindness and good humor brightening up the whole face, like heat-lightning in summer on the western sky; or the flashes of genius sparkling in the eyes with a splendor which the fires of no diamond can rival; or in the whole soul of intelligence, and noble thoughts, and heroic resolution, and strong and lofty passion glowing in the countenance,—there is a manifestation of creative power, of divine skill, unrivalled in any spot or portion of the works of God.

The extraordinary personal beauty of these Galway women was not mere imagination on my part, nor the result of any undue susceptibility I said to the coachman, as we passed through this part of the country, that I never saw a handsomer people. "That," said he, "travellers always remark;" and when I left the country, in casting my eyes over a recent book of travels in Ireland, I found the author's impressions corresponded with my own. Tradition says that a colony of Milesians formerly settled in this part of the country, and the remains of this race, or the offspring of the intermixture of them with the native tribes, present these results. This is a remarkable fact, and not without its bearing, upon one great branch of agricultural improvement.—REV. HENRY COLMAN.

MORNING HYMN.

These are thy glorious works, Parent of good,
Almighty! thine this universal frame,
Thus wondrous fair! Thyself how wondrous, then!
Unspeakable! who sitt'st above these heavens,
To us invisible, or dimly seen
In these thy lowest works: yet these declare
Thy goodness beyond thought, and power divine.

Speak ye, who best can tell, ye sons of light,
Angels, for ye behold him, and with songs
And choral symphonies, day without night,
Circle his throne rejoicing. Ye in heaven:
On earth, join, all ye creatures, to extol,
Him first, him last, him midst, and without end!

Fairest of stars, last in the train of night,
If better thou belong not to the dawn,
Sure pledge of day, that crown'st the smiling morn
With thy bright circlet, praise him in thy sphere,
While day arises, that sweet hour of prime
Thou Sun, of this great world both eye and soul,

Acknowledge him thy greater; sound his praise
In thy eternal course, both when thou climb'st,
And when high noon hast gained, and when thou fall'st
Moon, that now meet'st the orient sun, now fliest
With the fixed stars, fixed in their orb, that flies;
And ye five other wandering fires that move
In mystic dance, not without song; resound;
His praise, who out of darkness called up light.

Air, and ye elements, the eldest birth
Of nature's womb, that in quaternion run
Perpetual circle, multiform, and mix,
And nourish all things; let your ceaseless change
Vary to our great Maker still new praise.
Ye mists and exhalations, that now rise
From hill or steaming lake, dusky or gray,
Till the sun paint your fleecy skirts with gold,
In honor to the world's great Author rise,
Whether to deck with clouds the uncolored sky,
Or wet the thirsty earth with falling showers;
Rising or falling, still advance his praise.

His praise, ye winds, that from four quarters blow
Breathe soft or loud; and wave your tops, ye pines,
With every plant, in sign of worship wave.
Fountains, and ye that warble as ye flow,
Melodious murmurs, warbling tune his praise.
Join voices, all ye living souls! ye birds
That, singing, up to heaven's gate ascend,
Bear on your wings and in your notes his praise.

Ye that in waters glide, and ye that walk
The earth, and stately tread, or lowly creep,
Witness if I be silent, morn or even,
To hill or valley, fountain or fresh shade,
Made vocal by my song, and taught his praise.
Hail, universal Lord! be bounteous still
To give us only good: and if the night
Have gathered ought of evil, or concealed,
Disperse it, as now light dispels the dark!

MILTON.

THE MOTHER'S JEWEL.

Jewel most precious the mother to deck,
Clinging so fast by the chain on my neck,
Locking, thy little white fingers, to hold
Closer, and closer, the circles of gold.—
Stronger than these are the links that confine
Near my fond bosom this treasure of mine?
Gift from thy Maker, so pure and so dear,
Almost I hold thee with trembling and fear.

Whence is this gladness so holy and new,
Felt as I clasp thee, or have thee in view?
What is the noose that slips over my mind,
Drawing it back, if I leave thee behind?
Soft is the bondage, but strong is the knot,
Oh! when the mother her babe has forgot
Ceasing from joy in so sacred a trust,
Dark should her eye be, and closed for the dust.

Spirit immortal, with light from above,
Over this new-opened fountain of love,
Forth from my heart as it gushes so free,
Sparkling, and playing, and leaping to thee,
Painting the rainbow of hopes till they seem
Brighter than reason,—too true for a dream!
What shall I call thee? My glory? My sun?
These cannot name thee, thou Beautiful one!

Brilliant! celestial! so priceless in worth,
How shall I keep thee unspotted from earth?
How shall I save thee from ruin by crime,
Dimmed not by sorrow, untarnished by time?
Where, from the thief and the robber who stray
Over life's path, shall I hide thee away?
Fair is the setting, but richer the gem,
Oh! thou'lt be coveted,—sought for by them!

I must devote thee to One who is pure,
Touched by whose brightness thine own will be sure;
Borne in His bosom, no vapor can dim,
Nothing can win, or can pluck thee from Him;
Seamless and holy the garment he folds
Over his jewels, that closely he holds.
Hence, unto him be my little one given!
Yea, "for of such is the kingdom of Heaven!"

BEST LONG WOOLED BUCK AND EWE, over two years. Exhibited at the Fair of the New York State Agricultural Society, 1851, and owned by J. McDONALD and WILLIAM RATHBONE.

ANIMAL PHYSIOLOGY.

What is to be understood by animal physiology?

Physiology literally means the science of nature, and is synonymous with the term, physics or natural philosophy When applied to the vegetable creation, it is called vegetable physiology. When applied to the structure of man, it is called human anatomy, or human physiology. And when applied to the animal kingdom generally, it is called animal physiology.

Why are vegetables and animals called organic bodies?

Because they each possess the principle of life and certain organs or members active in the maintenance of this principle. For instance, plants possess roots and leaves which perform certain offices in collecting, assimilating, and perfecting the various elements of vegetable life. In like manner the heart and the stomach of an animal are called organs, because their functions are to circulate the blood and digest the food requisite in the animal economy.

What are called inorganic bodies?

Those bodies without vitality which are unable to perform the functions of living bodies, such as water, air, the various gases, rocks, and metallic substances.

How are living bodies described?

They pass through a certain routine of existence, from what may be called their birth to their death. Being in the possession of the quality called life, they take in nourishment from food and air, by virtue of which they grow to maturity; are afterwards, by the same means, supported for a certain period; and, at last, when the purpose of their existence in the world is accomplished, they cease to live, and are resolved into the elements from which they were formed. This is true of vegetables as well as animals.

How in this respect do animals differ from vegetables?

In addition to these peculiarities, common to animals and plants, animals possess properties which give them what is called sensibility, and which enables them to fulfil certain functions which do not belong to plants. In organic bodies, wherever there is sensation or voluntary motion, we have an animal; where these are wanting in a vegetable. In addition to this, animals have a will or a spiritual essence which gives them a measure of control over their physical organs, enabling them to move from place to place, to supply themselves with the means of subsistence, and also to indulge themselves in what may be called the social habits of organic life; not the least important in these habits is their own reproduction.

Why is animal physiology particularly interesting?

Because the animal structure is the most wonderful production in the whole range of the material creation; the complicated machinery of a watch or a steam engine bears but a faint resemblance, in contrivance, to the animal frame; but, the wisdom displayed in the mechanical perfection of an animal, scarcely deserves to be named—is a mere bagatelle—compared with that which is manifest in what is called animal sensibility or sensation—the mental attributes, which control the physical organs.

What branches of animal physiology are most deserving our attention and critical study?

Human physiology, or that which relates to man first deserves our regard. On a knowledge of the mechanism in the human structure; of the laws by which it is operated; of the agencies which may disturb its power of action; and, especially of the means for preserving and augmenting its vigor, depend what we esteem most valuable in this world. Physically as well as mentally the most valuable study is that which enables man to know himself—the mysteries of his own complex organization.

Without this knowledge what might be the consequence?

Suppose any intricate and complex machinery, for instance, an extensive factory for spinning and weaving, or for any results of mechanical labor, was placed in the care and under the management of an individual wholly destitute of all mechanical skill and taste, would it not speedily become deranged and useless? So likewise, if we are ignorant of the curious mechanism in our own bodies, and of the means for preserving its efficiency, the chances are the whole will become deranged and inoperative, the same as would a watch if placed for repair in the hands of a stevedore or any other common laborer.

But, independent of any such motives of interest for acquiring a knowledge of human physiology, what other rational motive is there for it?

An ignorance of and an indifference to this study, is not

very dissimilar to one's ignorance of his own local geography. With the existing instrumentalities for acquiring geographical knowledge, we can scarcely realize how any one with the least pretension to mental endowment, can willingly remain destitute of this knowledge. But, for an individual to continue ignorant of the geography of his own country and state, and especially of his own town, denotes a degree of stupidity as well as ignorance, absolutely disgraceful, and almost giving him rank among the brutes. The same may be said of the individual who is ignorant of himself; of the organization of his own body; and of the laws which control the preservation of his own life.

What other motives are there, especially in moral life, for the thorough and critical study of animal physiology?

It is apparent, that the animals upon a farm cannot be reared, fed, and fattened, to the best advantage, without a familiar acquaintance with their general habits, and particularly with the elements in their feed contributing most to a full and rapid development of their physical proportions and strength. Otherwise, the farmer can never make this branch of his business in the highest degree satisfactory or lucrative. A large proportion of his expenditure for feed will be essentially lost.

How can this be made to appear?

Some breeds of horses, horned cattle, swine, and sheep, it is well known, grow larger, acquire more symmetrical proportions, and in cows give more milk, on the same outlay in feed, than other breeds. So it is likewise in the labor of oxen. The husbandman must, therefore, by crossings and otherwise, procure the best kinds of farm animals, and be able to know the particular descriptions of nutriment for strength in those for work; also, for the production of fat in those designed for slaughter; and not less for milk in those on which the dairy products depend.

What other advantages to the farmer result from a particular and full knowledge of animal physiology?

A knowledge of the anatomy of farm animals enables one to judge better of their strength for labor. Besides, they are always liable more or less, to disease and broken or deranged limbs. If their owner has no knowledge of their physical organization how can he prescribe for fractured bones or for any diseased organs? A knowledge of the dis eases of animals, of the remedies for these diseases, and of the mode of treatment in case of severe accident, may frequently save the life of one of great value. This knowledge is as needful as skill in a medical practitioner when those of the human species are suffering from fever or dislocated joints, so far as restoration in both cases is of parallel importance. Without this knowledge, many a profitable cow, many a noble horse, many a powerful ox has been lost to the owner.

Are there any other motives for an acquaintance with the subject of animal physiology?

There are; motives of a sympathetic character. There is certainly a great pleasure felt in understanding and being able to converse on the attributes and points of all objects constantly about us, and on the profits of which we depend for our prosperity and comfort. And to be able to relieve a distressed or sick dumb animal must be a pleasure to every person of humane and kind sensibilities. A person has but a frail claim to a kind and Christian heart who is indifferent to the sufferings of the brute creation.

How many bones are in the human skeleton?

In the body of an adult there are two hundred and eight principal bones, besides eight very small ones in the thumb and great toe, making in all two hundred and sixteen. And if the teeth be added, thirty-two, when the set is completed, the whole number will be two hundred and forty-eight. Some, by distinguishing parts of the same bones, have reckoned the whole at two hundred and fifty.

Of what elementary substance are the bones and teeth composed?

Mostly of lime, which is received into the system from such articles of food as contain it. Hence, it is particularly needful that young animals, whose bones are maturing, should be liberally furnished with those articles.

What scientific advantage results from a knowledge of the particular conformation of bones?

When some prominent bone of an animal is found imbedded in the earth, the articulations of it denote the species of animal to which it belonged. Thus, if the radius of the human skeleton were thus found, which is the large or principal bone of the arm below the elbow, it would be clear from its articulations that it belonged to a moveable organ used for seizing objects. Such an one is called moveable, in distinction from those called fixed, which are to support the body, and are, instead of being connected with fingers or claws, connected with hoofs.

If such a prominent bone be found having had connection with hoofs, what inference is drawn?

The hoofs imply that the animal to which it belonged was a vegetable feeder, with grinding teeth, and a particular form of alimentary canal, and also a certain conformation of the spine. Hence, a pretty good idea would be had of the size, form, and general habits of the animal, simply from this bone.

How would a naturalist reason in such a case?

He would conclude that the animal had been put together, as in all cases with which we are familiar, upon rational and philosophical principles. Hence he concludes that no organ stands isolated, but that each has intimate relations with the rest, always forming a harmonious and symmetrical whole; and that, whether we examine the structure of the individual parts, or their relations, we must equally feel that all has proceeded from the hand of a being infinitely wise and good.

What practical illustration of this have we?

The following fact is stated by Dr. Buckland in his Bridgewater Treatise. Many years previous, a few bones of an extinct species of an animal were found. At that time nothing of the general form of the animal was known. But from these bones, Mr. Connybeare, a celebrated geologist, set himself to work to construct an animal, such as he supposed that to which these bones belonged. Some years afterwards, a complete skeleton of this singular animal, the Plesiosaurus, was discovered, with which Mr. Connybeare's drawing was found in a surprising degree to correspond.

Into what three classes are animals divided, when considered in relation to their food?

Carnivorous, or those which feed on flesh; herbivorous, or those which feed on herbage or vegetables; and omnivorous, or those which feed on flesh and herbage.

How are these different classes known by their teeth?

It is said that the teeth give us decisive indications of an animal's habits, of his bodily conformation, and of his general attributes. The general distinctive classification of teeth is into those called cutting and grinding; or incisors and molars; the former thin and sharp edged, and the latter double. In the human species and in herbivorous animals, the molars, or double teeth, are essentially flat, so as to operate, in reducing food to a fine state, like the stones of a mill, while in carnivorous animals, the lion and all the cat tribe, all of them are sharp-pointed except two; the ferret and polecats have four not pointed, and dogs have eight not pointed.

What remark is made respecting the design of Providence in the variation of the structure of teeth in different animals?

It is quite impossible that the sharp teeth of the tiger and the shark should ever be used for grinding food, like those of the horse and the ox, or that the teeth of the latter could be intended for seizing and tearing flesh; and this becomes still more striking, when we observe how the jaws in which the teeth are set have been articulated. In the carnivora, or carnivorous animals, as the teeth merely cut, the jaws are pointed like the blades of a pair of scissors, but the jaws of the herbivora, or herbivorous animals, and of man, allow of a grinding motion. This motion of the cow and horse is from side to side, or nearly circular, whereas in the rodential family this motion is rapidly performed in a longitudinal direction.

What is said of the provision in animals for teeth?

So long as animals are young and subsist on milk or pulpy substances, the teeth, which would then be an inconvenience, do not appear; but, in the jaws even before birth, provision for them may be seen, so that, as the wants of the animal require, the teeth begin to develop themselves—not all at once, but those most needful first, and then, at varying intervals with different species, the others in due succession. In the human species the set of teeth is not completed till the individual arrives at maturity—say, eighteen years of age, or in that neighborhood.

Of what are the teeth composed?

The human teeth are composed principally of two substances, the *enamel*, and the *ivory*, or bone. The enamel is placed externally and on the body of the tooth, and forms only a thin layer, and is so hard as to strike fire with steel; but when any part of it is destroyed, it is never restored. The internal part of all teeth is much of the nature of bone, and is produced by the lime contained in the food furnished for the animal. The teeth, bones, shells, fish-scales, cartilages, and coral are all compounds of the same elements combined in different proportions, and rendered harder or softer as they possess a larger or smaller quantity of calcareous salts; ivory and the enamel of teeth possessing the largest quantity, and consisting almost exclusively of phosphate of lime, with a small proportion of animal matter.

What is one of the most extraordinary facts in the animal economy that strikes the attention?

It is the law of nature, that animals feed on each other, or on vegetables, that seem to be produced for that end. This law applies to the greatest and to the least; the strongest and the weakest; in the sea and on the land. Whole species of animals apparently exist for nothing else, but to become food for other species. This is remarkably the case with marine animals. Who can imagine the millions of these smaller tribes that are annually consumed in this manner in nurturing a single whale, that in the end furnishes man with the means of dispelling from his abode the darkness of midnight? For what purpose do the clouds of insects fill our atmosphere but to become food for larger races of the animal kingdom, who in their turn serve the same end to others larger than themselves?

Through what processes does the food pass before answering the purpose of animal nutrition?

In the first place it is masticated or reduced to a fine, pulpy consistence; then, it passes into the stomach, where it is subjected to the action of what is called the gastric juice; by this action, digestion, or the decomposition of its elementary constituents, ensues; and, finally the nutritious portions mingle with the blood. The parts of the food not nutritious pass off through the alimentary canal, and are voided under the name of excrements.

To what has the process of digestion been compared?

To the making of soap. The grease used for the purpose is mixed in ley from ashes, or with potash, which is ley boiled down to a concrete substance. By this means the grease is dissolved and combines with the ley. If, however, too much grease is used for the alkali, portions of it will remain solid, that will not become soap. In a similar way the food is digested or dissolved by the gastric juice in the stomach; but, if a person eats too much for the gastric juice to dissolve, a portion of his food will remain in solid or undigested lumps, analagous to the lumps of undissolved grease in the soap barrel. This undigested food in the stomach not only produces the most disagreeable sensations, seemingly like what would be felt from having swallowed whole raw potatoes or small stones; but, it destroys the healthy action of the stomach, and leads to general derangement in the whole system. Hence intemperance in eating is nearly as detrimental to health as intemperance in the use of drink.

Is the food of all animals masticated in the mouth?

There are some exceptions. Some animals swallow their food first and then comminute it afterwards. This is the case with the lobster, the grasshopper, and some others, which have their teeth immediately connected with the stomach. One species of the grasshopper has no fewer than two hundred and seventy teeth. The same purpose is served by the gizzards of granivorous birds, only that the grain is ground between two hard, horny surfaces, which act like millstones, their effect being increased by numerous small stones swallowed instinctively by the animal. As many as two thousand of these stones have been counted in the gizzard of a single goose.

How is the sensation of hunger produced?

When there is no food in the stomach on which the gastric juice can operate, it acts on the inward coats of the stomach; if not to their destruction, yet to the production of a most disagreeable and painful sensation.

What is said of the power of the gastric juice?

In dogs it is so great that it is known to dissolve ivory and the enamel of teeth. In hens it will dissolve the hardest stones. And some years since, upon examining the stomach and intestinal tube of a man who had died in one of the hospitals of London, and who had in his lifetime out of hardihood swallowed quite a number of clasp-knives, their handles were found digested, and their blades blunted, though he had not been able to discharge them from the body.

What is said of the circulation of the blood?

The circulation of the blood is carried on by two distinct sets of vessels, the arteries and the veins. The arteries take the blood from the heart, loaded or impregnated with the nutrition from the food, and of a bright scarlet color, and carry it to every part of the animal system. It then returns to the heart, in the veins, receiving, as it courses along, the impurities of the system, particularly the wasted carbon, so that before reaching the lungs it has changed to a dark purple color. Here, by being spread over the lungs, in a beautiful, fine net work of air vessels, every particle of it is brought in contact with the pure, fresh gases just inhaled. Here the carbon is thrown off, not only from the blood, but from the system, by respiration. The blood being again restored to its scarlet hue, passes again to the heart, and is again freighted with nutritious aliment, to perform a second circulation, and so on in perpetuity to the end of life.

At what rate of time does the blood circulate?

It moves in its circulation with astonishing rapidity. Did it flow at an equal rate in a straight line, it would run in one minute through one hundred and fifty feet, or thereabouts. This swiftness, however, exists only in the larger vessels near the heart; the farther the blood recedes from the heart the slower its motion becomes. At the above rate, the blood rushes forward in its appointed channels more than forty miles in twenty-four hours, a greater distance than a horse should travel, or can travel for a protracted period.

What is understood by animal heat?

It is that property of all animals, by means of which they preserve a certain temperature, wholly independent of that of the medium by which they are surrounded, and appears rather to be in proportion to the degree of sensibility possessed by them. In birds it is the greatest.

How is the study of animal heat important to the farmer?

It is well known, that farm animals kept warm require less food and fatten better than those which are not sheltered in cold weather. It is supposed that their fat is consumed in generating, when exposed to the cold, the internal warmth necessary to preserve life, the same as the oil in a lamp is consumed in generating light and heat—or, as wood and coal are consumed in producing heat to warm our ordinary dwellings.

PROGRESS OF AGRICULTURE.

Why is it, that the subject of agricultural improvement commands such general attention? Why has it become a favorite topic with the statesman and the philanthropist?—with the scientific man, as well as with the man of business?—with the deep-thinker, as well as the hardy operator? Is it alone for the pecuniary profits which may be secured? Or do more exalted considerations enter into the account? To increase the reward of honest labor—to ensure a just return for capital invested—are objects legitimate and worthy to be considered. But *the spirit of the age* is directed to higher and nobler ends! It aims at intellectual conquest—to elevate the immortal mind, and to establish its dominion over animate and inanimate matter! It is, then, a leading object of this enlightened effort, to improve not only the *cultivation* but the *cultivators* of the soil—to combine in their behalf the principles of science, with the strength and skill of labor; and so to direct their efforts as to secure a just elevation of character to the one, and a profitable return to the other.

Agriculture has been termed the primitive art. To the savage the earth yields a spontaneous but precarious and inadequate support. Hunger, necessity—the mother of invention, lead him to that rude culture which alleviates, first, the cravings of appetite. The very act awakens within him the nobler spirit and properties of his nature. This is the dawn of civilization. With it agriculture commences. With civilization agriculture advances, and recedes with it wherever civilization recedes. Agriculture has flourished most in the most enlightened ages and nations of the world, and has been most elevated and prosperous where the people have been most free. Her true friend and disciple may therefore exclaim, with the American sage and patriot,—

"Where Liberty dwells, there is my country."

In the palmy days of Greece and Rome, when originated those bright examples of patriotism and virtue which have been venerated, quoted, and deemed worthy of imitation, in all succeeding ages, the cultivation of the soil was held to be the most honorable of pursuits. The sages and heroes of that period, not only directed the operations of farming, but thought it no disparagement to till the soil with their own hands. From the plough they were called to deeds of valor and renown in the service and defence of their country; and after the performance of those deeds, they returned voluntarily to the peaceful walks of rural life. The names of some of the noblest families of Rome were derived from products of the soil in the cultivation of which their ancestors excelled. It was deemed to be the highest praise of a good man to call him "an agriculturist, and a good husbandman; and he was thought to be greatly honored (says Cato) who was thus praised." The writers upon the subject, of that era, the most eminent men of the day, both Greek and Roman, appear to have been well acquainted with the principles and practices of good tillage.

The Roman conquest introduced into Great Britain a better system of husbandry than then existed there, and which has since been greatly improved. To that old and wealthy portion of the world, we have been in the habit of looking for examples, and although self-dependence is our true policy, we may still learn much from her experience and practice. Comparatively free, enlightened and enterprising, Great Britain has outstripped almost every other country of the old world in agricultural as well as manufacturing skill. Although she possesses an extensive commerce—although numerous colonies in all parts of the globe are tributary to her—the great truth is recognized by her statesmen and intelligent subjects, that the true wealth and power of a nation consists in its soil.

And yet, while vast improvements have been there accomplished, in every branch of husbandry—while science and practice have brought forth abundant fruits of their united skill, doubling within the last twenty-five years the yield per acre of grain and hay; doubling in the same time the number of domestic animals, and enabling the farmers to keep twice as many on the same amount of food as they did—while the finest of cattle, horses and sheep, of improved breeds, are sprinkled in numerous herds over her fertile meadows—while the lordly and privileged owners of the soil, and the wealthy tenantry of large estates, have been greatly profited—the mass of the people, the actual operators—still labor in unrequited toil. It is not, however, to be denied, that within the last half century, through the spirit of agricultural improvement and the philanthropic zeal which has been exerted in behalf of popular education, both in Great Britain and on the continent of Europe, the condition of the toiling millions has been greatly ameliorated.

In France, the government has established experimental farms and agricultural schools in the various departments of husbandry, and is annually expending large amounts to promote agricultural improvement throughout its domains. One of the results of this policy, has been, to double the product of wheat in that kingdom, which now produces annually more wheat than is grown in Great Britain and the United States. Similar results have followed a similar policy in the German States. "Flemish husbandry" has long been acknowledged to surpass that of any other country of Europe. The soil of Flanders is naturally sterile; but by a systematic and skilful plan of cultivation, combining the plough with the spade husbandry, upon small farms of from forty to fifty acres each, nearly the whole face of that country has been made to present the features of garden cultivation.

But it is not my intention to attempt a history of agriculture, ancient or modern, past or present. I have stated these facts by way of illustration, to show that, long centuries ago, great, patriotic and intelligent men appreciated the pursuit, and understood the essential theories and practices of the art, of cultivating the soil—that the spirit of improvement, which has lingered through ages, is now abroad and active in other lands, speaking to us in the voice of instruction, of encouragement and admonition; and that, though much has been discovered and accomplished, much remains to be developed and matured, in this great and beneficial field of enterprise.

And upon whom, let me ask, within the wide compass now occupied by civilized man, rest stronger and deeper obligations to come up to this great work with energy and effect, than upon the people of these United States of America? With advantages, such as no other people ever possessed—with a variety of soil and climate known to no other division of the habitable globe; with institutions more wise and more free than were ever enjoyed by man—we are under obligations, in gratitude to Providence, in justice to our fellow men, to set an example not only of political but of moral greatness, and to excel all other nations in the arts and pursuits of civilized life—in the production, the extension, and the enjoyment of happiness. Proclaiming to the world the broad doctrines, that all men were created equal, and that man, even in the imperfection of his nature, is competent to self-government,—it becomes us to elevate, not only theoretically, but practically; not only politically, but morally—the laboring and productive classes. Honest labor, to be deemed honorable, should only be required to be virtuous and intelligent.

And, fellow-citizens, I lay it down as a position not to be controverted, that the cultivators of the soil, the agricultural class, constitute in this country the basis of society—the foundation of our social and political system. From this basis spring all other classes, productive or non-productive. On this broad and deep foundation rest the prosperity and liberties of our country. Trace the records of our history, past and present! Cast your eyes around, among the multitudes of men! Can you name a patriot, a hero, or a statesman, who could not or cannot trace his genealogy, directly or indirectly, to the cultivators of the soil? Can you name a man who is distinguished in any profession,—on the bench, at the bar—in our halls of legislation—as a poet, as an orator, as a man of literature or science—who is not himself a farmer, the son of a farmer, or the son of a farmer's son? So with our merchants, proud and wealthy as they may have become—so with our mechanics, who constitute, with the farmers, the bone and sinew of our country. And so it must continue to be while our free institutions endure.

In our own country the tillers are the owners of the land, or may become so by industry and economy. Here, all is diffusive. The elements of our social and political order are continually active: Freedom, enterprise, the equality of rights, and the equal protection of the laws, stir up and intermingle the masses of which it is composed. High and humble, rich and poor, are terms which distinguish but the position of a day. Wealth and honors, possessions and pursuits, undergo continual changes. The professional man, the merchant, the mechanic, deriving directly or remotely their existence from the tillers of the soil, return, themselves or their descendents, in whole or in part, to that pursuit with which their earliest and dearest associations are connected—thus filling the places of those whom the allurements of fame or fortune have drawn from its peaceful paths Indeed, the business of agriculture, ever productive, honorable, untrammelled, open and free to all —furnishes a refuge to those who are driven, by the force of circumstances, by excessive competition, by the substitution of machinery for manual labor, and other causes, from other employments. It affords an asylum to disappointed hope, to exhausted energies and to satiated ambition, in all other pursuits. They return, like the dove to the ark, to that refuge where peace, where competence, and independence, are ever the fruits of honest toil and well-directed efforts!

But, who can estimate the importance of agriculture, in a national point of view, as controlling the character, the prosperity and independence of our country? Upon it, all classes depend for subsistence. It furnishes the elements of commerce and the materials of manufactures. It aids in the development of those mineral resources which supply the useful arts, constitute the essentials of household goods and structures and the useful implements of trade, and which set the mighty power of steam in motion. Who can estimate, who conceive, the value in dollars and cents of those unnumbered acres which constitute the surface of this vast republic?—of this immense *terra firma*, which remains, with all its productive powers, while all things else perish?—which now nourishes upon its bosom twenty millions of human beings, and is capable of sustaining innumerable millions yet unborn? Who can look forward, through the long vista of time to come, and foretell what vast improvements future ages are destined to witness—what, upon this great theatre of action, freedom, and enterprise, science and industry are yet to accomplish?

Agriculture, while it offers reasonable inducements for the investment of capital, affords a competence, which is better than wealth, to all who pursue it with becoming skill and energy. But let no one hope for success in that pursuit, without labor, or at least without vigilant application and strict economy. The declaration made by God to man—"by the sweat of thy brow shalt thou eat bread," was rather a gracious promise than a threat or denunciation. It was a guaranty to him, of a full reward for labor bestowed—a sweet return for honest toil: and the abundant harvests which crown the labors of the industrious and skilful husbandman, are consummations of the Almighty promise!

The pursuit of agriculture, is as favorable to intellectual as to physical improvement. It promotes the strength of the mind, as well as that of the body. Surrounded by scenes the most beautiful and sublime in creation—"looking from Nature up to Nature's God,"—the mind of the intelligent farmer must become imbued with the most exalted conceptions. Men of genius, in every department of literature and science, have written amidst the scenes of rural life; and the best of poets, in all ages, have from thence derived their inspirations!

Health, too, without whose presence there is no enjoyment, holds her favorite seat amidst fields and forests. She sports on the mountains and gambols in the valleys. She loves the grove and the fountain. She loves to inhale the fragrance of the new-mown hay! She loves to follow the ploughman, whistling at his plough; and listen at early dawn to the music of the birds:

"Often by the sound of tabor,
She the rustic's care beguiles,
On the brow of honest labor
Are bestowed her richest smiles."

Woman, "Heaven's last, best gift to man," was placed by his side in Eden, to help him in dressing and tilling the garden. She is "at home" amidst the charms of rural life, and clusters around the farmer's fire side his choicest and richest blessings. Relieved by mechanical skill from the toil of carding and spinning and weaving,—exempt by the just estimate of civilization, from the labors of the field—woman finds her appropriate sphere in domestic duties—in cultivating the minds of her children—in the pursuits of household economy—in spreading the neat and frugal meal, and in cheering and softening the cares of labor to him upon whom its heaviest burthen falls. She finds, too, her delight, as in the days of primeval innocence, among the fruits and flowers: and for her sake, and for his own most rational and healthful enjoyment, the grateful husbandman will cherish the orchard and the garden!—*An Address before the Tompkins County (N. Y.) Agricultural and Horticultural Society, by* HON. EBENEZER MACK.

THE FORMATION OF SOILS.

The soils which now exist upon the face of our earth, have been produced by a variety of agencies; the chief of these have been the gradual decomposition and crumbling down of the rocks themselves, and deposition by water. We know that the external outline of the earth has undergone most extensive changes. In some places it has sunk, in others risen. Sometimes it is evident from the present conformation of surface, that violent currents of water have swept across strata of rocks, wearing away the uppermost, and transporting their ruins to fill up depressions elsewhere. We often find strata upheaved and dislocated by action from below, and in many cases see the inferior rock presenting itself on the surface, having burst upwards in a state of fusion, in despite of every obstacle. Scarcely a region can be found which does not present striking evidence of the throes, convulsions, and changes, which took place before man became an inhabitant of this planet.

It is for geologists to decide, if they can, how long a time was occupied in these changes; suffice it for our present purpose that they have taken place, and that they seem to have been especially ordered for our benefit. Had the stratum last deposited or formed, continued unbroken and unchanged around the whole earth, we should have had none of the beautiful variety of scenery which now greets our eyes on every side; no alternation of hill and dale, mountain. plain and valley, with the attendant variations of climate and production, which now so often remind us of perfection itself.

The soil would have been identical in composition over vast districts, if not over the whole earth, being all formed from at least allied species of rocks. Now as few rocks contain all the material for a good soil, this soil would doubtless have been imperfectly fitted to sustain most of the plants necessary for our existence and comfort. When exhausted too, we should have had no stores of mineral substances in forms convenient for supplying the deficiency. The convulsions of nature, however, have been directed for our good, and they seem to have continued in a very long series before this earth was deemed fit for the abode of man.

Geological researches have shown us the existence of races of animals, that lived and died, and succeeded each other in countless myriads, through long and indefinite periods of time. We find them all changed to stone, entombed in rocky sepulchres. Sometimes the appearance of the rock denotes that it was deposited from a calm and quiet sea, where the animals died naturally, and in consequence seldom remain whole or unharmed. In other cases life and its functions seem to have been suspended by some sudden change, so that we find large fish with smaller ones in their mouth, but half swallowed, and others with their thorny fins yet erect in the attitude of fear or rage with which they received their death shock, when that sudden mysterious destruction came upon them. In some of these periods also, upon that part of the land elevated above the water, there flourished a vegetation of exceeding luxuriance.

Internal fires have borne a decided part in all of these changes, if they have not been the chief agents. It is well known that even now, as we go towards the centre of the earth, for each foot in depth the heat increases, indicating interior combustion still active. In the earlier history of our globe these fires must have burst forth many times. The masses of melted matter may be plainly seen, penetrating the stratified rocks, filling cracks in their substance, flowing over their surfaces, or upheaving and contorting them.

But while some rocks were thrust upward, others sank into corresponding depressions; and vast currents of water produced by these convulsions, seas and lakes turned out of their beds, seem to have swept over the world; completing the scene of confusion by tearing away and grinding down strata, bearing the materials to other regions, there to form beds of sand, clay, or gravel, according to the nature of the original rock. The vegetation at such periods, seems to have been carried into hollows and buried deep by succeeding or continuing shocks, to form under enormous pressure and a high temperature, beds of coal for the advantage of beings yet to be created.

Thus all of these tremendous revulsions and changes of surface, seem to have been made with the great end of preparing the earth for the habitation of man, making its resources more available to him. In such a view, the globe appears to have been a vast manufactory for our benefit. Its beds of limestone, of marl, of gypsum, are dispersed in every direction, that they may be accessible to all; the various composition of its rocks, produces soils capable of growing every necessary plant; its ores are abundant in proportion as they are the more indispensable for the formation of necessary implements; while on the walls of our coal mines, we may still trace the forms of a gigantic vegetation which flourished long ages ago, and was then stored for our use.

It is not to be supposed that the present surface assumed its present shape, in every place at the same time. Some regions, without doubt, became tranquil long before others, but all must at first have presented a strange naked aspect. There was of course no soil, except in the tract of some former current where matter in suspension had been deposited. This appearance of absolute ruggedness and sterility, could not have continued long unaltered. Atmospheric influences, heat and cold, moisture and dryness, worked surely then as now, and after a time the most enduring rocks began to crumble. As the decomposing fragments became minute, little patches of soil were formed here and there. If it were on the side of a hill, the fine particles had a tendency to descend into the hollows, being washed down by the rain. In ordinary circumstances, therefore, soil must have first appeared in the valleys, and in every little hollow of the hill sides. The durability of each particular species of rock, had of course much influence upon the readiness with which the soil formed. Thus most of the slates, many limestones and sand stones, soften and decay readily when exposed to the air; on these were to be seen soils at a comparatively early period, and such soils soon became deep. But the granites, and some of the harder limestones, remain almost unchanged for a long period of years, and we see even at this day that the soils upon those formations are thin, while at frequent intervals project masses of the naked rock, yet defying the influence of time.

Almost all of the deep, and apparently inexhaustible soils which occasionally occur in our own and other countries, seem to have been originally formed by depositions from water, either as a stream or a lake. The distance to which finely divided particles are carried by a rapid stream, is truly astonishing. The fine clay found in the bottom of some lakes in Holland, is known from its composition to have been brought down by the Rhine from its upper waters, in the mountains of Switzerland and Germany. The deposit even from waters which flow through a very inferior soil is of good quality, or at least much better than would be expected. I have recently had an opportunity of seeing this fact exemplified in New Haven, Ct. The Farmington Canal, which terminated at that place, ran for nearly thirty miles from the city through a very light sand, so light that it was a long time after its completion before its banks could be made to hold water at all. This canal is now abandoned, and in cleaning out one or two basins near the city, a deposit of nearly a foot in depth was found, having quite a clayey character, baking hard, and cracking when dry. This deposit has proved worth nearly or quite as much as manure, on the light sandy soils of that neighborhood.

Soils formed in this way by water, are common in every country, and there are also large tracts covered by some of those terrible ancient floods of which I have spoken. This may all have been done at the period of the deluge, but however that may be, the original formation is covered sometimes to a vast depth by the debris of others. In all of these cases of superficial deposit, the character of the underlying rock has of course little or nothing to do with that of the soil: but in most situations it has a controlling influence, and a study of the one will give us a general idea of the other, beside leading to important practical results.

The variations in the composition of different rocks, are far greater than is ordinarily supposed. It might be thought by many, for instance, that the soils of limestone countries would, as a general rule, be nearly identical in composition, but this is by no means true. The pure limestones contain as high as ninety-five per cent. of carbonate of lime, but there are many which contain impurities to the amount of much more than half their weight. Then, too, there is a large class of limestones in which magnesia is found in greater or less proportion. The soils produced by these last, when the magnesia is in large quantity, are frequently very poor and cold; differing extremely from those formed by a limestone in which little or no magnesia is present. An unpracticed eye would be unable to distinguish between the two kinds of stone, and a farmer who had lived upon a good limestone soil, might be miserably deceived when he thought he had settled upon another of the same character.—*From an Address at the Annual Cattle Show of the N. York State Agricultural Society, Buffalo,* 1848, *by* PROF. JOHN P. NORTON, A. M., *of Yale College.*

AGRICULTURE FAVORABLE TO KNOWLEDGE.

There is a prevailing impression, especially among intelligent young men, that the pursuit of agriculture is unfavorable to the pursuit of knowledge, and the general cultivation of the mind—that the life of a farmer is a life of drudgery and toil, without any stimulus or opportunity for intellectual improvement—and that if a farmer is intelligent, he is so in spite of the earthly, degrading tendency of his occupation. We maintain just the opposite view—that the occupation of the farmer is favorable to the pursuit of knowledge—favorable to intellectual health, activity, and vigor of mind, so that if a young man has a taste for knowledge, he should for this very reason be a farmer, because he can thus gratify this taste for knowledge better than in any other calling.

The life of the farmer is favorable to the pursuit of knowledge, because it is favorable for health. The farmer, who breathes the fresh air, and listens to the songs of birds, and sees so much in nature to interest him, is seldom troubled with hypochondria, dyspepsia, and indigestion, which are as injurious to the pursuit of knowledge as to happiness and health. Can there be any doubt that the occupation which gives such health and cheer to the farmer, is favorable to the development of the mind, and the pursuit of knowledge, especially when we consider the intimate connection between health of body and health of mind, and how many minds are necessarily feeble, stinted, and *sickly*, because dwelling in a feeble and sickly body?

The farmer has *leisure* for the pursuit of knowledge. Aside from the leisure which winter evenings, rainy days, and intervals between hurrying seasons of labor afford; he can, almost every day, snatch a few moments, or an hour for reading, *if he has a desire for improvement.* If the farmer chooses to spend his leisure at the stores and taverns, or in idle vacancy, dreaming and dozing away his life, working like his ox, and like his ox only eating and sleeping, he can do so—but let him not blame his occupation, for if he only has thirst for knowledge, he can gratify it. No laborer has more leisure for improvement than the farmer.

And besides, the leisure of a farmer is worth more to him, in the pursuit of knowledge, than that of other laborers, not only because from his good health and spirits, he is better prepared to improve this leisure, but because it will furnish him with food for thought, reflection, and inquiry, during the day; his work, much of it, being of such a nature as to afford opportunity for digesting what he has read, especially if it relates to agriculture. The reason many farmers are no more intelligent is, *not because they have no leisure*, but because they *do not improve their leisure*. The most ignorant farmers are by no means the most industrious. Some of the most industious, efficient farmers of my acquaintance, are the most intelligent also. Nor does their intelligence make them lazy, but rather stimulates them to labor. They take hold of labor, too, with more zeal and interest, and feel less tired at the close of the day, than the mere *drudge*, whose vacant mind is uninterested in what he sees and does. The man who is to work on a compost heap will not do less, but more work, if he spends a few moments in reading an essay or lecture on manures, so that he may labor *intelligently*.

Agricultural pursuits have a healthy influence on the mind, and thus favor the pursuit of knowledge. The farmer is free, on the one hand, from the tormenting excitement, anxiety, and perplexity of the merchant and trader, on the other hand, from the dullness and monotony of the day laborer, or the mechanic, who does one thing the year round. The influences that surround the farmer are as favorable to health of mind as health of body; hence, if a man has a taste for knowledge, he may choose the life of a farmer, as being well adapted to gratify his taste.

The occupation of the farmer affords him an opportunity to cultivate an acquaintance with the natural sciences, and is thus favorable to the pursuit of knowledge. The shoemaker, or the blacksmith, may be interested in the study of meteorology, but his daily occupation does not, like that of the farmer, give him an opportunity to observe the weather, the wind, clouds and storms, and their influence on vegetable and animal life. The book of nature is constantly open before him, inviting him to read her laws. The investigation of the laws of nature affords a pure and exalted source of happiness; but who is so favorably situated to investigate these laws—*while pursuing his appointed labor*—as the farmer? Who can so well learn the laws of vegetable life, as he who is constantly experimenting on those laws? Who can so well observe flowers, grasses, plants, grains, and trees, and their habits, as the farmer, whose business it is to cultivate them, and bring them to perfection?

The practical advantage to be derived by the farmer from an acquaintance with science, renders his occupation favorable to the pursuit of knowledge. The natural sciences, botany, geology, chemistry, and many others, are not only interesting in themselves, but intimately connected with the cultivation of the farm. It is by the aid of these sciences that the great improvements in agriculture have been made the past few years, and that we may expect improvements hereafter. If the farmer will not study science because it is interesting, he must study it because it is *useful*—because it is necessary to the successful cultivation of his land. However interesting science may be, the great mass of laborers, having little leisure, and no particular taste for science, do not pursue it. Even professional men do not. They have no stimulus to pursue it, as the farmer has.

The farmer, on the contrary, has extraordinary facilities and motives to become acquainted with science, for almost every science aids him in his work, gives him skill and power, as well as pleasure and profit. He can read the theory, and then test the theory by his observation and experiments. Science comes not only to please but to profit; not only to enrich his mind with knowledge, but to enrich his farm—to improve his fruit and stock—to fill his barns and granaries. Formerly, it was thought a farmer had no use for knowledge. Now it is found that no laborer has more use for knowledge.

The pursuit of agriculture is favorable to the general development and cultivation of the mind. It furnishes a home for the farmer and his family, a pleasant, *rural home*—one of the most essential means of moral, social, and intellectual improvement. The farmer and his children are free from many temptations to vice, intemperance, idleness, and extravagance, which are the bane of intellectual improvement. His life is adapted to develop self-reliance, energy, manly independence, as well as habits of observation, comparison and reasoning. In the rotation of crops, the application of manures, the cultivation of fruit and raising stock, and in planning the work of a farm, as well as in buying and selling, there is abundant exercise for the judgment of the farmer. The business of the merchant is said to be favorable to developing the judgment, but we submit whether the occupation of the farmer does not afford a more enlarged and healthy sphere for the exercise of the judgment, than that of the merchant and mechanic.

If the farmer, therefore, remains ignorant and stupid, it cannot be for want of opportunity for improvement. He is a workman, an experimenter in the great laboratory of Nature, where all he sees and hears invites him to observe, and inquire, and learn; where he can employ in his daily labors whatever knowledge he may possess, and find motives to obtain more knowledge. The means of knowledge, too, are within his reach, so that his life need not be a life of drudgery and toil, unless he chooses to make it so. To be sure, the farmer must *work*, and work hard, and therefore he needs the stimulus of knowledge; for knowledge will stimulate and encourage him to work, so that he can not only do more work, but do it also to better advantage. *Intelligent labor is the most successful labor.* Many men who find no stimulus to labor, when it is a mere exercise of physical strength, will labor with zeal and enthusiasm, if the *mind* is only interested, as it may be, in almost all the work of the farmer.—JAMES TUFTS, *in the Albany Cultivator.*

THE RICHES OF A POOR MAN.

Others in pompous wealth their thoughts may please,
And I am rich in wishing none of these.
For say, which happiness would you beg first,
Still to have drink, or never to have thirst?
No servants on my beck attendant stand,
Yet are my passions all at my command;
Reason within me shall sole ruler be,
And every sense shall wear her livery,
Lord of myself in chief; when they that have
More wealth, make that their lord, which is my slave.
Yet I as well as they, with more content,
Have in myself a household government.
My intellectual soul hath there possest
The steward's place to govern all the rest,
When I go forth, my eyes two ushers are,
And dutifully walk before me bare.

My legs run footmen by me. Go or stand,
My ready arms wait close on either hand;
My lips are porters to the dangerous door:
And either ear a trusty auditor.
And when abroad I go, Fancy shall be
My skilful coachman, and shall hurry me
Through heaven and earth, and Neptune's watery plain,
And in a moment drive me back again.
The charge of all my cellar, Thirst, is thine;
Thou butler art, and yeomen of my wine.
Stomach the cook, whose dishes best delight,
Because their only sauce is appetite;
My other cook, Digestion, where to me,
Teeth carve, and palate will the taster be.
 And the two eye-lids, when I go to sleep,
Like careful grooms my silent chamber keep.
Say then, thou man of wealth, in what degree
May thy proud fortunes over-balance me?
Thy many barks plough the rough ocean's back,
And I am never frighted with a wreck.
Thy flocks of sheep are numberless to tell,
And with one fleece I can be cloth'd as well;
Thou hast a thousand several farms to let,
And I do feed on ne'er a tenant's sweat.
Thou hast the commons to enclosure brought;
And I have fixt a bound to my vast thought.
Thou hast thy landscapes, and the painters try
With all their skill to please thy wanton eye:
Here shadowy groves, and craggy mountains there;
Here rivers headlong fall, there springs run clear;
The heaven's bright rays through clouds must azure show
Circled about with Iris' gaudy bow.
And what of this? I real heavens do see,
True springs, true groves; whilst yours but shadows be.

RANDOLPH.

A COUNTRY CLERGYMAN.

Near yonder copse, where once the garden smiled,
And still where many a garden flower grows wild;
There, where a few torn shrubs the place disclose,

The village preacher's modest mansion rose.
A man he was to all the country dear,
And passing rich with forty pounds a year;
Remote from towns he ran his godly race,
Nor e'er had chang'd, nor wished to change his place;
Unskilful he to fawn, or seek for power,
By doctrines fashion'd to the varying hour;
Far other aims his heart had learn'd to prize,
More skill'd to raise the wretched than to rise.
His house was known to all the vagrant train,
He chid their wand'rings, but reliev'd their pain:
The long remember'd beggar was his guest,
Whose beard descending swept his aged breast:
The ruin'd spendthrift, now no longer proud,
Claim'd kindred there, and had his claims allow'd:
The broken soldier, kindly bade to stay,
Sate by his fire, and talk'd the night away;
Wept o'er his wounds; or, tales of sorrow done,
Shoulder'd his crutch, and show'd how fields were won
Pleas'd with his guests, the good man learn'd to glow,
And quite forgot their vices in their woe;
Careless their merits, or their faults to scan,
His pity gave, ere charity began.
  Thus to relieve the wretched was his pride,
And e'en his failings lean'd to virtue's side:
But in his duty prompt at every call,
He watch'd and wept, he pray'd and felt for all.
And, as a bird each fond endearment tries,
To tempt its new-fledg'd offspring to the skies;
He tried each art, reproved each dull delay,
Allur'd to brighter worlds, and led the way.
  Beside the bed, where parting life was laid,
And sorrow, guilt, and pain by turns dismay'd,
The rev'rend champion stood: at his control
Despair and anguish fled the struggling soul;
Comfort came down, the trembling wretch to raise,
And his last faltering accents whisper'd praise.
  At church, with meek and unaffected grace,
His looks adorn'd the venerable place;
Truth from his lips prevail'd with double sway,
And fools, who came to scoff, remain'd to pray.
The service past, around the pious man

With ready zeal each honest rustic ran:
Ev'n children follow'd with endearing wile,
And pluck'd his gown, to share the good man's smile.
His ready smile a parent's warmth express'd,
Their welfare pleas'd him, and their cares distress'd;
To them his heart, his love, his griefs were giv'n,
But all his serious thoughts had rest in heav'n:
As some tall cliff that lifts its awful form,
Swells from the vale, and midway leaves the storm,
Though round its breast the rolling clouds are spread,
Eternal sunshine settles on its head. GOLDSMITH.

ODE TO THE GLOW-WORM.

Bright stranger, welcome to my field;
Here feed in safety, here thy radiance yield;
  To me, oh! nightly be thy splendor given!
O, could a wish of mine the skies command,
How would I gem thy leaf with lib'ral hand,
  With every sweetest dew of heav'n!

Say, dost thou kindly light the fairy train,
Amid their gambols on the stilly plain,
  Hanging thy lamp upon the moisten'd blade?
What lamp so fit, so pure as thine,
Amid the gentle elfin band to shine,
  And chase the horrors of the midnight shade?

Oh! may no feather'd foe disturb thy bow'r,
And with barbarian beak thy life devour!
  Oh! may no ruthless torrent of the sky,
O'erwhelming, force thee from thy dewy seat;
Nor tempest tear thee from thy green retreat,
  And bid thee 'mid the humming myriads die!

Queen of the insect world, what leaves delight?
  Of such these willing hands a bow'r shall form,
To guard thee from the rushing rains of night,
  And hide thee from the wild wing of the storm.

Sweet child of stillness! 'mid the awful calm
  Of pausing Nature thou art pleas'd to dwell;

In happy silence to enjoy thy balm,
  And shed through life a lustre round thy cell.

How diff'rent man, the imp of noise and strife,
Who courts the storm that tears and darkens life;
  Blest when the passions wild the soul invade!
How nobler far to bid those whirlwinds cease;
To taste, like thee, the luxury of peace,
  And silent shine in solitude and shade!

WOLCOT.

HISTORY OF THE OX.

How is the term cattle used?

In the United States it is commonly applied to the bovine family; or, quadrupeds that bellow, have teats and horns, and are employed by man for labor and for food. Primarily it included not only these, but horses, camels, asses, sheep of all kinds, goats, and perhaps hogs. In the latter sense it is always used in the Bible. In England it now includes horses. In Great Britain, also, farm animals are arranged in two general divisions—*black cattle*, including bulls, oxen, cows and their young; and *small cattle*, including sheep of all kinds, and goats.

What is said of the ox or bovine family?

It is that class of animals to which naturalists have assigned this general name, comprehending eight species, the principal ones being the buffalo, the bison, and our domesticated ox. Of these three the last named is by far the most generally diffused, as well as the most important.

How was the ox esteemed by the ancients?

Their devotion to rural pursuits led them to consider him the most valuable of all domesticated animals. This animal constituted the chief wealth of the patriarchs, whose herds were so extensive that "the land could not bear them," and induced emigration to distant countries. So much dependence was placed upon his labors in the field, that Solomon says, "Where no oxen are, the crib is clean; but much increase is by the strength of the ox." The Egyptians regarded the cow with such respect and veneration, as to elevate her among their deities, and to this day she is worshipped by the natives of India and Hindostan. By the Greeks and Romans, the breeding of cattle was ever thought one of the most dignified and profitable occupations. In fact, in all ages of the world, and in every climate, the ox has received the attention which his excellent qualities so richly merit.

How are oxen now regarded in India and Africa?

In southern Africa they are as much the associate of the Caffres as the horse is of the Arab. They share his toils, and assist him in tending his herds; they are even trained to battle, in which they become fierce and courageous. In central Africa the proudest ebony beauties are to be seen riding on their backs. They have drawn the plough in all ages; in Spain they still trample out the corn; and in India they raise the water from the deepest wells to irrigate the thirsty soils of Bengal.

How long has the ox been known in England?

From the earliest times, and Cæsar, upon his invasion of the island, found the riches of the people consisted in their large flocks and herds, which have contributed in no small degree to her opulence and position at the present period. They have been greatly improved by skilful breeding, and maintain an undisputed superiority over the cattle of all other countries.

Are the varieties in England numerous?

They are; the ox breeds freely, and his character is easily changed by the influence of climate, food, or cultivation. He always accommodates himself to the situation in which he may be placed. In some countries he attains an immense frame, while in others he barely exceeds the size of a large sheep. Thus in England, where the climate and soil are both favorable to his development, much credit is due to the breeder, who, by judicious crossings, has corrected faults and perpetuated the good qualities of the animal.

Which are the most celebrated breeds of Great Britain?

The Durham, the Devon, the Hereford, the West Highland, the Ayrshire, the Galloway, and the Alderney. There are others which maintain a good reputation in their native districts, but have not as yet been brought into general notice.

What is said of Robert Bakewell?

He was an eminent farmer who died in the year 1795. He distinguished himself by his success in the breeding of

SHORT HORNED BULL, KING CHARLES II., imported by James Lenox, New York.

cattle and sheep. One of the most noted of his efforts was the improvement of the *Long-horned* cattle of Yorkshire. By perseverance he established a disposition to fatten, together with an early maturity, and corrected the other capital defects of the parent stock. At the time of his death no breed was in equal favor, but it has now lost its reputation, and is rapidly giving way to the later improved varieties.

How long have the Durhams been known?

The counties of York and Durham have for ages been celebrated for the possession of a most valuable race of cattle, noted for their extraordinary milking qualities. Somewhere about the year 1750, they were crossed with some fine stock from Holland, and the good points of the progeny thus obtained speedily attracted attention. Two brothers, Charles and Robert Colling, commenced the business of breeding, and prosecuted it with so much success that their names have become identified with the breed. At the sale of their stock, in 1810, forty-seven animals sold for over thirty-five thousand dollars, and a famous bull by the name of *Comet* brought one thousand guineas, or about five thousand dollars.

For what are the Durhams distinguished?

A large and symmetrical form; particularly agreeable to the eye; and their excellence for milkers is undisputed. They feed well, and mature at an early age. Their flesh is excellent, and always meets a ready sale. Of this breed was the noted ox, called Spottiswoode, probably the largest ever exhibited, which in 1812 measured as follows—height of shoulder six feet and ten inches; girth behind the shoulder, ten feet and two inches; and was computed to weigh 4480 pounds. It is certain, however, that the Durhams are inferior to some others for use as working cattle, as well as in the richness of the milk. Durhams are frequently called *Short-horns.*

To what are the Devons indebted for their popularity?

A beautiful form, strength of limb, hardihood of constitution, the richness of their milk, together with the superior flavor of their flesh. The oxen are the best known for agricultural purposes, and the milk of the cows, when compared with that from Durham cows, makes up in richness for any deficiency in quantity. Both sexes are gentle, intelligent and docile; and no animals afford a better return for the attention bestowed on them. The breed is a very ancient one, although largely indebted to the skill of the modern breeder.

What is the character of the Herefords?

They are peculiarly adapted to the wants of the grazier, but are not much used in husbandry, although their size and strength fit them for heavy work—and they have all the honesty and docility of the Devon, if not his activity. The cows have generally been considered inferior to other breeds for the dairy, but from the statements of such men as Corning and Sotham of our own country, it would seem that this opinion is not well founded.

What is said of the West Highlanders?

They are small black cattle, bred in the highlands of Scotland, and on the northern islands of Great Britain. They are very hardy, and will thrive in almost any locality. Although of moderate size, they will attain a very respectable weight, and their meat is of such superior quality that it commands the highest price in the markets. They are brought annually in immense droves to the southern part of the kingdom, and there fattened for sale. They are thought to afford a better profit than any other breed.

How is the Ayrshire breed esteemed?

This stock prevails principally in Ayrshire, Scotland, and are thought by many good judges to be indebted for some of their good traits to a remote intermixture with the Durham breed. They are highly esteemed in their native haunts for the dairy. The milk is abundant, and noted for its richness. But when intended for the butcher, the Ayrshires are not considered equal to the Herefords or Durhams. There is not, however, entire uniformity of opinion in regard to this breed. Mr. Tennant, of Scotland, says that these cows, in the best part of the season, will yield fifteen quarts of milk on an average per day ; and, that the annual average for butter is one hundred and seventy pounds. Another person says that a good cow of this breed, on the best pasture, will yield four thousand quarts of milk in the year. But Mr. Cushing, of Massachusetts, is of the opinion that the Ayrshires are not better than our own native breeds, unless crossed with the Durhams.

Why are the Galloways esteemed?

They have some excellent points—such as a compact body, aptitude to fatten, and docility of disposition. They are mostly black, and without horns. The cows are indifferent milkers ; but large numbers of this breed are annually fattened for market.

For what in particular are the Alderneys valuable?

They are chiefly valuable for their dairy qualities. The quantity of their milk is small, but it has an extraordinary degree of richness, affording a large proportion of cream, and butter of a beautiful color and fine flavor. The reputation of the breed is so well established, that a few of the cows are kept by nearly all the nobility of England, and even by friends of the other varieties. Apart from this, they do not possess a high rank in the estimation of farmers; yet, it has been made evident that they are susceptible of improvement.

Why is it that these English breeds of cattle deserve such particular notice?

Because the breeding of cattle has for centuries been a principal department of English husbandry, and, owing to a variety of causes, has been prosecuted with unparalleled success. In no other country have such results been obtained. The history of these improvements is important to American farmers, for the reason that our stock has been derived from England, and that from England must we make our future importations, whenever they become necessary.

At what period was the domesticated ox brought to this country?

At the time of its settlement. The greatest number of animals was brought by the English colonists, and they have been most extensively diffused. In certain sections, we can perceive traces of the stock brought by the French, Dutch, and Spanish emigrants.

Have these animals been bred with care?

They have not, except in a few instances. Little or no effort has been made for perpetuating valuable points, or the peculiar characteristics of a breed. Within the past few years, however, a great change has taken place; and, partly in consequence of the formation of Agricultural Societies, an increased interest has been manifest in the improvement of horned stock.

In what part of our country do we find the best native stock?

Probably in New England. The cattle here are supposed to have been brought from Devonshire, England, as the early settlers were mostly emigrants from that district. We can thus account for the remarkable similarity in the best native animals to the Devons of the present day.

Their general characters are alike; having much the same features, the same ready disposition to acquire flesh, the same excellence for labor, and the same value for the dairy. Without doubt, the mixture of some good English Devons with the best of this native stock, would be attended with the happiest results.

Have the Short-horns been introduced into the United States?

They are said to have laid the foundation of the fine stock, known as the *milk*, or *beef*, breed, which existed in Virginia about the close of the last century. The first authentic importations, however, were made by Mr. Heaton of New-York state, in the years 1791 and 1796. Many subsequent importations have been made from time to time, and the Short-horns appear to be more extensively diffused than any other breed.

Are the Durhams much esteemed in the United States?

They are in certain localities, as in the fertile regions of the west. It is generally well understood, that they cannot flourish on poor lands. Thus far they have been mostly used for breeding, and only a small number fattened for the shambles, yet their merits in this respect are fully appreciated. At one time, they sold for extraordinary prices—sometimes for a thousand dollars a piece,—and now they have not recovered from the reduction in value, which was occasioned by the commercial revulsion that passed over the country a few years since.

What other breeds have been introduced into the United States?

The Devon, the Hereford, the Ayrshire, the Alderney, together with many others of less note. In some instances these importations have been made by societies, but as a general thing, by the enterprise of a few spirited individuals.

Have these importations been of much benefit?

Occasionally we come across herds where the blood has been preserved pure and uncontaminated by mixture with animals of different breeds, but the greatest good has been effected by crosses upon our native stock. The grade animals thus produced have become very popular among farmers, and are evidently considered as possessing superior qualifications. In the course of a few years, when the number of high-blooded cattle is much increased, the improvement of the native breeds will be more decided.

Why have cattle in so many instances deteriorated in value?

It appears to be a law of nature, that progress, improvement, or advancement in excellence, is the result of care and labor, and not of neglect. Hence, in addition to necessary tillage, vegetables flourish best, if seeds are changed every few years; so also, animals must change their locality or be crossed, every now and then, to maintain their vigor and full development. If this is neglected, they will usually degenerate in character.

What other cause has led to a deterioration in American cattle?

In numerous cases farmers have raised their poorest, and not their best calves. The latter would be sold for the shambles, because they would command a high price; and, for a contrary reason, or because not fit for the shambles, or would only bring a low price, the former would be kept alive to perpetuate the stock. As absurd as this is, it has often been done. The best calves and the best only should be raised. Inferior ones should not be suffered to remain on a farm any more than the seeds of pernicious weeds should be sown in company with wheat.

Why should so much interest be cherished in regard to the improvement of the breeds of cattle?

Because no small portion of the farmer's available property is in them. He depends on them and his crops for his ready cash. Hence, if his cows eat more than the milk is worth; if his oxen eat more than their labor is worth; and if both eat more than their flesh is worth for slaughter, it is evident they impoverish rather than enrich him. They are like shares in a bank or railroad which pays no dividend. But, if his cows give milk annually double in value of their feed; if the labor of his oxen is annually worth double of their feed; and if the flesh of both is worth double in the shambles of what it cost to raise and fatten them, it is clear that he is enriched by his cattle.

For what different purposes is the ox family valuable?

The male is used for draught, and, in some countries as a beast of burden. The female supplies us with milk, that affords cream, butter, and cheese. Their flesh is a favorite article of food; their tallow is made into candles and soap; their skins are tanned for leather; their hair enters into the composition of mortar; their horns are manufactured into various ornamental articles; the bones make cheap substitutes for ivory, and are used as a manure; their blood is employed in the manufacture of Prussian blue; and the refuse of the skin, hoofs and shanks, is made into glue.

How many horned cattle may there be in the United States?

Although a census is made once in ten years, it is impossible to tell, for the number is subject to constant variation. With the increase of our population the number necessarily increases too. It is fair to conclude that there are three cows, two working oxen, and five young or beef cattle to each family. Then, if we have at the present time three millions of families, there will be in all thirty millions of horned cattle, which, at twenty dollars per head, will make this department of our national wealth amount to six hundred millions of dollars—an amount showing its importance and the occasion for improving our breeds as well as our general culture in rearing them.

What is said of the buffalo?

Although by naturalists, from some general resemblance, assigned to the ox family, it is evident that the alliance is more fanciful than real. For if the two species are kept under the same roof and fed on the same meadows, they manifest no sympathy for each other and refuse to couple together. And such is the antipathy between them, that the cows of neither will suckle the calves of the other. Besides, the period of gestation with the buffalo is twelve months, whereas with our cow but nine. The buffalo is a native of India, but is found in various hot countries.

In what consists the value of the buffalo?

In hot countries almost all the cheese is made from the buffalo's milk, which is more abundant than from our cow, although not so good. As these animals are in general much larger and stronger than our oxen, when domesticated, they are very serviceable in the plough; they draw well, but do not carry burdens; they are led by a ring drawn through their nose. Two buffalos, harnessed, or rather chained, to a wagon, will draw as much as four strong horses. Their hides are valuable, and their tongues are saved to be eaten, but the other flesh is not. In Africa and India great numbers of them run wild.

What is said of the bison?

The bison, or the American bison, usually called the

American buffalo, wanders constantly from place to place for food. He is particularly fond of the tender grass on our western prairies, which springs up after a fire has been over them. The bulls and cows live in separate herds, for the greater part of the year, but at all seasons, one or two old bulls generally accompany a herd of cows. The flesh of the bison, in good condition, is very juicy and well flavored; much resembling that of well-fed beef. And the tongue is considered a delicacy, and may be cured so as to surpass in flavor the tongue of an English cow.

What is said of the form and size of the bison?

The form is better perceived from a picture of the animal than any verbal description. The head, shoulders, and upper part of the anterior extremities, are covered with a long, brownish, woolly hair. The tail is tufted and black. The horns are black, and turned laterally upwards. His length is about eight feet.

CHANGE OF MATERIAL SUBSTANCES.

It will be seen that a vast series of changes of position and form termed *physical;* of composition termed *chemical;* and if in connection with vegetable or animal life, termed *vital*, is incessantly going on; all concurring and harmonizing with one another, to bring about the great result—the final end of creation—the fitting of the earth to yield subsistence to man in return for labor; allowing the agricultural laborer, like the strong man, "to rejoice in his strength;" the intellectual farmer to find his happiness in the search after truth, and impress upon all a deep conviction of the wise and benevolent arrangement of creation.

Neither the written word of God, or the material universe, his created revelation, affords any ground for the belief that there ever has been any *second creation* of matter, or that even a single particle of new matter has ever been called into existence since the beginning, when all was made. Of this matter when created, it is equally certain that not an atom has been lost or destroyed. What we see going on around us is change, incessant change; change of place or combination, and change only.

Besides the myriads of worlds in the firmament, visible to

the naked eye, the telescope brings into view numberless others, all moving on their axes and in their orbits, with a rapidity that defies conception, and in perfect harmony. The loss of a single atom would destroy the harmony of their movements. On the other hand, the microscope reveals movements in small masses of matter equally astonishing. The transparent web of a frog's foot, small enough to be covered by the cut end of a fine hair, exhibits in its circulating particles a degree of intestine movement which may be compared to that within the Exchange room in Wall-street at its busiest hour, to that of a legislative hall at the time of dividing the house, or to a bee-hive preparatory to swarming. These are physical movements; changes of *place* merely.

Changes in the *composition* of the bodies called chemical changes, are going on, alike wonderful both in manner and degree. Limestone, for instance, is quarried to be burned, to build houses and to enrich fields. In this process an air is separated which is food for plants. Besides this source of carbonic acid, for so this air is called, millions of tons of coal are raised from mines, and being burned for steam engines, for culinary purposes, and our social hearths, and also for artificial light; are thus fitted to become a component part of plants and next of animals. Large masses of wood and other vegetable matters are rotting on the surface of the earth. Animal substances are also undergoing decompositon, and rocks and earths are being dissolved. Thus carbonic acid, the wood gas, and ammonia, or the animal gas, as it may be termed, and the mineral earths, are supplied to become again parts of plants and animals, and each kingdom of nature gets back what it has given up. A tree is burned; its particles are either dissipated in the form of air, or remain behind as ashes. The ashes return to the earth from which they were extracted; the rest of the wood has been dissolved into gases, and both are fitted to become the food of future forests. The grazing animal while yet alive, supplies food to the herbage he feeds on; when he perishes, plants and animals alike feed upon his body.

When Major Andre was disinterred, nearly half a century after his execution and burial, " the roots of a peach tree," so says the official account of the British Consul, " were found entwined among the bones of his skull." Of those here present, there may be some who have eaten of the fruit of that tree, and in whose bodies have thus become incorporated particles of matter that once formed a portion of the living frame of that gallant and accomplished soldier. Thousands of tons of bones from the battle-fields of Europe, have been used to enrich the soil of England; and in our own country, many hundred Germans are employed in gathering bones for our corn lands.

The cast-off bark and leaves of trees become decomposed, and afford nourishment to their roots. Every part of the animal body is in a state of perpetual change. The organic and mineral excretions an individual casts off, even his breath, may return to him no less than to others, and again become a part of his living body. The same matter forms the bodies of successive, and even of the same generations and individuals of the animal and vegetable kingdoms. Were it otherwise, the existence here of all living things would be limited. The materials necessary to their support would ultimately be all consumed.

The results of *vital chemistry* are yet more wonderful. We inspire vital air, one source of which is the leaves of plants under the influence of the sun; we expire another air, carbonic acid, which is their food. Animals are kept warm by a fire ever burning within them. The fat of their bodies is burned by the air inhaled into their lungs. The product of this combustion is the same as that of fat burned in a candle, viz., carbonic acid and water. If a man is struck down by a blow, or a poison, or an apoplexy, so that he hardly breathes, his body becomes cold; the fire is almost extinguished. If he exercises or contracts a fever, he breathes more rapidly than usual, he becomes heated; the internal fire is increased.

If an animal is kept cold, it gets poor; its fat is consumed to keep it warm. You all know, gentlemen, the advantages of warm stables for your stock, in winter. Perhaps some of you may not know the reason why large lunged animals are fattened with difficulty. It is that they burn up their fat. Yet they are capable, with proper feeding, of becoming even fatter than others, because they keep their health and good digestion longer.

All living things are compounds of four elements. FIRST, *oxygen*, or vital air, a part of the atmosphere which supports combustion, rots wood, rusts metals, and turns cider sour

SECOND, *hydrogen*, which with some of its compounds, with carbon, is a gas, and which burns with a flame, explodes in mines, forms burning springs, and with oxygen becomes water. THIRD, *nitrogen*, a part of the atmosphere, and a principal constituent of manure, and with hydrogen forming ammonia, otherwise called hartshorn. This is the product of animal decomposition, whether by burning or putrefaction, and it holds a relation to animal decomposition, not unlike that which carbonic acid does to vegetable decomposition. FOURTH, *carbon*, the residuum of wood where the hydrogen gas and its compounds have been burned away. Carbonic acid, carbon and oxygen, is the gas which gives briskness to cider and to water, kills those who descend into wells in limestone localities, who sleep in close rooms where charcoal fire is burning.

Mineral springs and hundreds of volcanoes, some of them apparently extinct, and all breathing animals, are constantly casting forth carbonic acid gas. It enters largely into the composition of plants, and is discharged from them in the process of fermentation, of rotting, and of combustion. In fine, it is wood dissolved in air. But there are some parts of wood and some parts of animals, which will not dissolve in air. There are what are termed *mineral* portions, in contradistinction to the former, which are called the *organic* portions. A log of wood is burned, some of it is dissipated in air, chiefly into carbonic acid gas; a smaller portion, the ash, remains, differing somewhat in its nature according to the species of wood or vegetable consumed. When an animal substance is burned, or left to decay, the organic portion in like manner is dissipated in the air, in the form of hartshorn, together with hydrogen, holding in solution sulphur and phosphorus. The other mineral or earthy parts remain. These are the bones, which are formed chiefly of lime and phosphoric acid, the same which we smell after burning a loco-foco match.

The composition of plants and animals is the food they receive, deducting that which they cast off. No plant or animal creates any element that enter into its structures. It only combines them into new forms. There is, however this great difference between plants and animals: plants collect unorganized atoms, matter not arranged in definite regular forms, and build up from them their beautiful structures. Animals, on the contrary, receive their food from matter already organized; the herbivorous animals directly, the carnivorous through the medium of the herbivorous.

The sea furnishes fish and guano, and thus renders back some of the nutriment washed into it, in rain water falling upon it. What immense quantities must be washed into the sea by the Nile, the Ganges, the Mississippi, and the great rivers of South America, not of vegetable matters only, but of animal and mineral. These are the food of the animal and vegetable inhabitants of the sea. Some of these matters come back. In a certain sense, fishing boats and whale ships are manure boats, no less truly than those which bring guano. Coal mines, volcanoes, and limestone quarries, are immense manure beds. Thus the sea and the earth give back what they have received.

The changes of matter are three-fold: first, from earth and air to the condition of plants; next, to the animal creation; and lastly, back again to earth and air. In the various processes by which these materials are rendered available, there is a striking analogy. The rotting of manure in the ground may be described as a low, smothered, imperfect burning, much of the smoke or coal remaining behind. The smothered fire of a charcoal bed leaves still more of the coal unconsumed. In the burning of peat in smothered fires, some ash is formed, and the quantity of coal-like residuum is still less. When oil is burned, or rapidly rotted (if I may so say,) in a clear burning lamp, no soot, smoke or lampblack, (which is powdered carbon,) is left; all is consumed—that is—all is converted into carbonic acid gas and water.

The blacksmith sprinkles water on his coal fire, and stirs it up to make the coal rot faster. So, too, the farmer stirs up the earth to promote decomposition. The ship carpenter keeps his timber under water until he wants to use it, to prevent it from decay by securing it from the air, or he puts it under cover to keep it dry. The presence of both air and water is necessary for the rotting or slow combustion. When the old London bridge was being constructed, the remains of a former wooden structure, built by the ancient Romans, were found unrotted. The farmer wets his manure heap, and turns it over to make it rot; if it rots too fast, he puts on more water to moderate the combustion, and prevent the carbonic acid gas, ammonia, sulphur, and phosphorus from escaping. Deeply buried manure rots slowly, so do compact bodies of manure. The more slowly vegetable substances are decomposed, the greater the carbonaceous, coaly, or sooty residuum.

In preparing manure, heating and rapid decomposition should be carefully avoided, for by it is lost not only carbonic acid, the wood-forming gas; but also ammonia, sulphur, and phosphorus, essential elements of animals and vegetables. The aim of the agriculturist in the application of manure, should be, to make the period of its rotting coincide with the period when the crop wants the nutriment it furnishes in the act of rotting. Thus half rotted manure is preferred for potatoes, its greatest action being wanted for the tubers and not for the haulm.

Our social fires, whether of wood or coal, manure the atmosphere; but it is wasteful to keep burning when no one is present to enjoy them. It is equally wasteful to let manure rot in the ground, or be washed away, when there is no crop to feed upon its gases or its salts. As manure begins to decompose the first moment it is dropped from the animal, so of course it immediately begins to lose some of its valuable qualities. It might be supposed, therefore, and Sir H. Davy inclined to the opinion that it should be used immediately, either by putting under the earth, or on the surface. But general experience is rather against the practice, at any rate the rule is not absolute. It is a question to be settled by careful observation, and not by scientific reasoning.

As the subject has been discussed theoretically, it may be proper to remark, that dung is valuable as manure: first, from the mineral ingredients contained in it; second, from the ammoniacal compounds. Now these are mostly formed in the manure by fermentation, and so far as respects them, the droppings of animals are no more manure, than an apple is cider, or barley is beer. Certain chemical changes must first take place before the organic portion of the droppings is available. The case seems to stand thus:—If a farmer keep his manure, ammonia is escaping, and he suffers a loss; the rain is washing away the mineral salts from it. If he put it on the surface of his ground fresh, it dries, and the chemical change on which the formation of ammonia depends, is partly prevented, and of the ammonia that is formed, much escapes in atmosphere, instead of combining with the soil

If he ploughs it in fresh, it may be so separated that fermentation does not take place, for want of the requisite warmth, before the rain washes away the urea or principle upon which the formation of ammonia depends.

The mineral food, or ash of plants, comes from the earth, rocks and stones; which are only solid earth, as earth is only crumbled, crushed or dissolved rocks and stones, and all are rusty metals, chiefly distinguished from metals properly so called, in the facility with which they rust. They are slightly soluble in rain water, which combining with carbonic acid and ammonia, in its descent first purifies the atmosphere, and then enriches the soil. Who among us has not been struck with the grateful purity of the air after a fall of rain or snow. This is another striking illustration of the ordering of all things for our well being. The presence of organic matter undergoing decomposition, constitutes the difference between soil and earth. Earth, when first brought to the surface, is barren, incapable of supporting vegetation. A large portion of the globe was originally, or in remote periods, in this condition.

In removing rocks from his fields, the farmer often brings to the surface masses of earth. These are barren, and remain so for years, unless manure is applied, and that, too, liberally. Such earth is said to be hungry; hungry it is, for the organic food of plants; for left to itself, its first act, if I may so express it, is to acquire carbonic acid from rain water, and ammonia from the same source, and from the atmosphere. These are the organic food of plants. It combines also with fœtid gases. These gases often contain phosphorus and sulphur, which are mineral food. Thus, if old clothes are fœtid with ammonia, sulphur, and phosphorus, are buried in the earth, they are sweetened. The earth combines with these substances. So, too, if foul water from a wash tub is passed through earth, (and still more through charcoal,) it is purified, and the earth is enriched. Here is now another kind provision for the purification of the air we breathe, and the water we drink.

As many animals live entirely upon plants, the materials of flesh and bone and whatever enters into the composition of their bodies, must be contained in such food. How do plants get the materials by which they form flesh, bones, and fat? When an animal dies, and is left on the ground, if not consumed by other animals, and the air be not (as in some parts of South America) so dry as to convert it into a mummy, the soft parts of the body rise in the form of certain offensive gases, consisting of hartshorn with phosphorus and sulphur dissolved in hydrogen. It is this combination which gives to stale eggs and decaying teeth, their intolerable fœtor, and which blackens silver. All these matters are capable of, and actually do combine with water, and of course come back to the earth in rain water. The bones remain, and after a great length of time crumble into dust. This is in fact, animal ashes, the residuum after slow combustion, and like the ashes of vegetables, came originally from the earth.

The property in earth to combine with gases, is exhibited in the fact that animals buried therein are no longer offensive to the senses. The gases evolved from them (and into which, except the mineral part, they are resolved,) are not imprisoned, but combined with the earth. But the earth may become saturated, and then the gases escape, some portions go down with the water, rendering springs foul, and others rise and mixing with common air, render it unhealthy, unless plants are present to appropriate them. Hence the propriety of having grass or trees in grave-yards. A tree and a fountain are also appropriately placed in connection, the fountain nourishes the tree, and the tree purifies the fountain.

During a season of unusual health, a valued connection of mine lost his life by a fever, contracted in overlooking a swampy field, long cleared, but then for the first time ploughed up. The coarse grasses had before been sufficient to combine with the gases which the mass of vegetable matter covered by it, and protected from the sun and rain, had afforded. But under the double influence of the destruction of the herbage, and the increased supply of gases consequent to tillage, the stirring up of the slow fire and closing the chimney flue, the noxious gases or malaria, as they are termed, being diffused in the atmosphere and respired, a fatal fever followed.

These gases, when they rise from the earth, combine with water in whatever form they meet with it. Hence the proverbial fatality of inhaling the moist air of the night in unhealthy localities, especially at the period of the first dew, for the heat of the sun by increasing the exhalation of the gases from the surface, renders that more highly charged. If

FARM HOUSE, copied from L. F. ALLEN's valuable work on RURAL ARCHITECTURE, page 103.

we could suppose plants to be imitable like animals when they are taking their repast, one would imagine they would warn us to keep out of their way, to stay within doors, not to interrupt or rob them of any portion of their supper, after they had borne the heat and burden of the day, and were thirsting for the rich dew of the evening.—*From an Address before the State Agricultural Society, members of the Legislature, and of the Medical Society of the State of New-York, at the Capitol in Albany*, 1848, *by* ALEX. H. STEVENS, M. D.

RURAL LIFE IN ENGLAND.

"Oh! friendly to the best pursuits of man,
Friendly to thought, to virtue, and to peace,
Domestic life in rural pleasures past!"—COWPER.

The stranger who would form a correct opinion of the English character must not confine his observations to the metropolis. He must go forth into the country; he must sojourn in villages and hamlets; he must visit castles, villas, farm-houses, cottages; he must wander through parks and gardens, along hedges and green lanes; he must loiter about country churches; attend wakes and fairs, and other rural festivals; and cope with the people in all their conditions, and all their habits and humors.

In some countries the large cities absorb the wealth and fashion of the nation; they are the only fixed abodes of elegant and intelligent society, and the country is inhabited almost entirely by boorish peasantry. In England, on the contrary, the metropolis is a mere gathering-place, or general rendezvous, of the polite classes, where they devote a small portion of the year to a hurry of gaiety and dissipation, and, having indulged this kind of carnival, return again to the apparently more congenial habits of rural life. The various orders of society are therefore diffused over the whole surface of the kingdom, and the most retired neighborhoods afford specimens of the different ranks.

The English, in fact, are strongly gifted with the rural feeling. They possess a quick sensibility to the beauties of nature, and a keen relish for the pleasures and employments of the country. This passion seems inherent in them. Even the inhabitants of cities, born and brought up among brick walls and bustling streets, enter with facility into rural habits, and evince a tact for rural occupation. The merchant has his snug retreat in the vicinity of the metropolis, where he often displays as much pride and zeal in the cultivation of his flower-garden, and the maturing of his fruits, as he does in the conduct of his business, and the success of a commercial enterprise. Even those less fortunate individuals, who are doomed to pass their lives in the midst of din and traffic, contrive to have something that shall remind them of the green aspect of nature. In the most dark and dingy quarters of the city, the drawing-room window resembles frequently a bank of flowers; every spot capable of vegetation has its grass-plat and flower-bed; and every square its mimic park, laid out with picturesque taste, and gleaming with refreshing verdure.

Those who see the Englishman only in town are apt to form an unfavorable opinion of his social character. He is either absorbed in business, or distracted by the thousand engagements that dissipate time, thought and feeling, in this huge metropolis. He has, therefore, too commonly a look of hurry and abstraction. Wherever he happens to be, he is on the point of getting somewhere else; at the moment he is talking on one subject, his mind is wandering to another; and while paying a friendly visit, he is calculating how he shall economize time so as to pay the other visits allotted in the morning. An immense metropolis, like London, is calculated to make men selfish and uninteresting. In their casual and transient meetings, they can but deal briefly in commonplaces. They present but the cold superficies of character—its rich and genial qualities have no time to be warmed into a flow.

It is in the country that the Englishman gives scope to his natural feelings. He breaks loose gladly from the cold formalities and negative civilities of town; throws off his habits of shy reserve, and becomes joyous and free-hearted. He manages to collect round him all the conveniences and elegancies of polite life, and to banish its restraints. His country-seat abounds with every requisite, either for studious retirement, tasteful gratification, or rural exercise. Books, paintings, music, horses, dogs, and sporting implements of all kinds are at hand. He puts no constraint either upon his guests or himself, but in the true spirit of hospitality provides the means of enjoyment, and leaves every one to partake according to his inclination.

The taste of the English in the cultivation of land, and in what is called landscape gardening, is unrivaled. They have studied nature intently, and discover an exquisite sense of her beautiful forms and harmonious combinations. Those charms, which in other countries she lavishes in wild solitudes, are here assembled round the haunts of domestic life. They seem to have caught her coy and furtive graces, and spread them, like witchery, about their rural abodes.

Nothing can be more imposing than the magnificence of English park scenery. Vast lawns that extend like sheets of vivid green, with here and there clumps of gigantic trees, heaping up rich piles of foliage: the solemn pomp of groves and woodland glades, with the deer trooping in silent herds across them; the hare bounding away to the covert; or the pheasant, suddenly bursting upon the wing; the brook, taught to wind in natural meanderings, or expand into a glassy lake: the sequestered pool, reflecting the quivering trees, with the yellow leaf sleeping on its bosom, and the trout roaming fearlessly about its limpid waters; while some rustic temple or sylvan statue, grown green and dank with age, gives an air of classic sanctity to the seclusion.

These are but a few of the features of park scenery; but what most delights me, is the creative talent with which the English decorate the unostentatious abodes of middle life. The rudest habitation, the most unpromising and scanty portion of land, in the hands of an Englishman of taste, becomes a little paradise. With a nicely discriminating eye, he seizes at once upon its capabilities, and pictures in his mind the future landscape. The sterile spot grows into loveliness under his hand; and yet the operations of art which produce the effect are scarcely to be perceived. The cherishing and training of some tree; the cautious pruning of others; the nice distribution of flowers and plants of tender and graceful foliage; the introduction of a green slope of velvet turf; the partial opening to a peep of blue distance, or silver gleam of water: all these are managed with a delicate tact, a pervading yet quiet assiduity, like the magic touchings with which a painter finishes up a favorite picture.

The residence of people of fortune and refinement in the country has diffused a degree of taste and elegance in rural economy, that descends to the lowest class. The very laborer, with his thatched cottage and narrow slip of ground, attends to their embellishment. The trim hedge, the grass-plot before the door, the little flower-bed bordered with snug box, the woodbine trained against the wall, and hanging its blossoms about the lattice, the pot of flowers in the window, the holly, providentially planted about the house, to cheat winter of its dreariness, and to throw in a semblance of green summer to cheer the fireside : all these bespeak the influence of taste, flowing down from high sources, and pervading the lowest levels of the public mind. If ever Love, as poets sing, delights to visit a cottage, it must be the cottage of an English peasant.

The fondness for rural life among the higher classes of the English has had a great and salutary effect upon the national character. I do not know a finer race of men than the English gentlemen. Instead of the softness and effeminacy which characterize the men of rank in most countries, they exhibit a union of elegance and strength, a robustness of frame and freshness of complexion, which I am inclined to attribute to their living so much in the open air, and pursuing so eagerly the invigorating recreations of the country. These hardy exercises produce also a healthful tone of mind and spirits, and a manliness and simplicity of manners, which even the follies and dissipations of the town cannot easily pervert, and can never entirely destroy. In the country, too, the different orders of society seem to approach more freely, to be more disposed to blend and operate favorably upon each other. The distinctions between them do not appear to be so marked and impassable as in the cities. The manner in which property has been distributed into small estates and farms has established a regular gradation from the nobleman, through the classes of gentry, small landed proprietors, and substantial farmers, down to the laboring peasantry; and while it has thus banded the extremes of society together, has infused into each intermediate rank a spirit of independence. This, it must be confessed, is not so universally the case at present as it was formerly; the larger estates having, in late years of distress, absorbed the smaller, and, in some parts of the country, almost annihilated the sturdy race of small farmers. These, however, I believe, are but casual breaks in the general system I have mentioned.

In rural occupation there is nothing mean and debasing. It leads a man forth among scenes of natural grandeur and beauty; it leaves him to the workings of his own mind, operated upon by the surest and most elevating of external influences. Such a man may be simple and rough, but he cannot be vulgar. The man of refinement, therefore, finds nothing revolting in an intercourse with the lower orders in rural life, as he does when he casually mingles with the lower orders of cities. He lays aside his distance and reserve, and is glad to waive the distinctions of rank, and to enter into the honest, heartfelt enjoyments of common life. Indeed the very amusements of the country bring men more and more together; and the sound of hound and horn blend all feelings into harmony. I believe this is one great reason why the nobility and gentry are more popular among the inferior orders in England than they are in any other country; and why the latter have endured so many excessive pressures and extremities, without repining more generally at the unequal distribution of fortune and privilege.

To this mingling of cultivated and rustic society may also be attributed the rural feeling that runs through British literature; the frequent use of illustrations from rural life; those incomparable descriptions of nature that abound in the British poets, that have continued down from "the Flower and the Leaf" of Chaucer, and have brought into our closets all the freshness and fragrance of the dewy landscape. The pastoral writers of other countries appear as if they had paid nature an occasional visit, and become acquainted with her general charms; but the British poets have lived and resided with her, they have wooed her in her most secret haunts, they have watched her minutest caprices. A spray could not tremble in the breeze—a leaf could not rustle to the ground—a diamond drop could not patter in the stream—a fragrance could not exhale from the humble violet—nor a daisy unfold its crimson tints to the morning, but it has been noticed by these impassioned and delicate observers, and wrought up into some beautiful morality.

The effect of this devotion of elegant minds to rural occupations has been wonderful on the face of the country. A great part of the island is rather level, and would be monotonous, were it not for the charms of culture; but it is studded and gemmed, as it were, with castles and palaces, and embroidered with parks and gardens It does not abound in grand and sublime prospects, but rather in little home scenes of rural repose and sheltered quiet. Every antique farm-house and moss-grown cottage is a picture: and as the roads are continually winding, and the view is shut in by groves and hedges, the eye is delighted by a continual succession of small landscapes of captivating loveliness.

The great charm, however, of English scenery, is the moral feeling that seems to pervade it. It is associated in the mind with ideas of order, of quiet, of sober well-established principles, of hoary usage and reverend custom. Everything seems to be the growth of ages of regular and peaceful existence. The old church of remote architecture, with its low, massive portal; its gothic tower; its windows rich with tracery and painted glass, in scrupulous preservation; its stately monuments of warriors and worthies of the olden time, ancestors of the present lords of the soil; its tombstones, recording successive generations of sturdy yeomanry, whose progeny still plough the same fields, and kneel at the same altar—the parsonage, a quaint irregular pile, partly antiquated, but repaired and altered in the tastes of various ages and occupants—the stile and footpath leading from the church-yard, across pleasant fields, and along shady hedgerows, according to an immemorial right of way—the neighboring village, with its venerable cottages, its public green, sheltered by trees, under which the forefathers of the present race have sported—the antique family mansion, standing apart in some little rural domain, but looking down with a protecting air on the surrounding scene: all these common features of English landscape evince a calm and settled security, and hereditary transmission of homebred virtues and local attachments, that speak deeply and touchingly for the moral character of the nation.

It is a pleasing sight of a Sunday morning, when the bell is sending its sober melody across the quiet fields, to behold the peasantry in their best finery, with ruddy faces and modest cheerfulness, thronging tranquilly along the green lanes to church; but it is still more pleasing to see them in the evenings, gathering about their cottage doors, and appearing to exult in the humble comforts and embellishments which their own hands have spread around them.

It is this sweet home-feeling, this settled repose of affection in the domestic scene, that is, after all, the parent of the steadiest virtues and purest enjoyments; and I cannot close these desultory remarks better, than by quoting the words of a modern English poet, who has depicted it with remarkable felicity:—

Through each gradation, from the castled hall,
The city dome, the villa crown'd with shade,
But chief from modest mansions numberless,
In town or hamlet, shelt'ring middle life,
Down to the cottaged vale, and straw-roof'd shed;
This western isle hath long been famed for scenes
Where bliss domestic finds a dwelling-place;
Domestic bliss, that, like a harmless dove,
(Honor and sweet endearment keeping guard,)
Can centre in a little quiet nest
All that desire would fly for through the earth;
That can, the world eluding, be itself
A world enjoy'd; that wants no witnesses
But its own sharers, and approving heaven;
That, like a flower deep hid in rocky cleft,
Smiles, though 'tis looking only at the sky.

WASHINGTON IRVING.

THE DEER'S INSTINCT.

A large deer was running at full speed, closely pursued by a panther. The chase had already been a long one, for, as they came nearer, I could perceive both their long parched tongues hanging out of their mouths, and their bounding, though powerful, was no longer so elastic as usual. The deer having discovered in the distance a large black bear, playing with her cubs, stopped a moment to snuff the air; then coming nearer he made a bound, with his head extended, to ascertain if Bruin kept her position. As the panther was closing with him, the deer wheeled sharp around, and turning back almost upon his own trail, passed within thirty yards of his pursuer, who, not being able at once to stop his career, gave an angry growl, and followed the deer again, but at the distance of some hundred yards. Hearing the growl, the bear drew her body half out of the bushes, remaining quietly on the look out. Soon the deer again appeared, but his speed was much reduced—and as he approached towards the spot where the bear lay concealed, it was evident that the animal was calculating the distance with admirable precision.

The panther, now expecting easily to seize his prey, followed about thirty yards behind, his eyes so intently fixed on the deer, that he did not see Bruin at all. Not so the bear. She was aware of the close vicinity of her wicked enemy, and she cleared the briars, and squared herself for action, when the deer, with a beautiful and powerful spring, passed clean over the bear's head and disappeared. At the moment he took the leap, the panther was just balancing himself for a spring, when he perceived, to his astonishment, that he was faced by a formidable adversary. Not the least disposed to fly, he crouched, lashing his flanks with his long tail, while the bear about five yards from him, remained like a statue, looking at the panther with her fierce, glaring eyes.

For a minute they remained thus—the panther's sides heaving with exertion, agitated, and apparently undecided—the bear perfectly calm and motionless. Gradually the panther crawled backwards till at the right distance for a spring, when, throwing all his weight upon his hind parts, to increase his power, he darted upon the bear like lightning, and forced his claws into her back. The bear, with irresistible force, seized the panther with her two fore paws, pressing him with the weight of her body, and rolling over it. I heard a heavy grunt, a plaintive howl, a crashing of bones, and the panther was dead. The cub of the bear came to ascertain what was going on, and after a few minutes' examination of the victim, it strutted down the slope of the hill, followed by its mother, who was apparently unhurt. I did not attempt to prevent their retreat, for among real hunters in the wilds there is a feeling which restrains them from attacking an animal which has just undergone a deadly strife.

This is a common practice of the deer, when chased by the panther—that of leading him to the haunt of the bear; I have often witnessed it, although I never knew the deer to return as in this instance.—*From* The Reformer.

TO THE WILD BROOK.

Unheeded emblem of the mind!
  When weeping twilight's shadows close,
I wander where thy mazes wind,
  And watch thy current as it flows:
Now dimpling, silent, calm, and even;
Now brawling, as in anger driven;
Now ruffled, foaming, madly wild;
Like the vex'd sense of Sorrow's hopeless child!

Beside thy surface now I see,
  Reflected in thy placid breast,
Hush'd summer's painted progeny
  In smiles and sweets redundant drest;
They flaunt their forms of varying dye,
To greet thee as thou passest by;
And bending up thy ample wave,
They in its lucid lap their bosoms lave.

While on thy tranquil breast appears
  No fretting gale, no passing storm,
The sun-beam's vivid lustre cheers,
  And seems thy silvery bed to warm:
The thronging birds with am'rous play,
Sweep with their wings thy glittering way;
And o'er thy banks fond zephyr blows,
To dress with sweets the smallest flower that grows.

But when destroying blasts arise,
  And clouds o'ershade thy withering bounds,
When swift the eddying foliage flies,
  And loud the ruthless torrent sounds,
Thy dimpling charms are seen no more,
Thy minstrel's caroll'd praise is o'er;
While not a flow'ret, sunny drest,
Courts the chill current of thy alter'd breast.

Such is the human mind! Serene
  When Fortune's gloomy hour appears!
And lovely, as thy margin green,
  Are buds of Hope, which Fancy rears:

Then adulation, like the flow'r,
  Bends as it greets us on our way;
But, in the dark and stormy hour,
  Leaves us unmark'd, to trace our troubled way.

Mrs. Robinson.

THE WINTER WALK AT NOON.

  Here, unmolested, through whatever sign
The sun proceeds, I wander. Neither mist,
Nor freezing sky nor sultry, checking me,
Nor stranger intermeddling with my joy.
Ev'n in the spring and play-time of the year,
That calls th' unwonted villager abroad
With all her little ones, a sportive train,
To gather king-cups in the yellow mead,
And prink their hair with daisies, or to pick
A cheap and wholesome salad from the brook:
These shades are all my own. The tim'rous hare,
Grown so familiar with her frequent guest,
Scarce shuns me: and the stock-dove, unalarm'd,
Sits cooing in the pine-tree, nor suspends
His long love-ditty for my near approach.
Drawn from his refuge in some lonely elm
That age or injury has hollow'd deep,
Where, on his bed of wool and matted leaves,
He has outslept the winter, ventures forth
To frisk awhile, and bask in the warm sun,
The squirrel, flippant, pert, and full of play:
He sees me, and at once, swift as a bird,
Ascends the neighb'ring beech; there whisks his brush,
And perks his ears, and stamps and scolds aloud,
With all the prettiness of feign'd alarm,
And anger insignificantly fierce.
  The heart is hard in Nature, and unfit
For human fellowship, as being void
Of sympathy, and therefore dead alike
To love and friendship both, that is not pleas'd
With sight of animals enjoying life,
Nor feels their happiness augment his own.
The bounding fawn that darts across the glade

When none pursues, through mere delight of heart,
And spirits buoyant with excess of glee;
The horse as wanton, and almost as fleet,
That skims the spacious meadow at full speed,
Then stops and snorts, and throwing high his heels,
Starts to the voluntary race again—
These, and a thousand images of bliss,
With which kind Nature graces every scene,
Where cruel man defeats not her design,
Impart to the benevolent, who wish
All that are capable of pleasure, pleas'd,
A far superior happiness to theirs,
The comfort of a reasonable joy.

COWPER.

MILK, BUTTER, AND CHEESE.

On what account is the cow particularly interesting?

Because of the importance of her products in domestic economy. Her milk is of great commercial value; and it is so necessary as an article of food in every family that we could not well do without it. To the poor it may so far become a substitute for other food, that half the meat and bread will suffice that would be requisite without it. The butter made from it is both a luxury and an ordinary ingredient in the culinary process; and, with many there would be the greatest reluctance in relinquishing the use of cheese.

By what characteristics is milk scientifically known?

It is well known that the milk of most animals is nearly white and opaque. Its specific gravity is about three per cent. heavier than water, having an agreeable sweetish taste, but a somewhat peculiar odor, especially when it is warmed. The milk of the cow is slightly alkaline, when newly taken from the animal, but in a short time, if exposed to the atmosphere, this property becomes imperceptible.

Into what substances may milk be converted?

When left at rest a few hours, it separates into two portions, the lighter part rising to the surface and is called cream. If the entire milk, or the cream alone, be put into a churn and successively agitated, the temperature of the liquid is moderately increased, the liquid becoming sour, and the fatty matter separates from the remainder in the form of butter. If a little acid, such as vinegar, diluted muriatic acid, or rennet, be diffused in the milk, raised to the temperature of one hundred degrees, it immediately coagulates and separates into two parts, a solid and a liquid—the curd and the whey.

How is the quality of milk changed by the age of the cow?

Dairymen think that milk of the best quality is yielded by cows that have had three or four calves. Such cows will continue to give excellent milk till ten or twelve years of age, and have had eight or ten calves. Some will remain as profitable milkers till fifteen years old, provided their teeth do not become impaired.

What effect does the season of the year have on the quality of the milk?

In the spring the milk is not only more abundant, but of finer quality than in other seasons. Moist and temperate climates are favorable to the production of milk in a large quantity. In hot countries and in dry seasons, the quantity is less, but the quality is richer. Cool weather favors the production of cheese, and also of sugar, which is a component of milk; while hot weather increases the oily substance of which butter is to be made.

How is the milk changed by the time of milking?

If the cow is milked only once a day, the milk will yield a seventh part more butter than an equal quantity of that which is obtained by two milkings in a day. When the milk is drawn three times a day, it is more abundant, but less rich. It is also universally remarked, that the morning's milk is of better quality than that obtained in the evening.

How do different portions of the milk drawn at the same milking compare in quality?

That which is drawn off first is thin and poor, and gives little cream. That which is drawn last—sometimes by dairywomen called "strippings"—is rich in quality and yields much cream. Compared with the first milk, the same measure of the last will give at least eight times, and even more, as much cream. The quality of the cream also, and of the milk when skimmed, is much better in the latter than in the earlier drawn portions of the milk.

How is the milk changed by the kind of food given the cow?

It is familiar to every dairy farmer that the taste and color of his milk and cream are materially affected by the plants on which his cows feed, and by the food he gives them in the stall. The taste of the wild onion and of the turnip, when eaten by the cow is often perceptible, both in the milk and in the butter. If madder is given to cows, the milk is tinged with red : if they eat saffron, it becomes yellow. Hence, if it is an object to increase the quantity of butter, feed cows with substances rich in oily matter; and, if the milk is to be made into cheese, let them be fed with substances containing the materials for curd—as peas and beans.

How does the milk of sheep compare with that of the cow?

In appearance they are much alike; but the milk of the ewe is generally more dense, and yields a pale yellow butter, which is soft, and soon becomes rancid. The curd is separated from this milk with greater difficulty than from the milk of the cow.

How does the milk of the goat compare with that of the cow?

It is richer in butter and sugar; is considered to be very wholesome; and is often recommended to invalids. The butter is white and hard, and keeps long fresh. Yet the milk of the goat generally possesses a characteristic unpleasant odor and taste, which is said to be less palpable in animals of white color, or that are destitute of horns.

How does the milk of the ass compare with that of the cow?

The milk of the ass has less of butter or cheese in it, than that of the cow, goat, or ewe; but, in sugar is more rich than that of either of them, ranking next in this respect to the milk of the human species. On account of such peculiarity, it is often recommended to invalids, as a light and easily digested drink. The little butter that can be made from it, is white, and soon becomes rancid; and, from the large amount of sugar in it, the process of fermentation soon begins.

What is there in the form of a cow to show that she will be good for milk?

Mr. Youatt says the milk cow should have a long and thin head, with a brisk but placid eye; should be thin and hollow in the neck, narrow in the breast and point of the shoulder, and altogether light in the fore quarter; but wide in the loins, with little dewlap, and neither too full fleshed along the chine, nor showing in any part an inclination to put on much fat. The udder should especially be large, round, and full, with the milk veins protruding, yet thin skinned, but not hanging loose or tending far behind. The teats should also stand square, all pointing out at equal distances, and of the same size, though neither very large nor thick towards the udder, yet long and tapering towards a point.

How should the size of the cow be regarded in reference to her value for milk?

The smaller breeds of cattle yield, as is to be expected, a smaller daily produce of milk, though from the same weight of food, they occasionally give even a greater volume of milk than the larger breeds. Simply, therefore, for the profit of the milk after deducting the cost of keeping, a large cow is not always to be chosen in preference of one of moderate size. The profits of a cow depend on other characteristics, rather than her size.

How much milk ought a farmer to expect from his cows?

It is impossible to fix on any number of quarts that a cow should yield at any specified time, or for the year; for it will cost fifty per cent. more to keep some cows than others No cow, however, deserves to be kept for milk or dairy purposes, unless she yields milk in the year, double the value, at wholesale prices, or upon the premises, that will pay the cost of feeding her; for instance, if it cost twenty-five dollars to keep a cow for the year, and milk is worth two cents per quart, she ought at least to yield twenty-five hundred quarts, which is only ten quarts per day on an average, for two hundred and fifty days. Prime cows, if well fed, will do far more than this.

What is the early history of butter?

Beekman, in his history of inventions, comes to the conclusion that butter is not of Grecian or Roman origin; but, that the Greeks received it from the Scythians, Thracians, and Phrygians, and the Romans derived it from the people of Germany, and used it as a medicine rather than as a culinary luxury. The ancients appear to have been wholly deficient in the art of giving it consistency. The word *chameah*, translated *butter*, in the English version of the Bible, means some liquid preparation of milk or cream.

In what places is butter mostly used for culinary purposes?

The European countries, in which oil and butter is used, says Malte Brun, may be separated by a line extending along the Pyrenees, the Cevennes, the Alps, and Mount Hæmus. Accordingly in warm countries, the place of butter, for the most part, is supplied by oil. In Italy, Spain, Portugal, and the south of France, butter is to be purchased in the shops of apothecaries. Here the olive-groves supersede the use of butter by that of oil.

How do the Hindoos prepare butter?

They make use of what they call *ghee*, which means butter clarified by boiling. They boil the milk two or three hours, which, when cool, is fermented with curdled milk, left to sour, churned, and when it is sufficiently rancid, is boiled with salt, or betel-leaf and neddle, to improve its taste and color.

How does cream differ from milk?

Milk is a kind of natural emulsion, in which the fatty matter exists in the state of very minute globules, suspended in a solution of casein and sugar. Cream is a similar emulsion, differing from milk chiefly in containing a greater number of oily globules, and a much smaller portion of water. In milk and cream these globules appear to be surrounded with a thin white shell or covering, probably of casein, by which they are prevented from running into one another, and collected into larger oily drops.

By what means is the fatty matter in these globules, of which butter is made, separated from the other substances with which it is united?

When the cream is heated for a length of time, these globules, by their lightness, rise to the surface, press nearer to each other, break through their coverings, and unite in a film of melted fat. In like manner, when milk and cream are strongly agitated by any mechanical means, the temperature is found to rise, the covering of the globules are broken or separated, and the fatty matter unites into small grains, and finally into lumps, which form our ordinary butter.

What is said of the butter obtained by heating the cream?

The cream is to be heated nearly to boiling, and kept some time at that temperature, the butter will then gradually rise to the surface, where it may be collected in the form of oil. On its being cooled, this oil becomes solid. In this way the fatty substance of the milk is procured in a purer state than by churning. It may hence be kept for a long period without salt, and without becoming rancid, but it has neither the agreeable flavor, nor the consistence of churned butter. It is scarcely known in our climate as an article of food

How is butter prepared by the Russians?

They take common churned butter and melt it, and then pour off the transparent liquid which floats upon the surface. This is the only form in which sweet butter is known in many parts of Russia. In warm weather it has the consistence of thick oil, is used instead of oil for many culinary purposes, and is denoted by the same Russian word as other oils.

When butter is to be procured by churning, how is the cream prepared?

It is usually allowed to become sour; and it ought to be at least one day old, and if the weather is cool it may be kept several days. If well freed from milk, it should be frequently stirred to keep it from curdling. If new sweet cream be put into the churn, more time and labor is required for the operation of churning; for, usually even then, the cream becomes sour from the agitation, before the butter is distinctly formed.

For what length of time should cream be churned?

The common idea is, that the quicker the butter is produced the better. Accordingly, so many efforts have been made in getting up patent churns, giving the cream a rapid agitation, and thus constantly keeping every part of it in contact with the atmosphere, so as to produce butter in ten or fifteen minutes. It is, however, pretty well ascertained, that the quicker the butter is produced, the paler, softer, and less rich it will be.

What is said of churning the milk and cream together?

Butter in many places, particularly in Scotland and Ireland, is thus prepared; but this is by far a more laborious process, from the difficulty of keeping in motion such a quantity of fluid. It is said, however, to have the advantage of giving a larger amount of butter. There can be no objection therefore to it, provided the skimmed milk is not wanted, and especially provided some other than human power is applied to the process of churning.

How long should the operation of churning continue?

It is said that cream may be safely churned from an hour to an hour and a half, while milk should be churned double that time. The agitation of the milk or cream should be regular, slower in warm weather, that the butter may not be soft and white, and quicker in winter, that the temperature may be kept up. It is not desirable that butter be produced in less than three quarters of an hour.

What are the best churns for general use?

The old-fashioned upright barrel churns, to be operated by hand, for a single cow, or a very small number, will not probably be wholly superseded. The horizontal churns, or upright ones, operated with a crank by hand, are evidently a very considerable improvement on the other; but for a large number of cows, whether the entire milk or the cream only is churned, what is called the dog power applied to the churn is an important saving of manual labor.

What other improvement has been made in churning?

There has recently been invented by F. G. Simpson of New Jersey, a churn to be operated by machinery with a weight applied, similar to the running of a clock. Nothing is required but to put the milk or cream into the churn, and then wind up the machinery, when the moderate and uniform agitation of the liquid is begun and continued till butter is produced, without the aid of any other power. Thus far this invention works well; and if no difficulty hereafter arises in its use, it will be generally adopted wherever the labor heretofore required in churning has been found a great burden.

What quantity of butter is annually made in the United States?

It is impossible to tell. There are no data from which anything like an accurate estimate can be formed. It might be supposed, that, on an average, each individual will consume twenty-five pounds in the year. If so, with our present population, the entire consumption will be annually over five hundred millions of pounds, the value of which cannot be far less than one hundred millions of dollars. And if each cow makes one hundred and fifty pounds, more than three millions of cows are required to furnish a supply. This shows the importance of this branch of rural economy.

How is butter kept from becoming rancid?

It should, in the first place, be well washed and worked in cold water, so as to be entirely relieved from the curd and watery substances with which it was combined. Then it should be sufficiently salted, and packed in a clean vessel, from which the action of the atmosphere is wholly excluded. A little sugar well worked into the butter, by some, is thought to improve its flavor.

What is the process of making cheese?

The common mode of separating the curd from the whey in milk is by the application of rennet, which is the stomach of young calves having been prepared for the purpose. Any acid combined with the milk will cause coagulation; but rennet is preferable. When the curd is collected, the whey being sufficiently excluded, it is made as solid as possible by the application of a powerful press.

On what does the quality of cheese depend?

It is obvious that whatever gives rise to natural differences in the quality of the milk must affect also that of the cheese prepared from it. If the milk be poor in butter, so must the cheese be. If the pasture be such as to give a milk rich in cream, the cheese will partake of the same quality. If the herbage or other food affect the taste of the milk or cream, it will also modify the flavor of the cheese. Hence the great difference in the price of different qualities of cheese.

How may the profits of cheese-making be estimated?

It takes from four to six quarts of milk, according to its quality, to make a pound of curd. Some cows are known to have yielded milk for four hundred weight of cheese in the year; but the average amount falls much below this. If cheese-making is to be rendered profitable, a first-rate article should be produced, which will always find a ready market and command a high price. Besides a good article improves in value for years, whereas a poor one from age becomes worthless.

EDUCATED FARMERS.

Why is it the fact—and a fact it is—that many of the best and most successful farmers in our country are those, who, bred to other pursuits, and toiled in them to middle age—and many far beyond it—till from inclination, or necessity, they have embraced agriculture as an occupation, with a determination to succeed? It is because investigation has been the habit of their lives. They do nothing without a good and satisfactory reason for doing it. They bend every faculty of the mind to acquire success in this, as they did in their previous pursuits; and the application of the same intelligence upon the farm that had there been exerted, produced the same results, although their early education and subsequent labors had kept them in profound ignorance of the simplest rules of practical agriculture. The most gratifying success has been thus accomplished, while he, who has from childhood tilled his paternal acres in obstinate and persevering ignorance of the true principles of his art, although scorning in the pride of his own fancied superiority, the more timid efforts of his thoughtful neighbor, delves on through life, a wretched and unsuccessful farmer, and in time leaves the world no better, so far as his own labors were concerned, than he found it; and is finally buried beneath a soil over which he plodded for three score years, and never knew a single part of its composition!

This, though perhaps an extreme, and certainly not a flattering picture, is still a type of agricultural life, in its way, existing in every one of our United States. In what profession throughout the length and breadth of our land is there so little progress—nay, such determined opposition to progress, as in the ranks of agriculture? I would not assert that numerous eminent examples of improvement have not existed among those of purely agricultural occupation. But they are rare, as compared with men of other pursuits when applied with all their research and intelligence to agriculture alone.

And it may well be inquired, why is this so? Agriculture occupies four-fifths of the laboring population of the land. From the agricultural ranks have sprung many of the most illustrious names whose services have adorned and honored their country. From its ranks, too, have perhaps a majority of the most successful among those engaged in the various other pursuits and occupations of life arisen. In short, there can be no class of our population which affords so sure a basis on which to rely for an infusion into all other pursuits to the durable prosperity of a state as the agricultural. Such is the gratifying truth; and it is to the health-giving influences of the soil itself; the free wild air of heaven that he breathes; cheerful exercise and occupation; contentment; and the full, unrestrained enjoyment of man's first estate bestowed by God himself, that thus constitutes in him who tills the soil, the full development of his faculties in all the admirable proportions of body and of mind that his Creator intended.

Notwithstanding all this, the question still recurs, and may be variously answered. The very ease and contentment of condition in the farmer, is one probable cause of his inactivity in improvement. The quietude of his avocations prevents that constant attrition of mind inseparable from the bustling activity of most other pursuits; and the certainty with which the soil yields its annual tribute to his labor, dispels that spirit of investigation common to classes the result of whose labors is contingent or uncertain. Nor yet is the farmer an ignorant, or a slothful man. In the great responsibilities of life—in domestic duty—in love of country—in the orderly support of the institutions of the land—in stern watchfulness over the acts of those he has placed in authority, and in that exalted patriotism which is ever ready for the heaviest sacrifice to the benefit of his race, he, as a class, stands without a rival. And yet, possessed of all these qualities, and enjoying all these advantages, the absence of the spirit of association, leaves him in effect the least benefited at the hands of those he elects to govern him, of all others.

Who invents, improves, and perfects the plough, and all the nameless implements which alleviate his toil and accelerate his labor? Who analyzes his soils, instructs him in their various qualities, and teaches him how to mix and manure them for the most profitable cultivation? The mechanic—the chemist. Who, ascertaining that his seeds are imperfect and unprofitable, searches foreign lands for new or better ones, and introduces them to his notice? The commercial adventurer, or the travelled man of inquiry and observation. Who, on comparing the inferior domestic animals which he propagates, and in whose growth and fattening he loses half his toil and the food they consume, sends abroad, regardless of expense, and introduces the best breeds of horses, cattle, sheep and swine for his benefit? In nine cases out of ten these labors and benefactions—and their name is legion—are performed by those whose occupations have been chiefly in other channels, and whose agricultural tastes have led them into the spirit of improving it. And in how many examples have we witnessed the apathy, if not determined opposition with which the farmer proper—or at least he who claimed to be one—has set his face like flint against their adoption, even after their superiority had been demonstrated beyond a question!

So, too, with the farmer's education. They have been content that the resources and the bounty of the state should be lavished upon the higher seats of learning, where the more aspiring of our youth should receive their benefits, not caring even to inquire whether such youth should again return among them to reflect back the knowledge thus acquired. They have failed to demand from the common treasure of the state those necessary institutions which shall promote their own particular calling, and which every other pursuit and profession in the land has been most active to accomplish In all this the latter have progressed with railway speed; while the farming interest has stood still with folded arms, and done comparatively nothing; and what good has been forced upon it by others, even regarded with suspicion. It is not because we as farmers, compared with others, are either ignorant or stupid. We only neglect to assert our rights, and appropriate the share to which we are entitled in the common patronage of the state to the benefit of our own professions. It is for us to ask—to will—to do it. We hold the power of the state by our numbers. We can control the halls of legislation. We can so direct the laws that we may share equal advantages in our institutions with others. We desire nothing exclusively to our own advantage, but we do deserve an equal participation in those institutions established for the common benefit of all.

These remarks are not made in a querulous or fault-finding temper. It is right that we have colleges and academies for the few who aspire to the higher walks of professional or scientific life, as well as common schools for the million. No state can be well, or wisely constituted without them, and I would not abate one jot or tittle from the wholesome support which a broad and liberal system of education demands. But we should claim, and insist, that departments devoted to agricultural teaching, or to the development of agricultural science, should be established, either as branches of our seats of learning, or as independent institutions. Why should not the farmer be educated to the top of his faculties, as well as those who select what are termed the learned professions as their pursuit? If our sons cannot be taught the education they seek in the colleges—and there are well grounded doubts of this fact from the moral malaria too often existing within and around them—institutions for their sole education should be aided, or erected, and endowed by the state.

We are a growing people; not in population alone, but in wealth, and in resources. Our whole country is comparatively new, and wealth is accumulated with us as with no other people of which history gives an example. I speak of substantial, enduring wealth; that which adds to the enjoyment, the happiness, and the truly elevated condition of man. Of all this wealth and prosperity, agriculture is the basis—the indispensable support. Yet, in defiance of this reiterated truth, as an occupation, agriculture of itself, is degraded. Let politicians, or demagogues chant their pæans to the tillers of the soil as they may, and tell them of the honor, and the dignity of their estate; yet, practically, simple farming is considered by those who assume to give tone and opinion in social and political life, an inferior occupation, fit only for dull, unthinking, and uneducated men.

Were it not so, why are the agricultural ranks so continually deserted by our active and aspiring youth for the more worldly popular pursuits, under the belief that they are more advantageous? Look at our great, bustling cities, and towns. See on all sides our professions crowded to excess; with, among the masses which throng them, but a comparatively few who are successful either in fame or fortune. View our merchants, and shopkeepers, overrun and undermined in competition with one another; and clerks, and shopboys, plentier and cheaper on their hands than the wares they hold on sale; and all the motley congregations which are drawn about them by the spirit of adventure and of novelty; while the petty political offices of the day are held up like lottery tickets, to an unscrupulous and indiscriminate scramble;—all for the possession of a fancied prize in the great raffling match of adventure; while the shop of the mechanic, or the artizan, which holds out a safe and durable reward to honorable labor, is hard pressed to find apprentices; and the broad, inviting acres of the farmer, are lying sterile or unproductive, for lack of cultivation.

Among the benefits arising from well directed agricultural education, aside from spreading the requisite learning and intelligence applicable to the chief pursuit of our people, deep and broad among them, the retention of that portion of active capital, acquired by the industry of our agricultural population, among themselves, would be one important consequence. In place of the prevailing and mistaken notion that monied capital invested in agriculture is either unproductive, or less so than in other pursuits, our farmers would be taught that, coupled with the knowledge to direct it, no branch of our national industry is so steadily remunerating as that connected with the soil—a fact now practically disbelieved; or why would such amounts of monied capital be continually drawn from the agricultural districts to your commercial cities, to be embarked in hazardous enterprises, or doubtful investments?

The merchant or the speculator may fail—and fail he does, very often—and in his downfall are often buried the toils of a long life of patient industry. But who ever knew a good farmer, of prudent habits, to fail? Nay, who did not, with an exemption from extraordinary ills in life, ultimately grow rich, and discharge meantime, all the duties of a good citizen? I concede to you the many prominent cases which exist, of wealth rapidly accumulated by bold and successful speculation; of fortunate, perhaps accidental adventure; of hoards heaped up by a long course of perseverance in trade, directed by that intuitive sagacity of which but few among us are endowed, and which so dazzlingly invite our imitation. Yet these are but a few glaring instances, standing out in bold relief among the many who have sunk in the same career, perhaps with a ruined peace; happy afterwards to retire, were it in their power, upon the limited possession which they had thrown away, to commence their wasting strife upon the broad sea of adventure.

A second advantage would be, that it would invite, annually, a large class of educated men of capital from our cities, to invest a portion of their wealth in our farms, convinced by the knowledge acquired in a course of agricultural education, that husbandry was a good business, and intending to pursue it as the occupation of their lives, it would cause a reflux of that capital and population which had been drawn away from agriculture. Nor would such associations among us detract from the industrious habits of our farmers by their example. They, by the possession of larger estates than we enjoy, might give more of their time to leisure than we are accustomed to spend; but they must, if good farmers, attend to the daily routine of their affairs, as well as we. They would diffuse intelligence among us; introduce improved implements, seeds, and stock; and in time, surely exalt the character of our husbandry. They might not, indeed, work at the muck heap, nor guide the plough with their own hands; but they must be capable, from education, to direct the labor of both; for we must not forget that the merchant who, from his luxurious counting room, plans his voyages, and directs the course of his ships; or the engineer who projects the rail-way, or the ocean steamer, once performed the duties of a shop boy, or hammered at the anvil. And thus with the farmer: he should be capable of directing the cultivation of the soil to its greatest possible extent of production; and he will find that, in achieving such result, all the powers of his mind, and the knowledge with which it is stored, will be required.

This thought will bear a little examination. The farmer is apt to think that the professional man, or the merchant, lives an easy and luxurious life. In many instances their families may do so; but with the eminent and successful man of law, or science—the artizan, or merchant himself, such supposition is a great mistake. There are not, under heaven, a more laborious class of men than these. Labor of body, and of mind is theirs—and that incessant. See them early, late; in season, and out of season—their whole energies devoted to their several callings, without rest, or intermission—and far too frequently, to the premature wasting of life itself. It is no wonder that such industry, directed by good education, (and by this term I mean the entire training of the boy to manhood in its most extended sense,) and stimulated by a laudable ambition, should lead to success. Yet with all these appliances, the labors of such men are often disastrous; and if not so, after a life of anxiety, their toils too frequently end with but the means of a slender support. Compared with these, the toils of the farmer are light. Physical labor he endures, it is true, and oftentimes severe labor, but his mind is easy. He enjoys sound rest, and high health. He has much leisure; in many cases more than is for his good. He has abundant time to discuss politics, law, religion—everything, in fact, but what relates to his own profession, on which subject, I lament to say, his mind seems less exercised than on almost any other..

Now, let the same early education be given to the young farmer of an equally acute intellect that is given to him who chooses professional, mechanical, or mercantile pursuits—education each in his own line. Let them start fair. Apply the same thought, investigation, energy, and toil, each in his particular sphere, and beyond all question agriculture will, in the aggregate, have the advantage—and for this reason, if no other: there are few contingencies connected with agriculture. Its basis is the solid earth, stamped with the Divine promise, that while it remains, seed-time and harvest shall continue; while commerce, and trade; mechanics, and arts are liable to extraordinary and continual accident. Look at the devastations by flood, and fire—of ship, and cargo, upon ocean, lake, and sea, and river; conflagrations in your towns and cities; and the thousand other casualties which almost daily occur—all which are a dead sink upon labor and capital not agricultural, and the risks of the husbandman are scarce one to ten, in the comparison. Rely upon it, farmers, you are on the safe side.

But, I hear some one remark, "Why, if agriculture, through the improved education proposed, holds out such alluring advantages, all our young men will rush into it, and competition will destroy it." Not the slightest danger. Our young men are already running into the other trades and professions, where competition is ruinous; and all we ask, is the opportunity to get a share of them back again. Besides, there is no fear that the other avenues of industry will not be filled; for, in the constitution of our natures, there will always be enough unquiet spirits born into the world which the farm cannot hold, to keep the bustling part of it in motion.

Another, and a prominent advantage which we should receive from good agricultural education, would be, that of more stability of character in our farming population. It is proverbial among travelled foreigners in this country, and it would be a subject of wonder among our staid people at home—if an American could wonder at any thing—that we are the most changing people in the world. We, as a population, have few, scarce any, local attachments. This, to an extent, is a true, although a severe censure. It arises, no doubt—and naturally enough, too—from the wide extent of national domain of which we are the possessors, and from the natural sterility of much of the soil in our older communities, which cause an effort, and a laudable one, too, to better their condition in our rural population. But more, I imagine, from the low standard of agricultural improvement, and a mistaken estimate of the value of the soil, and its application to the products which properly belong to it.

But, no matter what the cause. The fact is so, and it is a defect in our national character. How many among us but will, with a slightly tempting offer, sell his homestead without remorse—break up the cherished association of his life—turn his back upon the graves of his kindred, and his children—his birth-spot—the old hearth-stone of his boyhood—his family altar, even the brave old trees, which have, life-long, waved their branches over his childish sports, and shadowed his innocent slumbers when weary of his play, all—all, pass out of his hands, like a plaything of yesterday, unwept and unregretted, for the fancied advantage of a fresh spot in a strange and a newer land.

I must, however, in justice, make some exceptions to this general propensity in American character. There are some among the descendants of the early New England Puritans, and the ancient Dutch settlers of this state, who have, with a pious regard to the memories of their ancestors, and a wise attachment to the spots of their birth, retained, and through the influences of a correct education, and well settled principle, bid fair to retain, the paternal acres which they have inherited—homes of plenty, contentment, and genuine hospitality; where retired virtues, like those practised by their fathers, have long hallowed them with a local habitation and a name. Such, stand out as strong landmarks in the fitful changes of place and name throughout our country and redeem, to some extent, the caustic remark of the late John Randolph, of Roanoke, who once declared, on the floor of Congress, that he scarce knew an American but would sell his very dog for money!

We are not slow in finding out when we are well off,

although all are not satisfied under such condition; but with these advantages around and among us, of which we feel the daily benefit, and of which, by removal, we should forever be deprived, their tendency would be to fix us more firmly to our homes, and lead us to examine the resources within our reach, which otherwise might never have been developed. Associations of an elevated character are among the most powerful in thus keeping us content; and institutions in which the farmer has a direct interest, would, more than almost any other, allay this tendency to change. Our resources, and our productive power, are thus retained, far beyond what can be acquired by the continued restlessness common to us. Such influences would certainly be most wholesome.—*From Address before the New-York State Agricultural Society, delivered at Albany, by* LEWIS F. ALLEN, ESQ., *of Black Rock.*

THEORY AND PRACTICE OF FARMING.

The bearing of agriculture upon the future prosperity and destiny of our free country, must be apparent to every reflecting mind. The farmers comprise a very large majority of the population of this country—its bone and sinew. Their suffrages decide the character of our rulers—as a general rule, it may with safety be affirmed, that the higher the improvements in agriculture, the more intelligent and well informed will be the character of those who direct them—and the more likely to maintain unsullied those principles which actuated our worthy forefathers, and secured through their agency the blessings of that free government and those liberal institutions we are permitted to enjoy.

If then you would perpetuate the blessed institutions which are so highly prized and so richly enjoyed, if you would transmit them to your posterity unimpaired, do all in your power to enlighten and elevate the farmer, to encourage his calling—cherish it as the most important—for in it you have the palladium of freedom, and while the agriculturist continues enlighted, intelligent and virtuous, your liberties will be secure. In whatever aspect then we view agriculture —whether as necessary to provide for our wants, or as suited to the development of our physical and mental resources, or in its influence upon our moral, social, and political relations, it presents itself, as entitled to our highest regard, and claims our most cordial support. How important, then, that we should be prepared rightly to improve in every respect this most noble employment.

In all the pursuits of life, practical knowledge is absolutely essential to perfection. It is quite a different thing to understand principles and to work them out; and he who is a good theoretical agriculturist, may be a very bad practical farmer. The two kinds of knowledge, theoretical and practical, require to a certain extent, somewhat of a different order of mind; but the excellent practical farmer can understand the principles on which his practice is founded, with very great advantage to his practical efforts. A knowledge of the principles of Agriculture, though not absolutely indispensable to the cultivation of the land, cannot fail of being in the highest degree useful. A purely practical farmer repeats certain acts, and necessarily follows the plans handed down to him by his forefathers; he tills his land at a certain season; he sows his seed, fallows his land, rotates his crops, uses manures in the same manner as his father or grandfather; and provided he remains on the same farm and soil, he succeeds to the same extent.

If you ask him *why* he does certain things, he answers that he had seen his father do so; but he can assign no better reason. Transport this excellent *practical* man to another locality, let him be placed on a different soil, and watch the result. Suppose, for example, such a man removes from a farm, the soil of which is a strong loam or clay, to one covered with a thin layer of sandy light soil, and not being acquainted with the difference of working such a soil, he may perchance, instead of looking around him, and observing what treatment is used by his neighbors, at once go on in his beaten track. Such a practice would inevitably destroy the fertility of his new farm. He would experience bitter disappointment, and have to recommence his education.

Here, it will be perceived, the purely practical man would be at fault; he would feel that mere practice, although excellent as regards one locality, will entirely fail in another; so that the necessity for some general principles will be forced on his mind. Hence the utility of such principles, and the necessity of such an education as will enable the farmer to comprehend and embrace principles, as well as mere practice, may be considered as demonstrated. A small portion however, of the principles which illustrate agriculture will answer the purpose of the practical farmer, and enable him to improve the processes he at present employs for the cultivation of the soil.

We need not cross the Atlantic to find evidence of the advantages resulting from a right application of practical skill and science to the advancement of agriculture. Some of the once fertile portions of our own State have been exhausted by bad husbandry, and have been abandoned—while others, sterile by nature, have been made fruitful and have become among our most productive and valuable lands. It is one of the beneficial results of agricultural associations, that improvements of this description have through them been fostered and encouraged. An impulse in many instances has been given which will not soon be forgotten.

The pursuit which most of you have chosen is a noble one. The father of our country, the immortal Washington, has left behind him an imperishable monument of his views on this pursuit, in his agricultural correspondence,—a work which should be in every farmer's dwelling, and which will be read around the fireside of the American farmer, with his farewell address to his countrymen, to the latest period of time. Most of the eminent men who succeeded him in the Presidential chair, have as they retired from the cares of the office, entered upon agricultural pursuits—and some of them still live to cheer on the farmer, by their precept and example in his noble pursuit.

The late Governor of our state, whose sudden and unexpected death has cast a gloom over our nation, and whose talents were respected every where, whatever diversity of opinion there was as to his views on political questions, had retired from the cares of office, to the cultivation of his farm; and I have the means of knowing, that in the active pursuits of the farmer, he enjoyed more satisfaction, than when clothed with the highest honors of the State, or when occupying a seat in one of the most distinguished bodies the world has ever known, the Senate of the United States.

Another distinguished statesman, still living to bless and adorn his country and the world by his superior talents, a few years since, when occupying a very important station, said, that the real luxury of life to him, was, when released from the duties and cares of office, and the toils of a most laborious profession, he was permitted to enjoy the sweet retreat at Marshfield, superintending his farm and attending to his herds, and enjoying for a season only those delights and blessings which may, and should ever cluster around the home of every American farmer.—Hon. B. P. Johnson, *Albany*.

THE BOY AND THE RAINBOW.

Declare, ye sages, if ye find
'Mongst animals of every kind,
Of each condition, sort and size,
From whales and elephants to flies,
A creature that mistakes his plan,
And errs so constantly as man?
Each kind pursues his proper good,
And seeks for pleasure, rest, and food,
As Nature points, and never errs
In what he chooses and prefers;
Man only blunders, though possest
Of talents, far above the rest.
  The happiness of human kind
Consists in rectitude of mind,
A will subdued to Reason's sway,
And passions practis'd to obey;
An open and a generous heart
Refin'd from selfishness and art;
Patience, which mocks at Fortune's pow'r,
And Wisdom, never sad nor sour.
In these consists our proper bliss
Else Plato reasons much amiss.
But foolish mortals still pursue
False happiness in place of true:
Ambition serves us for a guide,
Or lust, or avarice, or pride;
While reason no assent can gain,
And Revelation warns in vain.
Hence, thro' our lives, in ev'ry stage
From infancy itself to age,

A happiness we toil to find,
Which still avoids us like the wind;
Ev'n when we think the prize our own,
At once 'tis vanish'd, lost, and gone.
You'll ask me why I thus rehearse
All Epictetus in my verse,
And if I fondly hope to please
With dry reflections such as these,
So trite, so hackney'd, and so stale?—
I'll take the hint, and tell a tale.
One ev'ning, as a simple swain
His flock attended on the plain,
The shining bow he chanced to spy
That warns us when a show'r is nigh:
With brightest rays it seem'd to glow,
In distance eighty yards or so.
This bumpkin had, it seems, been told
The story of the cup of gold,
Which Fame reports is to be found
Just when the rainbow meets the ground;
He therefore felt a sudden itch
To seize the goblet and be rich!
Hoping (yet hopes are oft but vain)
No more to toil thro' wind and rain,
But sit indulgent by the fire,
'Midst ease and plenty, like a squire.
He mark'd the very spot of land
On which the rainbow seem'd to stand,
And stepping forwards at his leisure,
Expected to have found the treasure.
But as he mov'd, the color'd ray
Still chang'd his place, and slipt away,
As seeming his approach to shun;
From walking he began to run,
But all in vain, it still withdrew
As nimbly as he could pursue.
At last thro' many a bog and lake,
Rough craggy rock, and thorny brake,
It led the easy fool, till night
Approach'd, then vanish'd in his sight,
And left him to compute his gains,
With nought but labor for his pains.

WILKIE.

THE BENEFITS OF AGRICULTURE

Agriculture is the greatest among the arts, for it is first in supplying our necessities. It is the mother and nurse of all other arts. It favors and strengthens population; it creates and maintains manufactures; gives employment to navigation, and materials to commerce. It animates every species of industry, and opens to nations the surest channels of opulence. It is also the strongest bond of well-regulated society, the surest basis of internal peace, the natural associate of good morals.

We ought to count among the benefits of Agriculture the charm which the practice of it communicates to a country life. That charm which has made the country, in our own view the retreat of the hero, the asylum of the sage, and the temple of the historic muse. The strong desire, the longing after the country with which we find the bulk of mankind to be penetrated, points to it as the chosen abode of sublunary bliss. The sweet occupations of culture, with her varied products and attendant enjoyments are, at least, a relief from the stifling atmosphere of the city, the monotony of undivided employments, the anxious uncertainty of commerce, the vexations of ambition so often disappointed, of self-love so often mortified, of factitious pleasures and unsubstantial vanities.

Health, the first and best of all the blessings of life, is preserved and fortified by the practice of agriculture. That state of well-being which we feel and cannot define; that self-satisfied disposition which depends, perhaps, on the perfect equilibrium and easy play of the vital forces, turns the slightest acts to pleasure, and makes every exertion of our faculties a source of enjoyment; this inestimable state of our bodily functions is most vigorous in the country, and if lost elsewhere, it is in the country we expect to recover it.

The very theatre of agricultural avocations, gives them a value that is peculiar: for who can contemplate, without emotion, the magnificent spectacle of nature when, arrayed in vernal hues, she renews the scenery of the world! All things revive at her powerful voice; the meadow resumes its freshness and verdure; a living sap circulates through every budding tree; flowers spring to meet the warm caresses of Zephyr, and from their opening petals pour forth rich perfume. The songsters of the forest once more awake, and in tones of melody again salute the coming dawn; and again they deliver to the evening echo their strains of tenderness and hymns of love. Can man—rational, sensitive man—can he remain unmoved by the surrounding presence! and where else than in the country can he behold, where else can he feel this jubilee of nature, this universal joy?

Ennobled, indeed, must be the profession, whose proper abode is amidst the finest scenes of creation, and under the immediate influence of the celestial phenomena. The agriculturist stands in connexion with the agencies of the universe. The refreshing dews, the enriching rains, the winds, the snows, the frosts, all contribute to the results he prosecutes. When the sun shines, it is to ripen his harvests; when the clouds collect, it is to water his pastures; and if, from time to time, destructive meteors excite his fears, or disappoint his expectation, they also recall him to a sense of his dependence on Heaven, and they give an increased value to what they spare. He is reminded every moment that all the occupations, all the labors of agriculture tend to the good of society. The hands he puts in motion, the poor he preserves from idleness, the products of the earth which he multiplies, are all so many benefits which he confers on his country and his kind. He naturally becomes a better man in a vocation which is composed of useful actions. And what better guarantee is there of happiness, here or hereafter, than the daily practice of good works!—Mac Neven.

WINTER AT COPENHAGEN.

From frozen climes and endless tracts of snow,
From streams which northern winds forbid to flow;
What present shall the muse to Dorset bring,
Or, how, so near the pole, attempt to sing?
The hoary winter here conceals from sight
All pleasing objects which to verse invite.
The hills and dales, and the delightful woods,
The flowery plains, and silver-streaming floods,
By snow disguis'd, in bright confusion lie,
And with one dazzling waste fatigue the eye.
  No gentle breathing breeze prepares the spring,
No birds within the desert region sing.

The ships, unmov'd, the boisterous winds defy,
While rattling chariots o'er the ocean fly,
The vast Leviathan wants room to play,
And spouts his waters in the face of day.
The starving wolves along the main sea prowl,
And to the moon in icy valleys howl.
O'er many a shining league the level main
Here spreads itself into a glassy plain:
There solid billows of enormous size,
Alps of green ice, in wild disorder rise.
And yet, but lately have I seen, ev'n here,
The winter in a lovely dress appear.
Ere yet the clouds let fall the treasur'd snow,
Or winds began thro' hazy skies to blow;
At evening a keen eastern breeze arose,
And the descending rain unsullied froze.
Soon as the silent shades of night withdrew,
The ruddy morn disclos'd at once to view,
The face of Nature in a rich disguise,
And brighten'd every object to my eyes.
For every shrub, and every blade of grass,
And every pointed thorn seem'd wrought in glass;
In pearls and rubies rich the hawthorns show,
While through the ice the crimson berries glow.
The thick sprung reeds, which watery marshes yield,
Seem'd polished lances in a hostile field,
The stag, in limpid currents, with surprise,
Sees crystal branches on his forehead rise:
The spreading oak, the beech, the towering pine,
Glaz'd over, in the freezing ether shine.
The frighted birds the rattling branches shun,
Which wave and glitter in the distant sun.
When, if a sudden gust of wind arise,
The brittle forest into atoms flies;
The crackling wood beneath the tempest bends,
And in a spangled shower the prospect ends:
Or, if a southern gale the region warm,
And by degrees unbind the wintry charm,
The traveller a miry country sees,
And journeys sad beneath the dropping trees.
Like some deluded peasant, Merlin leads
Thro' fragrant bowers and thro' delicious meads,

While here enchanted gardens to him rise,
And airy fabrics there attract his eyes,
His wandering feet the magic paths pursue,
And, while he thinks the fair illusion true,
The trackless scenes disperse in fluid air,
And woods, and wilds, and thorny ways appear,
A tedious road the weary wretch returns,
And, as he goes, the transient vision mourns.

PHILLIPS.

THE HEART'S CHARITY.

A rich man walked abroad one day,
And a poor man walked the selfsame way,
When a pale and starving face came by,
With a pallid lip and a hopeless eye.
And that starving face presumed to stand,
And ask for bread from the rich man's hand;
But the rich man sullenly looked askance,
With a gathering frown and doubtful glance.
"I have nothing," said he, "to give to you,
Nor any such rogue of a canting crew;
Get work, Get work! I know full well
The whining lies that beggars can tell."
And he fastened his pocket and on he went,
With his soul untouched and his conscience content.

Now this great owner of golden store
Had built a church not long before,
As noble a fane as man could raise.
And the world had given him thanks and praise,
And all who beheld it lavished fame
On his Christian gift and godly name.

The poor man passed, and the white lips dared
To ask him if a mite could be spared;
The poor man gazed on the beggar's cheek,
And saw what the white lips could not speak.
He stood for a moment, but not to pause
On the truth of the tale, or the parish laws.
He was seeking, to give—though it was small,
For a penny, a single penny was all;
But he gave it with a kindly word,

While the warmest pulse in his breast was stirred;
'Twas a tiny seed his Charity shed,
But the white lips got a taste of bread,
The beggar's blessing hallowed the crust
That came like a spring in the desert dust.

The rich man and the poor man died,
As all of us must, and they were tried
At the sacred Judgment seat above,
For their thoughts of evil and deeds of love.
The balance of Justice *there* was true,
And fairly bestowed what fairly was due,
And the two fresh comers through Heaven's gate,
Stood there to learn their eternal fate.
The recording angel told of things
That fitted them both with kindred wings;
But as they stood in the crystal light,
The plumes of the rich man grew less bright,
The angels knew by the shadowy sign,
That the poor man's work had been most divine;
And they brought the unerring scales to see
What the rich man's falling off could be.
Full many deeds did the angels weigh,
But the balance kept an even sway;
And at last the church endowment laid
With its thousands promised and thousands paid,
With the thanks of prelates by its side,
In the stately words of pious pride,
And it weighed so much that the angels stood
To see how the poor man could balance such good.
A cherub came and took his place
By the empty scale with radiant grace,
And he dropped the penny that had fed
White starving lips with a crust of bread.
The church endowment went up with the beam,
And the whisper of the Great Supreme,
As he beckoned the poor man to his throne,
Was heard in this immortal tone—
"Blessed are they who from great gain
Give thousands with a reasoning brain,
But better still shall be their part
Who give one coin with a pitying heart."

ELIZA COOK.

LADY MESSENGER AND HER COLT, formerly owned by S. W. JEWETT, Weybridge, Vermont.

HISTORY OF THE HORSE.

What is the early history of the horse?

In his original state he was undoubtedly wild, but of the time of his domestication we have no knowledge. Arabia is generally considered his native country; but we know that he did not flourish there until after the birth of Mahomet, in 571 A. D.

What are some of the earliest records of the horse in the Bible?

The first time he is there mentioned is in the forty-seventh chapter of Genesis, where Joseph, during the seven years of famine, received horses in payment for corn. Before the time of Solomon, horses were not as common among the Hebrews as afterwards. Before his time, they were not mentioned in the armies of Israel; and the kings were cautioned against multiplying them. The judges and princes of that country generally rode on mules or asses. Solomon is the first king who had many; and he was accustomed to obtain them from Egypt.

In what estimation was the horse held among eastern nations in those early times?

It is well known that the sun was worshipped all over the East, and the horse was consecrated to this deity, who was represented as riding in a chariot drawn by the most beautiful and swiftest horses in the world, and performing every day his journey from the east to west, to enlighten mankind. Zenophon describes a solemn sacrifice of horses to the sun; all being of the finest breed, and were led with a white chariot, crowned, and consecrated to the same god. King Josiah is said to have removed from the temple the horses which his predecessors had consecrated—probably with a view thus to sacrifice them. The Jewish Rabbins tell us, that these horses were every morning harnessed to the chariots dedicated to the sun, and that the king, or some of his officers, got up and rode to meet the sun at his rising, as far as from the eastern gate of the temple to the suburbs of Jerusalem.

What is said of wild horses?

They are found in various countries, and it is impossible to ascertain whether those on the eastern hemisphere belong to the original stock, or are descended from such animals as have escaped from servitude. Those on the western plains of our own continent are derived from the animals brought over by the Spaniards. Wild horses associate in large droves, as well with a regard for their mutual protection, as from an attachment to each other. They are not naturally ferocious, and are, indeed, timid, although gay and high spirited. They are seldom known to make an attack, but when attacked by other animals, they either disdain their enemies, or trample them to death. Sentinels are stationed on the elevated grounds, while the remainder of the herd are feeding on the plains below, and as soon as any impending danger is discovered, they all seek safety in flight. Indians capture many of these wild horses by means of the lasso.

What is said of the value of the horse?

The horse is known to most nations as the most useful and manageable of those animals that live under the sway of man. In gracefulness of form, and dignity of carriage, he is superior to almost every other quadruped; he is lively and high spirited, yet gentle and tractable; keen and ardent in his exertions, yet firm and persevering. The horse is qualified for all the various purposes in which man has employed him; he works steadily and patiently in the loaded wagon, or at the plough; becomes as much excited as his master on the race course, and appears to rejoice in the chase.

For what other purposes is the horse valuable?

From the milk of the mare the Calmucs and other Tartars prepare a spirituous drink of considerable strength. And besides the valuable services of the horse when alive, after death his skin is used for a variety of purposes, and the hair of the mane and tail is used for chair bottoms and mattresses. His flesh, although rejected by civilised nations, is eaten by several rude tribes.

Have climate and soil any influence upon horses?

Nearly every country possesses a breed of horses peculiar to itself, occasioned by the difference in soil and climate. In wild regions, the animal is small, compactly built, and having a very hardy constitution; where the forage is scant, he has a light frame, with a rapid gait, so that he can travel easily in search of food; while in a more favorable situation, he combines speed and power of endurance, with beauty of form and elegance of action. The diminutive Shetland pony, when transported to England, in the course of a few generations, loses this peculiarity of form, and approaches the native horse in bulk and general characteristics. The noble Arabian degenerates in England, and the English dray horse, noted for its heavy limbs, if carried to Arabia, acquires, in time, the symmetry, grace, and speed of the native.

In what country was the breeding of horses first practiced?

In Egypt, which became distinguished for their number and excellence. Joseph, when he carried his father's corpse into Canaan, was accompanied by horses and chariots, and we are told that the hand of the Lord was upon all the cattle of the Egyptians, and that the horses and chariots of Pharaoh were overwhelmed in the Red Sea.

How many horses is Solomon said to have had?

On marrying a daughter of Pharaoh, he procured horses from Egypt, and they multiplied so exceedingly that he afterwards had four hundred stables, forty thousand horses, and twelve thousand horsemen.

What is the character of the Arabian horse?

For the combined excellencies of symmetry of form, beauty of action, speed, power of endurance, intelligence, docility, and attachment to its master, this horse has long maintained an unrivalled superiority. It is bred with the greatest care, and the Arabs believe that the breed originated with the steed of King Solomon, and exhibit the pedigrees of some mares reaching back near two thousand years. The poor Arab regards his horse with the greatest affection, feeds it from his own hand, and makes it an inmate of his own single tent. Arabian stallions have been brought from Arabia at different periods, and have given rise to all the improved breeds of Turkey, Persia, Barbary, Europe and America.

How is the Barb, or the Barbary horse, esteemed?

He is inferior to the pure-blooded Arabian alone, from which he has derived his good qualities. The Barbary horse was introduced into Spain by the Moors, and it appears he has given character to the horses of that country and Italy. He is highly esteemed, and has been frequently taken to England and America.

Where has the breeding of the horse been most successful?

In England. At the time of the invasion of Julius Cæsar, he was much pleased with the appearance of the horses used by the Britons in their war chariots. He carried some of them to Rome, where they were at once received into favor. The spirit of improvement upon this excellent native breed commenced at an early date. The first foreign stock introduced were German race horses and Spanish stallions. In 1121, the Arabian was introduced. By the judicious intermixture of the best foreign blood, the horse has apparently been brought to perfection in England. Some of the animals bred there, have surpassed in speed and endurance even the Arabian on his own soil. Flying Childers, a celebrated racer in the last century, could clear the space of eighty-two and a half feet in a second.

What are the kinds of horses bred in England?

The English do not seem to have what the Americans designate as "a horse of all work;" but on the contrary, breed the animal with especial reference to the kind of labor he will be called upon to perform. They have the racer, the hunter, the roadster, the carriage-horse, the pony, the dray-horse, and the farm-horse, intended for sporting, travel, and useful labor. Of the kinds most esteemed in agriculture, are the *Suffolk Punch*, which is large and serviceable—the *Cleaveland Bay*, of good frame and possessing great strength—and the *Clydesdale*, which is a favorite in the northern counties, and throughout the whole of Scotland.

When was the horse introduced into America?

At the time of its discovery by the Spaniards. The horse was regarded with great terror by the simple minded natives, and they even thought the rider a part of the animal. Their fear of the Spanish cavalry contributed in no slight degree to the ease of the early conquests. Some of these horses escaped, and laid the foundation of the immense droves which range through the wild region of both North and South America.

What is said of the Narraganset Pacers?

They were spirited animals, noted for speed and endurance, being common many years ago in Rhode Island. Tradition has ascribed their origin to an Arabian stallion, that escaped from a wreck, and associated with some mares that were allowed to roam in the woods. These Pacers were highly esteemed in the West Indies, whither they were shipped in large numbers. At the present day, they appear to be almost, if not quite extinct.

What is the history of the Barb Ranger?

Barb Ranger exercised so important an influence upon

the stock of this country, that his history is well worthy of remembrance. Not long before the Revolutionary War broke out, a beautiful white Barb was presented by the Emperor of Morocco to the commander of an English frigate. On the way home, the frigate stopped at a port, where the horse was taken ashore, and turned into a lumber-yard for exercise. In his gambols, he climbed a pile of boards, fell and got three legs broken. He was thought to be useless, and was presented to the captain of a New England vessel, then lying at anchor in the same harbor. This captain carried the horse to the ship, and placed him in slings, by which treatment he eventually recovered. He was taken to Pomfret, Conn., where he remained several years. During the war, the attention of Captain Lee, who commanded a famous troop of cavalry, was drawn to some eastern horses in his troop, and he sent a Captain Lindsey to make inquiry as to their origin. Lindsey purchased Ranger at a high price, and carried him to Virginia. In his new quarters he gained a high reputation, and was generally known as Lindsey's Arabian.

Have other Arabians been brought to the country?

Several have been, and although they were considered very superior animals, yet their introduction has not been followed by such decided benefit as in the case of Ranger. Breeders appear to think that further improvement in our stock will be due to England, to which country we are already indebted for much of our success.

Have many English horses been brought to this country?

They are almost without number, and many were of a high character. It would be difficult to mention even their names. Messenger, who was imported in the latter part of the last century, seems however, to merit a short notice. He stood for several years in the northern states, and obtained a distinguished reputation. His progeny are very widely spread, and are every where esteemed as roadsters.

What was the origin of the Morgan horse?

The Morgan horses are highly valued, particularly in Vermont. They originated from a two-year old colt which Mr. Justin Morgan, of Randolph, in that State, procured from Massachusetts in the year 1795. This colt is said to have obtained some of his good qualities from an English stallion, that was afterwards carried back to his native country. The Morgan horses are of medium size, strong, very hardy, spirited, easily kept, docile, and excellent for the road.

What is said of the Norman horse?

It is a cross of the Spanish and Flemish horses. It is remarkable for vigor of constitution, spirit, strength, and excellence in harness. Several of this breed have been introduced into this country, and they appear to be well adapted to the use of farmers.

Is the horse superior to all other animals in individual, physical, and moral attributes?

He is not. In sagacity he falls far short of the ponderous and drowsy elephant; in muscular development and grace of limb, he does not surpass the stag; in ardor and constancy of devotion, he can scarcely be said to equal his friendly companion and rival of his master's affections, the faithful dog; and his courage fails him at sight of a "lion in the way;" while, in the humbler qualities of patience and availability to the very last, even to the hair and the hoof, that unambitious drudge, the ox, may well assert his pretensions to comparison, if not to superiority.

Why then is the horse such an universal favorite with man?

It is the admirable *combination* of the several qualities which, taken simply, serve to confer distinction on other quadrupeds, that united in the horse, fit him for employments so various; giving him pre-eminence alike in the wagon or the plough—the coach or the battle-field.

What beautiful description does Job give of the horse?

Hast thou given the horse strength?
Hast thou clothed his neck with thunder?
Hast thou taught him to bound like the locust?
How terrible is the sound of his nostrils!
He paweth in the valley; he exulteth in his strength,
And rusheth into the midst of arms.
He laugheth at fear; he trembleth not,
And turneth not his back from the sword.
Against him rattleth the quiver,
The glittering spear, and the lance.
With rage and fury he devoureth the ground;
He standeth not still, when the trumpet soundeth.

How is it known that the horse existed before the flood?

The researches of geologists prove it. There is not a portion of Europe, nor scarcely any part of the globe, from the tropical plains of India, to the frozen regions of Siberia—from the extreme northern to the extreme southern limits of America, in which the fossil remains of the horse have not been found mingled with the bones of the hippopotamus, the elephant, the rhinoceros, the bear, the tiger, the deer, and various other animals, some of which, like the mastodon, have passed away.

What pecuniary profit have some of the most noted horses of England and this country brought to their owners?

The stud horse, King Herod, in nineteen years, earned more than a million of dollars. Marsk, won in matches upwards of eighty thousand dollars. Shark, beside a cup of the value of one hundred and twenty guineas, and eleven hogsheads of claret, won the vast amount of seventy-seven thousand dollars. Highflyer, won and received about fifty thousand dollars, in matches.

At what high prices have horses occasionally been sold?

Eclipse, at an advanced age, sold for the sum of ten thousand dollars; and at the age of twenty-seven years, he was in vigorous health, receiving in Kentucky, where he was kept, one hundred dollars for each of his colts. And one of the progeny of Eclipse, by Lady Lightfoot, was sold to a gentleman of Pennsylvania for ten thousand dollars. Others have sold at prices incredibly great to persons not skilled in the points that render the animal peculiarly valuable.

At what speed will the best American horses trot?

The cases are not unfrequent where they will trot from fifteen to seventeen miles an hour. A roan mare at Providence, R. I., called *Yankee Sal*, was reported to have trotted fifteen miles and a half in 48 minutes and 43 seconds. Lady Kate, a bay mare, trotted on the Canton Course, near Baltimore, sixteen miles in 56 minutes and 13 seconds. Mount Holly, on the Hunting Park Course, Penn., in 1836, trotted seventeen miles in 53 minutes and 18 seconds. In 1839, Tom Thumb, an American horse, was driven in England sixteen and a half miles in 56 minutes and 45 seconds. In 1833, Paul Pry, on the Union Course, L. I., won the match, performing eighteen miles in 58 minutes and 52 seconds. Numerous cases might be named of about the same speed.

What cases are on record of horses trotting long distances with extraordinary rapidity?

In 1831, Chancellor, on the Hunting Park Course,

Pa., ridden by a boy, performed thirty-two miles in one hour, fifty-eight minutes, and thirty-one seconds, and to save the bet, trotted the last mile in three minutes and seven seconds. In the same year, Whalebone, went the same distance in one hour, fifty-eight minutes, and five seconds. He commenced the match with a light sulky, which broke down on the fourteenth mile, and was replaced by one much heavier. In 1835, Black Joke was driven in match, on the Course at Providence, R. I., fifty miles in three hours and fifty-seven seconds. In 1837, Mischief, a grey mare, in harness, performed about eighty-four miles and a half in eight hours and thirty minutes. And in 1829, Tom Thumb, on Sunbury Common, England, performed one hundred miles in eight hours and a half.

What account is given of the horse, Tom Thumb?

Tom Thumb was an American horse, brought from beyond Missouri, and is reported to have been an Indian pony, caught wild and tamed. Others allowing him to have been thus domesticated, think him not to have been the full blood wild horse of the western prairies, but to have some cross of higher and purer blood. He was fourteen and a half hands high, of remarkable hardihood, and at the time of the above feat, was eleven years old. The whole time allowed for refreshments, during his great performance, amounted to but thirty-seven minutes. He did his work with great ease; and, at eleven miles the hour, seemed to be only playing, while other horses accompanying him labored hard.

What is said of the rearing of horses as a source of pecuniary advantage?

That it is an important department in rural economy, no one will deny. This is apparent from the great number of horses in our country, from the amount of capital invested in the animal, and from the fact that a family in the country cannot do without him, unless it be with the greatest inconvenience. The profit depends much on the goodness of the breeds, and the cheapness of feed where it is done. Near markets, where feed of every kind is at a high price, and especially if the animals be of inferior grade, the business would evidently be attended with loss. Farmers will always find that raising a poor horse will cost more than he is worth. But, remote from such markets, with superior animals, it is otherwise. There is not a more interesting spectacle, than to see on a farm, from fifty to an hundred, of all ages, from sucking colts upwards, of the best breeds of the animal; and it cannot be doubted, that where seen, the proprietor annually receives a handsome income, in addition to the pleasure experienced.

What are the evils of running and trotting horses on public courses?

It leads to the congregating of large numbers of persons—not a few of them idle and perhaps dissolute, and most of them neglecting needful labor and occupation; and in such a concourse, it is almost a matter of course there will be dissipation and gambling. The worst feature in the practice is, that many young men of doubtful tendency in morals and attention to useful occupation, will here receive an unfavorable bias, that in the end will fix their ruin. No other conclusion can be drawn. Nothing, seemingly, but a miracle can prevent such a result.

How is it supposed that these evils are overbalanced, or that the practice in question can be justified?

Moralists would say that the evils cannot be overbalanced; but it must be admitted, that the practice in question is connected with the improvements made in the breeding and rearing of horses. Had the practice never existed, no one can doubt that we should now be unable to find, in our country, those points of decided excellence prevalent in almost every section of it. Moreover, there may be persons engaged in the business of rearing horses for the race and trotting course, governed by motives as honorable and pure, as pertain to other departments of rural enterprise and advancement.

THE GREAT INTERESTS OF THE COUNTRY.

It is readily granted that an increase of domestic manufactures, to keep pace with the growth of the country, is good policy. Where there are many persons employed in manufactures, the fruits of the husbandman's labors will find a ready market and a higher price. But here may be an excess. If, instead of cultivating the earth, still greater numbers devote their labors to manufactures, the products of agriculture will not be sufficient for the wants and comforts of the people, and the advanced price of necessary articles of living will prove oppressive to those who are obliged to purchase them. The manufacturer may be well able to give the higher price, because his profits are greater. But those in other classes of society, and mechanics of other professions are burdened, without means to support or remedy the evil.

While, then, due protection is afforded by government to commerce and navigation, and also to domestic manufactures, the interests of agriculture should not be depressed nor neglected; and why should not the latter be favored by the general government, as well as the two former? That our country is susceptible of immense advances in agriculture, of very great progress, both in the quantity and kind of products from the soil, no one can doubt. And it is believed that Congress is constitutionally competent to aid and encourage agricultural pursuits as well as manufactures; "having power to promote the general welfare and the progress of the useful arts." Agriculture is truly an art or science as much as manufactures, and discoveries and improvements may be made in the former equally with inventions in mechanics. Agriculture has generally been admitted to be the most necessary of all occupations or pursuits. It is, indeed, essential to the comfort and welfare of mankind. The three great departments or branches of human labor are agriculture, manufactures, and commerce; but agriculture has the priority. As men become civilised, the two last are important, and will be encouraged as society becomes improved. What is ornamental and convenient, will be added to what is necessary.

A great and chief employment should be agriculture. It is favorable to health, and morals, and to republican liberty. And where one man or woman is now thus employed, there had better be ten. Where one bushel of wheat is now raised, there might and ought to be ten. We had better raise grain for exporting than cloths. The increase of agriculture will not prove detrimental to commerce; nor, indeed, necessarily, to manufactures. But is it not worthy of inquiry, whether less investment in manufacturing establishments, and fewer hands employed in them, and more in agriculture, would not be for the permanent prosperity of the nation, and much more favorable to morals, and therefore, to the peace and stability of our republic? Let manufactures be supported, and even extended, as the country increases in population; but let them not be considered the highest object, or worthy of the greatest anxiety to be enlarged.

The soil of a great part of the United States is favorable to agricultural pursuits, and great improvements may yet be made in the manner of cultivation, and the product to be raised. Most parts, even of the New England states, might be made to produce double and treble their present amount and value. Some states might easily increase their products sevenfold. And we need not fear a surplusage. Europe will afford a market. England and France, some years, may stand in need of ten times the amount which we now usually export.

Great Britain is as much indebted for her prosperity and wealth to agriculture, as to her commerce and manufactures, and the government there has long given protection to the interests of agriculture. We do not wish to have the agricultural interests predominate, certainly not to be cherished to the injury or diminution of manufactures and navigation. But the farming interests seem not to have the high estimation and comparative value which they deserve in such a country as this. These interests have been too much neglected by the government; and the attention of public spirited men is therefore invoked to the subject. If the federal government cannot agree to afford aid, or in what particular way to give it, let it furnish funds to each state for the purpose, and leave it with the legislatures of the several states to appropriate it in such a manner as shall be considered most useful to the whole people. We advocate not the policy of cramping commerce or manufactures; we plead only for a share of attention to agriculture, corresponding to its vast importance and essential value.

Has proper encouragement been given to the interests of agriculture by either the general or state governments? Is agriculture cherished as the support, and the permanent support, of commerce? And is it not essential to the prosperity of commerce, that agriculture be extended and improved in the United States, if our immense population, and it is constantly and rapidly increasing, cannot subsist, or cannot have all their wants supplied, but by a proportionably large commerce. The products of other climes, to a vast amount, are annually consumed or sought for among us. And how are these to be procured, except by means of commerce and navigation; and by the transportation of the surplus products of our country in exchange?

The exports of cotton from some of the southern states

are to a great amount, and some other states export tobacco and rice; but the middle and western states can only raise grain for exportation,—or, that is their principal dependence for it. And where they now raise one hundred bushels, they might raise two or five hundred. There should be encouragement for such a purpose, as well for the benefit of the merchant who exports, as of the farmer himself who grows it. Before the revolution, fish, and naval stores, and tobacco, were the great surplus staples of which our exports consisted. Now it is cotton and grain; and the latter will probably be of superior amount, unless there is some unexpected change in the state of Europe, or a mistaken indifference of the importance of agriculture should prevail in the country.

It would certainly be a most mistaken and unfortunate policy, to compel the citizens of the United States to withdraw from the ocean, and become exclusively an agricultural people. Commerce is indispensable to our growth and power as a nation. In this age of the world, a people situated as those of the United States are, would be little accounted of or respected, who were without commerce. And it is, therefore, worthy of deep consideration, even with the friends of commerce, whether, if agriculture is not encouraged and carried to a more prosperous state, the interests of foreign trade and navigation will not speedily and extensively suffer.

It cannot be said that manufactures are as favorable to the interests of navigation as agriculture. Their extension is undoubtedly for the general interest of the country; but they are not to be cherished for the sake of any direct or peculiar advantage to the mercantile portion of the nation. For many generations, we shall have little of manufactures to send to distant countries. Our population is increasing as already remarked, and will need all the goods and articles we can manufacture, unless agricultural pursuits are unusually neglected. It is true, that we export cotton goods and shoes, but they are all wanted in the country, and there is still a great importation, especially of the former. All trade with foreign nations will not, indeed, be at an end, for we cannot manufacture many articles so cheap as they can be imported from old countries, where they are made with ease and perfection.

But generally speaking, as manufactures increase among us, our navigation will be curtailed, except when the labors of our mechanics shall so far exceed those of other nations, that the articles here made shall be far superior to those of a similar kind in other countries. But the tendency of agriculture is different, and its surplus products will never be an evil, for they may always be conveyed with some profit to other regions of the globe. Unless the government should be unfriendly to commerce, and should adopt the narrow policy of preventing all intercourse with other nations, the enterprising spirit of the people will lead them to engage in trade to other countries, as their ancestors have done, and the abundance of our agricultural products will constitute their chief motive for it, as well as their profits from it.

If the subject is correctly considered, and with liberal and patriotic views, there will be no jealousy in the merchants towards the husbandman, but his success and prosperity will be just cause of satisfaction. For not only will the prosperous farmer furnish articles for transportation to other countries, but will consume a much greater amount of the articles imported in exchange from foreign parts, where they are more naturally or more easily produced. Thus, it will be perceived, that each helps the other; and, in no aspect is either one adverse to the other. They are both essential to the growth and permanent well-being of the country.

There is another consideration, which should lead us to appreciate more highly than we do, at present, the advantages of agriculture, and that is, to induce a greater portion of the rising generation to give their attention and labors to that important department of national wealth and prosperity. It is believed there is less fondness for cultivating the soil than there should be in such a country as ours ; far less than would conduce necessarily to a general supply and competence. As sincere patriots, we must desire to witness an increase of farmers, either great or small, with five hundred acres or fifty ; for these are the citizens who are truly attached to the welfare of the country, and who are too independent to be bribed or influenced by the hollow-hearted demagogue.

And as friends to commerce as well as to agriculture, we should like to see a less number crowding into shops, and engaging in mere traffic for a subsistence. To be either respectable or successful, there must be a due proportion of people concerned in merchandise and navigation. Espe cially is it necessary, to be an honorable and prosperous me chant, to have experience and knowledge of the course of trade in other countries. And yet many seek to become merchants, that they may suddenly acquire wealth, whose only claim to the honorable distinction is, that they are extravagant and injudicious speculators. A good merchant must have enterprise, but he must also have prudence and experience, or his enterprise will only be another name for recklessness and folly.

It is precisely because we desire to see our merchants honorable and respectable, as they have usually been, and the concerns of foreign commerce in the hands of upright and intelligent men, that we intimate a wish that the profession may not be crowded by persons eager only for wealth, and destitute of the education and probity proper for so elevated a rank in society. And all the present generation need to be admonished, of the pernicious results of that reckless spirit for speculation which now prevails in this country, beyond that of all others; and which aims to secure wealth or competence without steady labor and habits of industry. It would greatly promote the general prosperity, if a far larger portion of young men would become practical farmers; and in most cases, their worldly lot would thus be far more free from calamitous reverses, than by engaging in speculation, where the chance is three to one against success; or in trade, with an expectation of becoming rich or independent, without diligent application to business.—ALDEN BRADFORD, LL. D.

THE PRODUCER AND THE CONSUMER.

Let me say how greatly I am gratified at such an opportunity of meeting with the farmers of Norfolk, and their wives and daughters, and how sincerely I thank you, Mr. President, for presenting me to them. I am deeply sensible how poor a title I bring to their consideration or attention at such a festival as this. I can give them the results of but little agricultural observation abroad, and of still less agricultural experience at home. I am not sure, indeed, that it might not be said of me in regard to agriculture—I hope that it cannot truly be said of me in regard to any thing else—that I once put my hand to the plough, and looked back.

Certainly, sir, I was once the owner of a farm ; but now the only claim, the only pretence, I can set up to the favor of the farming interest, is that I believe I consume my share, or, as the common saying is, *eat my allowance* of the products of the soil.

Let me not seem, however, to disparage or depreciate this title ; though I confess that it is not quite peculiar enough to myself to be the subject of any great individual pretension. Seriously, sir, I am at a loss to know what better friend a farmer can find, the world over, than a good consumer. I know not what may be the influence of the prizes which you are presently about to distribute, but I am inclined to doubt whether, in the long run, there be any more effective premium for the promotion of agriculture, than a sure, steady, remunerative price for the crops which it produces. You may talk as you will about the strong bond of fellowship, the close tie of sympathy which exists between those who till the same soil, or wield the same scythe, or "feed the same flock, by fountain, shade, or rill," or together drive their teams a-field, "'ere the high lawns appear, under the opening eyelids of the morn ;" but I am inclined to think that not a few of the practical producers of Norfolk would rather fall in occasionally, and more especially about harvest time and on a market day, with those who have been accustomed to admire the operations of agriculture a little more at a distance. It is a very pleasant thing for any one of us to be able to say —I wish I could say it as well as my illustrious friend at my side, (Mr. Webster ;) "I raise my own potatoes and kill my own pork ;" but I think there are those around me who would rather hear us both say, that we are ready to *buy theirs.*

I must not be betrayed, however, Mr. President, into an essay on political economy, or into any further discussion of the relations between producers and consumers. Perhaps I may change my views if I am ever permitted to join the producing class myself; for to that complexion we all hope to "come at last." There is no more striking evidence of the estimation in which agriculture is held among the arts of life, than that all men, of all sorts and conditions, seem with one consent to look forward to it as the occupation of their later and better years. We rarely hear of a farmer coming down from the country to exchange his pure air, and clear skies, and ample elbow-room for the smoke, and dust, and din of a crowded city. The footsteps are all in the other direction. The mechanic at his bench, the merchant in his counting-room, the physician and the clergyman in their studies, the lawyer in his office, the statesman in the Senate chamber, all seem to indulge a common hope. At the end of the cherished vista of each one of them alike, may be seen a snug farm, a few trees, a strawberry bed, a flower-garden, a potato patch, and, above all, a quiet, independent, rural *home.* I need not say that it is among my own hopes that one day or other I may realize such an enjoyment, and then I shall at least have something more practical to speak of at a cattle show dinner.—*Extract from a Dinner-Table Speech at the Fair of the Norfolk Agricultural Society,* 1849, *in Massachusetts, by the* HON. ROBERT C. WINTHROP.

MUSIC OF NATURE IN NORWAY

Still as everything is to the eye, sometimes for a hundred miles together along these deep sea valleys, there is rarely silence. The ear is kept awake by a thousand voices. In the summer, there are cataracts leaping from ledge to ledge of the rocks, and there is the bleating of the kids that browse there, and the flap of the great eagle's wings, as it dashes abroad from its eyrie, and the cries of whole clouds of sea birds which inhabit the isles; and all these sounds are mingled and multiplied by the strong echoes, till they become a din as loud as that of a city.

Even at night, when the flocks are in the fold, and the birds at roost, and the echoes themselves seem to be asleep, there is occasionally a sweet music heard, too soft for even the listening ear to catch by day. Every breath of summer wind that steals through the pine forests, wakes this music as it goes. The stiff, spiny leaves of the fir and pine vibrate with the breeze, like the strings of a musical instrument, so that every breath of the night wind in a Norwegian forest wakens a myriad of tiny harps, and this gentle and mournful music may be heard in gushes the whole night through. This music, of course, ceases when each tree becomes laden with snow; but yet there is sound in the midst of the longest winter night. There is the rumble of some avalanche, as after a drifting storm a mass of snow, too heavy to keep its place, slides and tumbles from the mountain peak. There is also now and then a loud crack of the ice in the nearest glacier; and, as many declare, there is a crackling to be heard by those who listen when the northern lights are shooting and blazing across the sky.

Nor is this all. Wherever there is a nook between the rocks on the shore, where a man may build a house, and clear a field or two; wherever there is a platform beside the cataract, where the sawyer may plant his mill, and make a path for it to join some road,—there is a human habitation, and the sounds that belong to it Thence, in winter nights, come music and laughter, and the tread of dancers, and the hum of many voices. The Norwegians are a social and hospitable people; and they hold their gay meetings in defiance of their arctic climate, through every season of the year.

HARRIET MARTINEAU

SMALL FARMS.

The greatest obstacle to the improvement of agriculture in New England, is the propensity of the farmer,—the mania I might well call it,—to own more land than he can till to advantage. And it is thus that we see scattered over the country, large tracts of sterile, unproductive land, which under good cultivation would yield bountiful and valuable crops.

Not only the dictates of sound philosophy, but numerous facts, drawn from experience, are constantly and loudly calling upon the farmer, from every quarter, to occupy a small farm and cultivate it well. I wish that this admonition could be thundered into the ears of the agricultural population of New England, until a complete revolution should be produced in the farming system.

This great truth is already beginning to be understood in other countries, and is attended with corresponding advantages. The densest population in Europe may be found in Flanders and Lombardy, where the land is divided into small holdings, and, being thoroughly tilled, produces abundant food for the inhabitants. And the experience of a quarter of a century in France, under small-holding farmers, the land is producing one third more food, and supporting a population one third greater, than when it was possessed in large masses. The law is universal,—applying to every country,—that success in agriculture consists in the thorough cultivation of a small piece of ground, which, well-manured and well-worked, yields up its treasures in prodigal profusion.

In almost every part of New England, one capital error runs through the whole system of farming. A great deal of money is invested in land, and a very little employed in its cultivation. And it is sad to see the owner of a large farm pride himself on the number of acres which he possesses, and undertake to cultivate the soil without sufficient means. Such a man has been happily compared to a merchant, who expends all his capital in building for his own use a large, roomy store, and is afterwards seen gazing with complacency on his bare walls and empty shelves.

He has chalked out to himself a hard lot, and involuntarily enters into a state of servitude worse than Egyptian bondage. His work is never accomplished. He toils at all hours, and yet is never ahead of his work, and his work is never well done. He has not time to accomplish anything thoroughly. His house is out of repair,—his cattle poor,—his barn dilapidated,—his fences in ruins,—his pastures overrun with bushes,—and acres of land, which, under proper cultivation, might be made to yield a rich harvest, are but little removed from barrenness, or perhaps covered with noxious plants.

What a harrassed, unhappy being must be the owner of such a farm! He has no time for recreation or mental improvement. He is doomed to the tread-mill for life, with his spirits depressed, despondency stamped on his haggard lineaments, and the worm of discontent gnawing at his heart. For him there are no pleasant associations with the past, the present is full of anxiety, care and hard labor, and a cloud rests upon the future.

Such a man has little reason to pride himself on his extensive possessions; and, paradoxical as it would appear, he would, in nine cases out of ten, add to his riches as well as his enjoyment, by giving away at least one half. He is, in the true sense of the words, miserably poor,—in fact, a slave; and when his eyes are opened to his real condition, it is no wonder that he is glad to emancipate himself, by selling his farm for what he can get, and escape, post haste, to Texas or Iowa.—FARMERS' MONTHLY VISITOR.

OUR NATIVE LAND.

Our eastern borders behold the sun in all its splendor rising from the Atlantic, while the western shores are embraced in darkness by the billows of the Pacific. Our country has indeed a vast extent of territory, with the diversified climates of the globe. On the one hand, is the ever-smiling verdure of the beautiful and balmy south, and, on the other, the sterile hills and sombre pine forests of the dreary north; and, intermediate, the outstretched region where the chilling blasts of winter are succeeded by the zephyrs and the flowers of summer.

The snow-clad summits of the mountains look down upon the elemental war of the storm-clouds floating above the shrubless prairie, that realizes the obsolete notion of the earth being an immense plain, and toward the ocean on the east and the west, upon the broad, rich valleys where the father of waters, the "endless river," and the majestic Columbia, with its hundred branches, gently wind along, or rapidly rush on to mingle their waters with the waves of the Pacific, the gulf of Mexico, or the magnificent expanse of our northwestern Caspian seas.

Could the power of vision at once extend over our whole wide domain, what a grand, ennobling scene would be presented to a spectator standing upon one of the lofty peaks of the Rocky Mountains, or, as Washington Irving aptly denominates it, "the crest of the world." And then to take, upon a summer day, a bird's-eye view of all our roads, canals, railroads, lakes and rivers; the innumerable post-coaches whirling along over our one hundred and thirty thousand miles of post-road; or steamers gliding magically along our waters; our locomotives shooting off like the comet upon its track; our rapid intercourse between the seaboard and the inland maritime cities; and our ships approaching and departing, with the commerce of the world; with all the various, complicated movements of the country, town and city; and then, like Prior on Gronger Hill, to hear all the different musical and discordant sounds coming up to this "crest of the world," if they could comprehend the entire scene, from the bellowing of the buffalo, leading his shaggy hundreds over the prairie, to the roar of the cataract as it shakes the earth with its stupendous plunge;—with all this beneath the eye and upon the ear, well might the enraptured spectator exclaim, "What a sublime panorama!"

For variety, beauty, grandeur, and sublimity of scenery, what country can surpass our own; what country can equal the life-sustaining power that slumbers in her soil! With all her wealth, improvements and intelligence, still we have but just commenced the settlements of our country, and are only on the borders of the mighty wilderness. Her undeveloped resources are capable of sustaining a free population of more than one hundred millions. A century hence, if permitted to enjoy the blessings of peace, the United States of America, with fifty stars upon her banner, may number no less than one hundred and twenty millions of happy freemen. How exalted may then be the intelligence and virtue of the people. The success of our efforts in the improvement of our schools, and the general diffusion of knowledge, enable us to make an estimate of what our posterity of the third generation are likely to become.

Active must be the ardent imagination that can picture the scene at a glance. The ideal landscape cannot equal the reality, however lively may be the fancy. The idea of such a view as we have fancied may be taken from the mountain top, a hundred years hence, can never be conveyed by words; the picture must be painted by the wonder-working power of the pencil of ideality.

Our country! Such is thy physical greatness, and such the intellectual and moral power that now gives promise of a glorious destiny, far beyond all parallel in the annals of the world. For such a destiny may thy institutions be well sustained; and may a halo of glory play around the name of every man who honestly labors in behalf of his fellows and posterity, to uphold, purify, perpetuate and extend them.

SELECTED.

THE AMERICAN FOREST GIRL

Wildly and mournfully the Indian drum
  On the deep hush of moonlight forests broke;—
"Sing us a death-song, for thine hour is come,"—
  So the red warriors to their captive spoke.
Still, and amidst those dusky forms alone,
  A youth, a fair-haired youth of England stood,
Like a king's son; though from his cheek had flown,
  The mantling crimson of the island blood,
And his pressed lips looked marble. Fiercely bright,
And high around him, blazed the fires of night,
Rocking beneath the cedars to and fro,
As the wind passed, and with a fitful glow
Lighting the victim's face. But who could tell
Of what within his secret heart befell,
Known but to Heaven that hour?—Perchance a thought
Of his far home, then so intensely wrought
That its full image pictured to his eye
On the dark ground of mortal agony
Rose clear as day!—and he might see the band
Of his young sisters wandering hand in hand,
Where the laburnum drooped; or happily binding
The jasmine, up the door's low pillars winding;
Or, as day closed upon their gentle mirth,
Gathering with braided hair around the hearth
Where sat their mother;—and that mother's face
Its grave, sweet smile yet wearing in the place
Where so it ever smiled! Perchance the prayer
Learned at her knee came back on his despair;
The blessings from her voice, the very tone
Of her "Good-night," might breathe from boyhood gone!
He started and looked up: thick cypress boughs
  Full of strange sound, waved o'er him, darkly red
In the broad, stormy firelight; savage brows,
  With tall plumes crested and wild hues o'erspread
Girt him like feverish phantoms; and pale stars
Looked through the branches as through dungeon bars,
Shedding no hope. He knew, he felt his doom—
Oh! what a tale to shadow with his gloom
That happy hall in England! Idle fear!
Would the winds tell it? Who might dream or hear

The secret of the forest? To the stake
  They bound him; and that proud young soldier strove
His father's spirit in his breast to wake,
  Trusting to die in silence. He, the love
Of many hearts!—the fondly reared—the fair,
Gladdening all eyes to see! And fettered there
He stood beside his death-pyre, and the brand
Flamed up to light it in the chieftain's hand.
He thought upon his God. Hush! hark!—a cry
Breaks on the stern and dread solemnity—
A step hath pierced the ring! Who dares intrude
On the dark hunters in their vengeful mood?
A girl—a young, slight girl—a fawn-like child
Of green savannas and the leafy wild,
Springing unmarked till then, as some lone flower,
Happy because the sunshine is its dower;
Yet one that knew how early tears are shed,—
For hers had mourned a playmate brother dead.
She had sat gazing on the victim long,
Until the pity of her soul grew strong;
And, by its passion's deepening fervor swayed,
Even to the stake she rushed, and gently laid
His bright head on her bosom, and around
His form her slender arms to shield it wound
Like close Liannes; then raised her glittering eye
And clear-toned voice that said, "He shall not die!"
"He shall not die!"—the gloomy forest thrilled
  To that sweet sound. A sudden wonder fell
On the fierce throng; and heart and hand were stilled,
  Struck down, as by the whisper of a spell.
They gazed: their dark souls bowed before the maid,
She of the dancing step in wood and glade!
And, as her cheek flushed through its olive hue,
As her black tresses to the night-wind flew,
Something o'ermastered them from that young mien;
Something of heaven, in silence felt and seen;
And seeming, to their child-like faith, to token
That the great spirit by her voice had spoken,
They loosed the bonds that held their captive's breath;
From his pale lips they took the cup of death:
They quenched the brand beneath the cypress tree;
"Away," they cried, "young stranger, thou art free!"

HEMANS.

THE AMERICAN PIONEER.

Far away from the hillside, the lake, and the hamlet,
The rock and the brook, and yon meadow so gay;
From the footpath, that winds by the side of the streamlet,
From his hut and the grave of his friend far away;
He is gone where the footsteps of men never ventured,
Where the glooms of the wild tangled forest are centered,
Where no beam of the sun or the sweet moon has entered,
No bloodhound has roused up the deer with his bay.

He has left the green valley for paths where the bison
Roams through the prairies, or leaps o'er the flood;
Where the snake in the swamp sucks the deadliest poison,
And the cat of the mountains keeps watch for his food.
But the leaf shall be greener, the sky shall be purer,
The eye shall be clearer, the rifle be surer,
And stronger the arm of the fearless endurer,
That trusts naught but Heaven, in his way through the wood.

Light be the heart of the poor lonely wanderer,
Firm be his step through each wearisome mile,
Far from the cruel man, far from the plunderer,
Far from the track of the mean and the vile;
And when death, with the last of its terrors, assails him,
And all but the last throb of memory fails him,
He'll think of the friend, far away, that bewails him,
And light up the cold touch of death with a smile.

And there shall the dew shed its sweetness and lustre,
There for his pall shall the oak leaves be spread;
The sweetbriar shall bloom, and the wild grape shall cluster,
And o'er him the leaves of the ivy be shed.
There shall they mix with the fern and the heather,
There shall the young eagle shed its first feather,
The wolf with his wild cubs shall lie there together,
And moan o'er the spot where the hunter is laid.

BRAINARD.

HISTORY OF SHEEP.

Do we know the origin of the domesticated sheep?

Not with any degree of certainty. There appear to be four distinct kinds of wild sheep now known to naturalists, viz.: the *Argali*, found in the elevated ranges of northern and central Asia; the *Musmon*, belonging to the country which borders upon the Mediterranean; the *Bearded Sheep* of Barbary and Egypt; and the *Rocky Mountain Sheep* of our own continent. Many persons incline to the opinion that the Argali is the original stock, and that the variety found in North America came from Asia by crossing the ice at Behring's Straits. At first, it may seem strange that the domesticated animal could have sprung from either of these wild varieties, on account of the great dissimilarity in their appearance and habits; but, on examination, it will be found that the difference is not more stiking than that between the several distinct breeds now in our possession.

When was the sheep domesticated?

It was probably the first animal brought under the control of man, for Abel, we are told, was a "keeper of sheep." The wealth of the ancient patriarchs consisted in flocks and herds, while many of the most celebrated characters on the page of sacred history occupied the position of shepherds. The sheep of Job were fourteen thousand, and Mesha, king of Moab, was a sheep-master, and rendered unto the king of Israel one hundred thousand lambs and one hundred thousand rams with the wool. Solomon offered one hundred and twenty sheep at the dedication of the temple. Moses, the lawgiver of the Israelites, was a shepherd, while king David was brought from feeding ewes to "feed Jacob his people, and Israel his inheritance."

What is said of fat-rumped and broad-tailed sheep?

They are extended over the north and south of Asia, Palestine, and Russia, and almost entirely compose the flocks of the Calmucs and Tartars. They are thought to be the descendants from the flocks of the patriarchs. The sub-varieties are numerous, and from the wool of that in Thibet are the costly Cashmere shawls manufactured.

When was the keeping of sheep commenced in Europe?

It is generally understood that the animal, in its domestic state, was spread over western Asia at a very remote period,

VIEW OF FARM BUILDINGS AND MERINO SHEEP of DAVID HALL, Gaines, Albion county, N. Y.

but the date of its introduction into Europe is not known. The early Greeks had very choice breeds, which they probably obtained in the course of their commercial dealings with Egypt and Asia Minor. The business became one of importance in the hands of the Romans, who procured their stock from the Greeks. They endeavored to procure the best animals, and, according to Strabo, the price of rams in the time of Tiberius sometimes rose as high as a talent—worth about a thousand dollars of our money. Great care was taken with the flocks; the houses were daily washed and fumigated; the sheep were covered with skins and cloths; the wool was washed three or four times a year, frequently moistened with the finest oil, and when it became matted, drawn out and combed. Nevertheless, under their management, while the wool became remarkable for its fineness and beauty, the sheep gradually lost their ancient health and vigorous constitution.

What was the origin of the Spanish Merinoes?

It appears to be clearly established, that this valuable race originated from a mixture of the delicate and fine wooled sheep of Italy with the more robust animal of Spain. Columella, a Roman writer on agriculture, tells us that his uncle who lived in Bœtica, the present province of Estremadura, effected the cross, and that the progeny united the strength and color of the sire with the dam's softness of fleece. When Italy was overrun by the barbarians, the flocks there became extinct, while the new breed continued to flourish in their mountainous districts.

Was the excellent character of the Merinoes recognised?

It was at an early day, and the shepherds took especial care that the breed should not degenerate by intermixture with the coarser natives. The Saracens, with their luxurious habits, were attracted by the excellence of the wool, for articles of dress, and in the thirteenth century became renowned for their woolen manufactures, and Seville alone contained no less than sixteen thousand looms. After the expulsion of the Saracens, the weavers were driven from the kingdom, and the business of manufacturing became almost extinct. The government, in the course of time, saw its error, and came to look upon the business of breeding the Merino with favor, and placed all matters connected therewith under the exclusive jurisdiction of courts established for the purpose.

With a jealous watchfulness of their own interests, they strictly prohibited the exportation of the improved animals, and in a very few instances only were any allowed to be taken out of the country. But the manufactures could not be revived, and the wool was sent abroad.

When was the first importation of Merinoes into the United States?

Towards the close of the last century, some two or three animals were brought into New England, but it was not until the year 1801 that there was made any importation which resulted in the propagation of the pure breed. Large numbers were imported within the succeeding ten years, and much credit is due to the exertions of the Hon. William Jarvis, of Vermont, then United States Consul in Spain, who himself shipped thirty-six hundred from the most celebrated flocks in the kingdom, which were sold in consequence of the French invasion.

How were the Merinoes then valued in this country?

Their value for the production of wool soon became known, and the sheep commanded extraordinary prices, some rams having sold in 1811 for over one thousand dollars a-piece. Owing to a variety of circumstances, the excitement soon died away, and the Merinoes fell into undeserved neglect. But, by the sagacity of some breeders, a few flocks have been kept in the original purity, and, as within the past few years new and valuable importations have been made, it would seem that in a short period the breed can be extensively disseminated throughout the country.

What was the origin of the Saxon sheep?

It is known to be a descendant of the Spanish Merinoes carried into Saxony, in the years 1765 and 1768. These animals were selected with great care, and after their removal were regarded with an unusual degree of interest. They were placed under faithful and competent shepherds, and received the attention of the most distinguished farmers, as well as the nobility of the country.

How did this enterprise succeed?

Under this judicious management, prosecuted for a number of years, these flocks have attained a world-wide notoriety for the superior quality of their wool, being remarkable for the softness and fineness of its texture. Great pains are taken to perpetuate this character; and the Saxon shepherd has earned a high reputation for his skill. The sheep are the objects of his constant care, it being his duty to supply every want, and guard them against sudden changes of temperature. This has resulted in producing a beautiful fleece, but has somewhat weakened the natural vigor and strength of the Spanish Merino.

When were the first importations of Saxons to this country?

Four rams were brought over in the year 1823, and about eighty sheep were brought in the year following. These were received with so much favor, and sold so well, that speculators were induced to embark in the business. For three or four years after the first importation, the fever of speculation raged, and at public auction, some of the best sheep brought from four hundred to four hundred and fifty dollars a-piece. But the reaction was sudden, and the sale of several of the last lots was attended by serious loss. As may be supposed, the majority of these animals, selected with a view to commercial profit, were of very inferior character, and by being mixed with our native and pure Merino flocks, occasioned a serious injury. Hence the present dislike of the Saxon sheep in this country. But the importations of the late Henry D. Grove, of New-York State, Mr. Taintor, of Connecticut, and others, show that the pure blooded animal possesses a good constitution, and, for the quality of its wool, is well worthy of the attention of our sheep-masters.

What is said of the Rambouillet flock?

This was established as early as 1786, by the King of France, on his royal farm at Rambouillet. The stock was composed of the choicest Spanish Merinoes, and in their new situation they were bred with the greatest care. This has resulted in a marked improvement, and the sheep have acquired a larger carcase, with a much heavier and finer fleece. The very first Merinoes brought to the United States were taken from the flock at Rambouillet. Other importations have since been made, and the accounts of their value seem to be no ways exaggerated.

Have the Spanish breeds ever been introduced into Great Britain?

Henry VIII. is said to have imported three thousand Spanish sheep into England, but what became of them is unknown. Another lot of quite inferior quality was obtained in 1788, by George III., and a third of very valuable animals was sent to him in 1791. Although much interest was felt among the breeders at the time of these importations, yet it cannot be said that the Merino is held in very high favor on the English soil. The farmers evidently prefer those breeds which have there been brought into notoriety, alleging that the Spanish and Saxon varieties have not size enough to meet the demands of the market for mutton.

What are the favorite breeds of sheep in Great Britain?

The Leicester, the Cotswold, the South Down, and the Cheviot, rank the highest. Each one has its peculiar qualities, on account of which its friends claim for it a decided superiority to all the other breeds. It will be recollected that opinions upon such a subject are generally dictated by prejudice and self-interest. Nevertheless, these English breeds in general, have a deservedly high reputation for their beauty of form, fineness of bone, good quality of the fleece, disposition to fatten easily, and the excellent character of their mutton. This last is a very important recommendation, as it is well known that lamb and mutton compose a large portion of the food of the inhabitants.

Do the English breeds owe much to the skill of their breeders?

Very much; and it is gratifying to notice among the farmers of the present day a determination to secure still further improvement. It is not certain that the highest point of excellence has yet been attained. The Leicesters are chiefly indebted for their pre-eminence to the late Robert Bakewell. It is surprising to see how much he accomplished, and a high authority has said that he "seemed to have the power of modeling a sheep just as he liked, and then giving it life." One of his sheep attained the weight of 368 pounds when alive. In 1787 he let three rams for a single season for over six thousand dollars; in 1789, he let one for over five thousand dollars; and from others in the same year, he cleared more than thirty thousand dollars. What Bakewell did for the Leicesters, was done by the late John Ellman for the South Downs. Our agricultural papers give frequent illustrations of the success of modern breeders—among whom are numbered even the nobility of the land.

Have the English breeds been brought to this country?

They have by the enterprise of a few individuals, and

have already become known extensively throughout the Union. While some of their owners manifest a disposition to preserve their peculiar characteristics as near as possible, after a change of climate and soil, yet a desire to produce new varieties by crossing seems prevalent, and where conducted without skill or judgment, has undoubtedly been of little or no benefit.

What is understood by the term "Native Sheep?"

Neither North nor South America can boast of any variety of sheep that has been domesticated. The stock which we call native are derived from the animals brought over by the colonists, and are thus designated only to distinguish them from the Merino, the Cotswold, and other improved breeds. Being permitted to breed almost at will, and without any regard to improvement, they are of a very inferior character. Since public attention has been drawn to the capabilities of our country for the production of wool, more interest has been felt, and farmers in many sections have crossed their natives with the better varieties. The time will soon come when these old-fashioned natives will be extinct.

For what is the sheep valuable?

In eastern countries it is still employed by some of the tribes for carrying burdens, and by some nations its milk is used as a beverage and for being made into butter and cheese. The value of its fleece for the manufacture of clothing has been well understood from the earliest ages. The skin when partially tanned is used for parchment and the covers of books. The flesh is generally a favorite article of food, while the tallow, and even the horns and intestines, are converted into objects of utility. Add to this the fact of its being an inhabitant of every latitude, and that it can be reared in situations where other animals would starve, and we can readily see how justly it has been ranked in usefulness next to the cow.

What is said of the character of the sheep in its domestic state?

It is weak and defenceless, timid and inoffensive, so that it is altogether dependant upon the kindness of man. Its well known mildness of temper strongly recommend it to human affection, and have indicated it as a pattern of innocence, meekness, and submission. Hence the frequent allusions made to it by sacred writers. The destitute condition of the

Jews has been described as that of a flock "scattered upon the hills, as sheep that have not a shepherd." Our blessed Saviour was figuratively called "The Lamb of God," to signify his purity and quality as a sacrifice for the sins of the world; and in other places he is termed "The Good Shepherd, who gave his life for the sheep." And the sheep has been as much eulogised by profane writers, as by the Hebrew nation and ourselves.

What effect has climate upon the sheep?

It produces striking changes in the size and frame, and likewise affects the color and quality of the fleece. In cold regions we find the sheep protected by a heavy covering of wool, which is rather coarse upon the outside, but fine and soft near the skin. In the tropics, the wool gives place to hair, and it is in the temperate zones that the fleece is of that medium quality which is best adapted to manufacturing purposes. Under this wonderful ordinance of nature, when a soft-wooled sheep is carried to the equator, he speedily loses his unsuited coat, and it is replaced by a growth of coarse reddish hair.

Is the quality of the fleece influenced by the manner in which the sheep may be kept?

To a great degree—greater than would be generally supposed. Common sense indicates that the value of the wool, not only as regards its bulk but the price it will command in market—must be materially depreciated by the animal being exposed to inclement weather, when it should have the protection of warm, yet well ventilated buildings. No animal, and much less the tender sheep, thrives when exposed to the fury of the wind and storm. But, it is not so soon thought of, that there is a vast difference among articles of food in their wool-producing properties. This is a subject that has been as yet little investigated, and presents a wide field for experiment. Several trials have already been made, attended with astonishing results, but the matter is open to further research.

What is said of the Rocky Mountain sheep?

They dwell upon and about the mountains from which they receive their name. Although not found eastward of the declivity of these mountains, they frequent the elevated and craggy ridges with which the country is intersected between the great range and the Pacific. They collect in flocks, consisting of from three to thirty, the young rams and females herding together during the winter and spring, while the old rams form separate flocks. They are accustomed to pay daily visits to certain caves in the mountains, that are encrusted with a saline efflorescence, of which they are fond. It is said that the horns of the old rams attain a size so enormous, and curve so much forwards and downwards, that they effectually prevent the animal from feeding on level ground. Its flesh is represented by those who have fed on it, to be quite delicious when it is in season—far superior to that of any of the deer species which frequent the same quarter, and even exceeding in flavor the finest English mutton.

How is the Argali, or Wild Sheep, described?

It is an inhabitant of rocky and mountainous regions, and is principally found in the Alpine parts of Asia. These animals have large horns, arched semicircularly backward, and divergent at their tips, wrinkled on the upper surface, and flattened beneath; on the neck are two pendant hairy dewlaps. This creature is about the size of the fallow deer. It is of a grey ferruginous brown color above, and whitish beneath. The horns in the adult, or full grown animal, have much the appearance of those of the common ram. This animal has hair instead of wool, thus greatly differing from the general aspect of sheep; and the hair, which is close in summer, like the deer, becomes somewhat wavy, a little curled, and rough, consisting of a kind of wool intermixed with hair, and its roots concealed by a fine woolly down.

What description is given of the African sheep?

In temper it is extremely mild; but it is an uncouth looking creature. It is high on the legs, narrow in the loins, and its coat is rough and shaggy. Its horns are remarkably small, and within their curve the ears are enclosed. Whenever the ears escape from this seeming confinement, the animal exhibits much uneasiness; and difficult as it is for him to replace them, he never rests till it is accomplished. On his back and sides he is nearly black; the shoulders are reddish brown; the posterior parts of the body, the haunches, the hind legs, the tail, the nose, and also the ears, which are large, are white.

PHILOSOPHY OF PLOUGHING.

The chemist has ascertained, by analysis, what doubtless you all knew before but did not seriously consider, that every particle of the bodies of domestic animals—the bones, the muscles, the skin, the horns, the hoofs and the hair—are formed from the food they consume. He finds all the same elements in hay that he finds in any part of domestic animals, and in butter and cheese. It is impossible to calculate how much of the salts of the earth has been driven and carried from the country in cattle, and sheep, and horses, and hogs, and poultry, and not returned to the earth to preserve its fertility. But it is safe to say that if all that has gone from the county of Cheshire could be brought back into it and restored to our fields, every acre of our arable land would be made far more fertile than the richest fields of the West. And how much more of the salts of the earth, abstracted from its surface and never restored, has, in process of time, been buried deep in the grave?

Liebig, the German chemist, in his "Familiar Letters on Agricultural Chemistry," emphatically asks, "Can it be imagined that any country, however rich and fertile, with a flourishing commerce, which for centuries exports its produce in the shape of grain and cattle, will maintain its fertility, if the same commerce does not restore, in some form of manure, those elements which have been removed from the soil, and which cannot be replaced by the atmosphere?" Must not the same fate await every such country, which has actually befallen the once prolific soil of Virginia, now in many parts no longer able to raise its former stable productions—wheat and tobacco.

Elsewhere, Professor Liebig demonstrates the importance, if not necessity, of the use of manures containing phosphoric acid in combination with alkaline bases, such as potash and soda, and thence called phosphates. It has been well established that, without them, there can be neither blood, nor bones, nor brains. From the battle-field of Waterloo, bones enough have been transported to England to restore this essential element to the soil of many thousands of acres. Even twenty years ago, the estimated value of all the bones imported, in a single year, from the continent into England, was between half a million and a million of dollars.

May not the necessity that the phosphates, as well as other salts of the earth, should enter into the composition of the food of man, which chemistry demonstrates, account for many important revolutions or changes recorded in the history of the race? Existing in the smallest quantity, long and unskilful cultivation must inevitably exhaust them, so far at least as to render the supply at first deficient, and then more and more so, until the earth becomes unable to give perfection to its products. To what other cause can we attribute the decay of empires, and the depreciation in energy of the people of long inhabited countries? Why did Egypt, long ago, become less populous and less powerful than she had been? Why Assyria, Palestine, Greece and Rome? Why has eastern Virginia not only lost in population, but from the highest fell to the lowest rank among productive countries, and why has the rural population of New England become stationary? And why is it, on the other hand, that China, which has, according to her own annals, existed, as an industrious nation, many thousands of years, continues to be as populous and productive as ever, but because she exports nothing and wastes nothing that is derived from the earth?

Of all the principal articles of food, Indian corn contains the greatest proportion of the phosphates; next to corn, rye, and next to rye, wheat—rice and potatoes containing less than any others. Does not this remark lead you at once to contrast the tall and vigorous population of our own country, just opened to the occupation of civilized man, and of course yet abounding in all the fertilizing salts—a population vigorous in mind as well as in body—with the thoughtless, unambitious, imbecile people, who feed on rice and potatoes?

The art of agriculture consists in prolonging as well as in restoring the fertility of the soil. Ploughing, besides its utility for other purposes, is a mode of prolonging its fertility. The chemist tells us why it is beneficial; and to know the why and wherefore is an advantage even to him who knows what the actual result will be. Premising that all soil is disintegrated rock, which perhaps he learnt from the geologist, he informs us that all rocks, or nearly all rocks, contain, in greater or less proportions, the salts which are essential to the growth of plants; that, being contained in rocks, they exist also in their finely divided parts—sand, earth and soil;

that the roots of living plants, by the power which the principle of life gives them, can extract and imbibe whatever of these salts is near or adheres to the surfaces of these particles; that the plough, by its own operation, and by opening the earth to the operation of the weather and the atmosphere, carries the disintegration still farther, and exposes new surfaces to the roots.

But it is not only by this farther disintegration of the earth that ploughing is of service. By the experiments of the chemists it has been ascertained that at least nine-tenths in bulk of a plant consists of the constituents of the atmosphere, which enter by the roots as well as the leaves. Ploughing not only permits the gases to enter the earth more freely, but the roots to spread far and wide in pursuit of the appropriate nourishment of the plant; and it also promotes, by the free admission of the oxygen of the atmosphere, the decomposition of such vegetable matter, containing all the gases and fertilizing salts, as by chance may lie, or by design be placed, as manure, beneath the surface; and this remark is especially applicable to the most common of all manures, that taken from the barn-yard.

The salts which the chemist considers so important to agriculture exist, as he says, and as his analyses prove, in all earth as well as in all rock, even in the earth which lies far below the surface; they exist, it is true, in some localities in greater quantities and in more fit proportions than in others; and this is what constitutes the difference between good land and poor land; and when a primeval forest is first cut down and the land brought into use, they exist in greater quantity on the surface than below; for the trees have, by their roots, drawn these salts from a considerable depth; and when they die and decay the salts are left on the surface. But when the surface soil has been long in use, and the salts in it partly exhausted, the subsoil, as highly charged as ever, and perhaps more so, with similar salts, should be resorted to, and brought up by the plough. At first, it may seem to the eye unfavorable to vegetation; but if it be so, one year's exposure to the atmosphere will, in almost every instance, so change it as to render it fertile. The atmosphere and every thing with which it is charged, and all their influences, are favorable to the growth of plants. If to supply a sufficient quantity of manure has become, in this country, the first requisite of good farming, to plough frequently and to plough deep, where the subsoil is not coarse sand, is the second, so far as good farming depends on the management of the soil.—*From an Address before the Cheshire County Agricultural Society, N. H.*, 1848, *by the* HON. SALMA HALE.

SCIENCE AND PRACTICE.

An agricultural life, as it tends to locate men in fixed abodes, and give them permanent interests and associations, tends likewise to elevate them in the scale of being—awakens the higher and more ennobling sentiments of the soul, and thus conduces to the better development of our social and moral nature. The shepherd, who has a permanent possession in his flocks and herds, who makes it his care to provide for them food and drink, (they furnishing him in return with food and clothing,) is far removed, in respect to civilization, from the savage. Indeed we often speak of the patriarchal simplicity of the pastoral life, as if it were something more beautiful and desirable even, than our present state of society. But this is only the enchantment which distance lends to the view. If we really study this pastoral state of society in reference to its principles, or if we go and look upon it as it is still to be seen in the East, we may soon learn at what an immeasurable remove we are above such a state of existence.

Among these wandering shepherd tribes, there can be no settled institutions, except such as are of a very general nature. Men do not gather, and remain about any one point long enough to develop these institutions. The whole face of society is perpetually shifting so that you have no basis on which to build any thing great and lasting. Not only does such a state of things preclude the development of great public institutions, but how disastrous too is its influence in respect to individual character. There is an immense moral power in the fact that, in an agricultural state of society, a man is so fixed, that he must confront day by day and year by year, those who are in like manner fixed about him, as neighbors and fellow-citizens. He is so placed that his conduct must be thoroughly known, and his character well understood. He stands as in the presence of a permanent tribunal, before which he has the strongest motives to acquit himself well.

A wandering Arab lies, steals and murders, at one place, and then shoots off into another part of the country; and though as an old Roman poet has said, "They who pass over the sea have a change of skies but not of character"—that is they carry their characters with them wherever they go, yet in such a mixed and moving state of society, a man *seems* almost to escape from himself, because he perpetually slips away, before he is really confronted by a tribunal, which would search him out and reveal him to himself. Permanence of abode, permanence of associations, permanence of interests, are requisite, in order to the highest and best institutions, in order to the purest and noblest society. Yet even in reference to these things, there may be many degrees of perfection. The tillers of the soil, in England, for instance, are perhaps more fixed to one spot, than in this country. Many of them hardly ever go out of sight of their own cottages, and such a thing as a removal from one part of the land to another, would in vast multitudes of cases, be deemed utterly impracticable, and almost impossible. England yet retains the traces of that great feudal system, which so long prevailed throughout central and northern Europe, and which gathered men in clusters around a central baron or lord, as his vassals or retainers. The great difficulty under this arrangement is, that only a few persons comparatively are the real owners of the land, while the great multitude of those who till the soil lack that stimulus—that principle of elevation, which comes from actual possession. They have not that sense of personal independence and responsibility, which is one of the grandest elements of progress and growth.

In childhood and youth, we often have a burning desire to live in the city. Many of you, doubtless, have passed through this state of feeling and know its power. Many a boy at the age of twelve or fifteen, thinks there could be no higher bliss for him than to go and live in the city. The mind warms with the thought of being permitted to behold all its sights, and to catch all its sounds—to look upon its shows and processions—and mingle in its bustle and uproar. But as we grow older we are very apt to get over that feel ing. Our tastes become more natural. The city grows old and uninteresting. We weary of its artificial monotony—and are sick of its noise. But nature is ever fresh and young. We drink in health, amid the scenes which are open around us. The works and the operations of her hand never grow old. With every return of spring, many a dweller in the city pines for the old haunts, which he once frequented—dwells with longing upon the woods, fields and streams about which he once wandered. There distant scenes visit him amid the noise and uproar of the metropolis, amid the confusions of his business, and he would give much if he could return again to them. But no such pining for the city is found among the full grown men in the country. Nature asserts her supremacy in this, and maintains a dominion over the mind, which no artifice can supplant.

And the effect of this familiar and long continued converse with nature, amid the beauty and order of her works, is not lost upon the character. In whatever position of life we may be placed, we become indeed, after a a time, almost unconscious of the influence which is exerted upon us by surrounding scenes and objects. The dweller amid the Alps is not usually conscious of the effect which is wrought upon his mind by the presence of this grand and inspiring scenery. Nevertheless an influence is exerted, and if you take a people which have dwelt for generations amid such scenes, and compare them with another people, that have been living upon the open plains, and you perceive marked shades of difference which have been caused by the pressure of different outward objects. Precisely so with the husbandman who lives amid his fields, as compared with the man dwelling amid the associations of a city. The former is surrounded by influences far more genial and healthy—morally and physically healthy, and the result, looking at things on a wide scale, is visible in the characters of the two.

Still, we all have a timely sense of the trials, difficulties and hardships of our own particular lot, and we are often apt to fancy that there are fewer evils in some other conditions in life than our own. The sailor, when he has been a little while upon the land, tires of its monotony, and longs again to feel the motion of the sea, and share again in its excitements. But when he has been a little while at sea, tossed about with storms, and facing danger, he sighs for the quiet of home, and the comfort of rest, and thinks if he can ever reach the shore of his own land again, he shall be content to remain there. The farmer weary with his toil, sometimes thinks, that to do the work of a merchant or a professional man, would be no labor at all—mere play. The professional man, sitting down to a piece of writing which must be done, perhaps with aching head and thoughts confused, feels that if he had nothing to do but to go out in his garden or field, and use his hoe or spade—nothing to do but to work with his hands, it would be a very easy matter; but to work with his head is quite another thing. Allowing, however, for all these varieties of feeling—for this is a matter of feeling rather than of judgment—allowing for all these, we think it may safely be concluded, that no class of men occupy a more comfortable, desirable, and we may add, important place in society than the farmers.

Still, much may yet be done, to make their employment more inviting, and their lot in life more desirable. Not to dwell upon the various points that might here be touched, let me call your attention to one which you will certainly appreciate, as it is in keeping with the very object of this association. I allude to a knowledge of *agricultural science.* I well remember that there was an early prejudice among farmers against what is called scientific farming. There was a feeling that all such farming was necessarily of a slipshod and unprofitable character—and that what the farmer wanted was *practical* knowledge simply—information such as had come down to him in the natural course of things, from generation to generation. Now this practical knowledge is doubtless of the utmost importance, and one thing which has tended to keep alive this prejudice against science is, that many men who have set out upon a course of scientific farming, have been destitute of this kind of knowledge, and have accordingly failed.

I well remember that there came to my native place, when I was a boy, a retired merchant, who bought a farm, with the purpose of having everything done in true scientific style. He designed to show the ignorant natives about there the wonders of science, and deeply impress upon their minds their exceeding ignorance. Well, the farmers round about used to make themselves merry at watching his operations. Things went queerly. There was a magnificent outlay, but small results—great expenditures but small profits. They came to the conclusion that they managed quite as well, and a little better than he did. And so they did. The truth is, science and practical knowledge in this business, as in all others, must go hand in hand. Farming is, in a most important sense, a trade, in which a man must learn the use of his tools simply and only by practice. It is no more natural to a man to make a handsome hill of corn, than it is to make a nail for a horse-shoe—and ever so much instruction upon the point by mere words will not enable him to do it. He must come to the trial. He must arrive at the result step by step, and by slow degrees. And so on through all the details of agriculture. This practical knowledge of farming as a trade, which has been learned by years of experience, is something which cannot be dispensed with.

But in order to the highest pleasure, profit, and dignity of the farmer's life, he wants still another thing—and that is an accurate knowledge *scientifically* of the nature of plants and of soils—the uses of manure—the methods of vegetable growth—such information, in short, as will give him power to control the operations of his farm, and bring them into harmony with the laws of nature. Now here is a wide field, which has not yet, except in a few instances, comparatively, been entered upon, even by the intelligent farmers of New England. Indeed, it is but a little while since this field of study has really been rendered accessible to the common mind. But now the way is rapidly preparing, so that one can take up this branch of study without difficulty, and make himself, to a considerable extent, master of it. Liebig's work on agricultural chemistry, has been for some years before the people.

In the first place, science is always safe. It does not lie. It deals with fixed principles. It touches bottom at every point, so far as it goes. It does not send men off upon a wild goose chase after something that is not there. It does not simply leave them to arrive at results only after a long and expensive course of experimenting. It starts with fixed facts. It has analyzed things to their elements. Again, every branch of business becomes interesting and attractive to the mind, in proportion as we master its details and thoroughly understand it. That which is dry and barren on a mere superficial view, is full of life and meaning to him who has penetrated and comprehended it. When a farmer can penetrate the mere surface of agriculture, to the more hidden processes which are going on all about him, he dwells in a little world of his own, a most interesting world too, one which has within it abundant sources of thought and delight.

If farmers would thoroughly study the science of agriculture, and master it, they would by that very process, bring their minds in contact with many other sorts of knowledge, and the field of their intelligence would be greatly enlarged. All knowledge is connected, and it matters little what branch of study you pursue. If you take up any one, and follow it out thoroughly, you inevitably come in contact, in a merely incidental way, with all sorts of knowledge. The whole intellectual nature is quickened and expanded. It can be made profitable. It can be turned to great account in reference to the actual income of a farm. There is no manure so cheap as a wholesome application of science. It prevents useless outgoes and secures an income. In reference to this matter, a recent writer says :

"On the general subject of book-farming, we suppose that all the world knows there is a schism among farmers. The old conservative farmers don't believe in it. The young progressive farmers do. Here we have it, under shady orchards, along the forest furrows, among chattering birds, and in spring airs, the same controversies that are waged in politics and religion. Now we are on both sides; we are for book-farming, and we are against book-farming. If the farmer applies the book to the soil instead of to his own head, we are sure that his farming will be poor enough. But if books and papers are used on the farmer's mind—to give it habits of reflection and observation ; to possess it with the elements of Nature, with which he deals, and if then he directs that mind, thus stored and trained to his work, we are sure that he will benefit as much by education as any other man." There is of course at the present time much scientific information diffused abroad through agricultural papers, to be picked up here and there by piece meal. But this is not enough. Before the mind can be prepared to appropriate this so as to make it of much account, there must be the masters of the system, as a system, and then all these items will fall into their natural place, and help on the work.—*From an Address at Framingham, Mass.*, 1850, *by* REV. MR. TARBOX.

THE RISING OF THE WATERS.

About daybreak it began to rain, and continued to pour with increasing violence all the morning; no one thought of stirring abroad who could keep within shelter. My boys and I had for a task only to keep the fire at the door of the shanty brisk and blazing, and to notice that the pools which began to form around us did not become too large; for sometimes, besides the accumulation of the rain, little streams would suddenly break out, and, rushing towards us, would have extinguished our fire, had we not been vigilant. The site I had chosen for the shanty was near to a little brook, on the top of the main river's bank. In fine weather, no situation could be more beautiful; the brook was as clear as crystal, and fell with a small cascade into the river, which, broad and deep, ran beneath the bank with a swift and smooth current.

The forest up the river had not been explored above a mile or two; all beyond was the unknown wilderness. Some vague rumors of small lakes and beaver dams were circulated in the village, but no importance was attached to the information: save but for the occasional little torrents with which the rain sometimes hastily threatened to extinguish our fires, we had no cause to dread inundation.

The rain still continued to fall incessantly: the pools it formed in the hollows of the ground began, towards noon, to overflow their banks and to become united. By and by something like a slight current was observed passing from one to another; but, thinking only of preserving our fire, we no sooner noticed this than by occasionally running out of the shanty into the shower, and scraping a channel to let the water run off into the brook or the river.

It was hoped that about noon the rain would slacken; but in this we were disappointed. It continued to increase, and the ground began to be so flooded while the brook swelled to a river, that we thought it might become necessary to shift our tent to a higher part of the bank. To do this, however, we were reluctant, for it was impossible to encounter the deluge without being almost instantly soaked to the skin; and we had put the shanty up with more care and pains than usual, intending it should serve us for a home until our house was comfortably furnished.

About three o'clock the skies were dreadfully darkened and overcast. I had never seen such darkness while the sun was above the horizon, and still the rain continued to descend in cataracts, but at fits and intervals. No man, who had not seen the like, would credit the description. Suddenly a sharp flash of lightning, followed by an instantaneous thunder peal, lightened up all the forest; and almost in the same moment the rain came lavishing along as if the windows of heaven were opened; anon another flash, and a louder peal burst upon us, as if the whole forest was rending over and around us.

I drew my helpless and poor trembling little boys under the skirts of my great coat. Then there was another frantic flash, and the roar of the thunder was augmented by the riven trees that fell, cloven on all sides, in a whirlwind of splinters. But though the lightning was more terrible than scimitars, and the thunder roared as if the vaults of heaven were shaken to pieces and tumbling in, the irresistible rain was still more appalling than either. I have said it was as if the windows of heaven were opened. About sunset, the ground floods were as if the fountains of the great deep were breaking up.

I pressed my shivering children to my bosom, but I could not speak. At the common shanty, where there had been for some time an affectation of mirth and ribaldry, there was now silence; at last, as if with one accord, all the inhabitants rushed from beneath their miserable shed, tore it into pieces, and ran with the fragments to a higher ground, crying wildly, "The river is rising!"

I had seen it swelling for some time, but our shanty stood so far above the stream, that I had no fear it would reach us. Scarcely, however, had the axemen escaped from theirs, and planted themselves on the crown of a rising ground nearer to us, where they were hastily constructing another shed, when a tremendous crash and roar were heard at some distance in the woods, higher up the stream. It was so awful, I had almost said so omnipotent, in the sound, that I started on my feet, and shook my treasures from me. For a moment the Niagara of the river seemed almost to pause—it was but for a moment—for, instantly after, the noise of the rending of weighty trees, the crashing and the tearing of the rooted forest, rose around. The waters of the river, troubled and raging, came hurling with the wreck of the woods, sweeping with inconceivable fury every thing that stood within its scope; a lake had burst its banks.

The sudden rise of the waters soon, however, subsided; I saw it ebbing fast, and comforted my terrified boys. The rain also began to abate. Instead of those dreaded sheets of waves which fell upon us as if some vast ocean behind the forest was heaving over its spray, a thick continued small rain came on; and, about an hour after sunset, streaks and breaks in the clouds gave some token that the worst was over;—it was not, however, so, for about the same time a stream appeared in the hollow, between the rising ground to which the axemen had retired, and the little knoll on which our shanty stood; at the same time the waters in the river began to swell again. There was on this occasion no abrupt and bursting noise; but the night was fast closing upon us, and a hoarse muttering and angry sound of many waters grew louder and louder on every side.

The darkness and increasing rage of the river, which there was just twilight enough to show was rising above the brim of the bank, smote me with inexpressible terror. I snatched my children by the hand, and rushed forward to join the axemen; but the torrent between us rolled so violently that to pass was impossible, and the waters still continued to rise.

I called aloud to the axemen for assistance; and, when they heard my desperate cries, they came out of the shed, some with burning brands and others with their axes glittering in the flames; but they could render no help; at last, one man, a fearless back-woodsman, happened to observe, by the firelight, a tree on the bank of the torrent, which it in some degree overhung, and he called for others to join him in making a bridge. In the course of a few minutes the tree was laid across the stream, and we scrambled over, just as the river extinguished our fire and swept away our shanty.

This rescue was in itself so wonderful, and the scene had been so terrible, that it was some time after we were safe before I could rouse myself to believe I was not in the fangs of the nightmare. My poor boys clung to me as if still not assured of their security, and I wept upon their necks in the ecstasy of an unspeakable passion of anguish and joy.

About this time the mizzling rain began to fall softer; the dawn of the morning appearing through the upper branches of the forest, and here and there the stars looked out from their windows in the clouds. The storm was gone, and the deluge assuaged; the floods all around us gradually ebbed away, and the insolent and unknown waters which had so swelled the river shrunk within their banks, and long before the morning, had retired from the scene.

Need I say that anthems of deliverance were heard in our camp that night? Oh, surely no! The woods answered to our psalms, and waved their mighty arms; the green leaves clapped their hands; and the blessed moon, lifting the veil from her forehead, and looking down upon us through the boughs, gladdened our solemn rejoicing.—JOHN GALT, *in Canada*.

ANTIQUITY OF FREEDOM.

Here are old trees, tall oaks and gnarled pines,
That stream with gray-green mosses; here the ground
Was never touched by spades, and flowers spring up
Unsown, and die ungathered. It is sweet
To linger here, among the flitting birds
And leaping squirrels, wandering brooks, and winds
That shake the leaves, and scatter as they pass
A fragrance from the cedars thickly set
With pale blue berries. In these peaceful shades—
Peaceful, unpruned, immeasurably old—
My thoughts go up the long dim path of years,
Back to the earliest days of Liberty.

O FREEDOM! thou art not, as poets dream,
A fair young girl, with light and delicate limbs,
And wavy tresses gushing from the cap
With which the Roman master crowned his slave,
When he took off the gyves. A bearded man,
Armed to the teeth, art thou: one mailed hand
Grasps the broad shield, and one the sword; thy brow,
Glorious in beauty though it be, is scarred
With tokens of old wars; thy massive limbs
Are strong and struggling. Power at thee has launched

His bolts, and with his lightnings smitten thee;
They could not quench the life thou hast from Heaven.
Merciless Power has dug thy dungeon deep,
And his swart armorers, by a thousand fires,
Have forged thy chain; yet, while he deems thee bound,
The links are shivered, and the prison walls
Fall outward; terribly thou springest forth,
As springs the flame above a burning pile,
And shoutest to the nations, who return
Thy shoutings, while the pale oppressor flies.

Thy birthright was not given by human hands.
Thou wert twin-born with man. In pleasant fields,
While yet our race was few, thou sat'st with him,
To tend the quiet flock and watch the stars,
And teach the reed to utter simple airs.
Thou by his side, amid the tangled wood,
Didst war upon the panther and the wolf,
Thine only foes: and thou with him didst draw
The earliest furrows on the mountain side,
Soft with the Deluge. Tyranny himself,
Thy enemy, although of reverend look,
Hoary with many years, and far obeyed,
Is later born than thou; and as he meets
The grave defiance of thine elder eye,
The usurper trembles in his fastnesses.

Thou shalt wax stronger with the lapse of years,
But he shall fade into a feebler age;
Feebler, yet subtler. He shall weave his snares,
And spring them on thy careless steps, and clap
His withered hands, and from their ambush call
His hordes to fall upon thee. He shall send
Quaint maskers, forms of fair and gallant mien,
To catch thy gaze, and, uttering graceful words,
To charm thy ear; while his sly imps, by stealth,
Twine round thee threads of steel, light thread on thread,
That grow to fetters; or bind down thy arms
With chains concealed in chaplets. Oh! not yet
May'st thou unbrace thy corslet, or lay by
Thy sword! nor yet, O, Freedom! close thy lids
In slumber; for thine enemy never sleeps;

And thou must watch and combat, till the day
Of the new Earth and Heaven. But wouldst thou rest
A while from tumult and the frauds of men,
These old and friendly solitudes invite
Thy visit. They, while yet the forest trees
Were young upon the inviolated Earth,
And yet the moss-stains on the rock were new,
Beheld thy glorious childhood and rejoiced.

BRYANT.

NATURE'S NOBLEMAN.

Away with false fashion, so calm and so chill,
  Where pleasure itself cannot please;
Away with cold breeding, that faithlessly still
  Affects to be quite at its ease;
For the deepest in feeling is highest in rank,
  The freest is first in the band,
And Nature's own Nobleman, friendly and frank,
  Is a man with his heart in his hand!

Fearless in honesty, gentle yet just,
  He warmly can love—and can hate,
Nor will he bow down with his face in the dust
  To Fashion's intolerant state:
For best in good breeding and highest in rank,
  Though lowly or poor in the land,
Is Nature's own Nobleman, friendly and frank,
  The man with his heart in his hand!

His fashion is passion, sincere and intense,
  His impulses, simple and true,
Yet tempered by judgment, and taught by good sense,
  And cordial with me, and with you:
For the finest in manners, as highest in rank,
  It is *you*, man! or *you*, man! who stand
Nature's own nobleman, friendly and frank,
  A man with his heart in his hand!

TUPPER.

CHINESE HOGS, of the Hon. J. Delafield, exhibited at the Annual Fair of the New York State Agricultural Society, at Albany, 1851.

HISTORY OF THE HOG.

What is the early history of the hog?

This animal appears to have been known in the remotest ages, and yet its history, until within a few centuries, is clothed in obscurity. The circumstance of the Israelites being so strictly forbidden to eat its flesh, has been urged by some writers to show that they were familiar with its use.

How was the hog esteemed by the Israelites and Egyptians?

They both abstained from the flesh, regarding the animal as unclean. The Israelites held it in such detestation, that they would not even pronounce its name, alluding to it by some such title as "that beast," or "that thing." It is said that if an Egyptian touched a hog at any time, except on the feast-day of the moon, when he was allowed to eat of it, he was obliged to purify his person and clothing by plunging into the Nile. The swineherds formed an isolated race, considered as outcasts from society, being forbidden to enter a temple or marry into other families.

How was the hog esteemed by the Greeks and Romans?

Very highly by both nations, and with the latter the art of raising it became a study. It was their design to render the flesh as delicate as possible, and, to effect this, the poor beast was actually tormented to death, in various ways too horrible to be described. A celebrated dish, consisting of an entire carcase, stuffed with delicacies of every description, and bathed in wines and rich gravies, became so expensive that a sumptuary law was passed respecting it. Another great dish was a pig served whole, one side having been roasted while the other was boiled.

Whence originated our domestic hog?

It is supposed to have been derived from the wild boar, which is a native of Asia, Africa, and Europe. The wild boar is not found in America, and the large herds of wild swine that roam over certain districts, are only descendants of the stock brought to the continent at the time of its settlement. The general appearance of the wild boar and the domesticated animal are much alike. And it has been ascertained that the first named, when brought under the influence of man, gradually loses its peculiarities of form, and changes its habits of life.

How is the wild boar described?

He is generally of a dusky brown color, of large size, and possessing a remarkable degree of strength. He is active and very fierce. He is partial to thick woods, in the vicinity of a swamp, or of water, and makes his living principally of herbs, fruits, and roots. The females and their young are accustomed to herd together, but he wanders through the forest alone, conscious of his power, and neither seeking nor avoiding an enemy. He is supposed to live about thirty years. Hunting the wild boar, it is well known, has in all ages been a favorite sport.

In what manner is he hunted?

It is done by dogs, or else he is taken on surprise in the night by the light of the moon. As he runs slowly and leaves a strong scent behind him, he is easily pursued and overtaken. This is particularly the case with those which are old, the younger ones run swift and a great distance without stopping, so that they are taken with difficulty. During the day the wild boar usually hides himself in the thickest and most unfrequented part of the wood; and in the evening, and at night, he goes out to seek for food.

How long has the keeping of swine prevailed in France and Spain?

From time immemorial. By the accounts of ancient writers, it appears, that the people of those countries kept immense droves of swine, which supplied at one time nearly the whole of Italy with meat.

How long has it been practiced in England?

Our earliest accounts of the country speak of the large herds of swine, that constituted the wealth of the inhabitants, and chiefly supplied their tables. According to legendary history, the business was conducted on an extensive scale as far back as the year 863 B. C. The hot springs of Bath, so celebrated for their medicinal qualities, are said to have been discovered in that year, in consequence of the following circumstance. A young prince who was afflicted with the leprosy, disguised himself, and, at a distance from his father's court, performed the humble duties of swineherd. He noticed that the animals in his care were cured of a disease of the skin, by wallowing in a bed of warm mud; and, by adopting the same course of treatment, he was in a short time permanently restored to health.

Has the art of breeding been successfully conducted in England?

Under the care and skilful management of the English farmer, a very important change has taken place in the character of the swine of that country. The wild boar and the old English hog have both given way to the improved breeds, each of which has a character and a name. Nearly all of these improvements have been introduced to the American breeder.

What is said of the Chinese hog?

It appears to have produced all our modern improved breeds, by having been crossed either with the wild boar, or the old English hog. It was brought from China by ship-masters on their return voyages, although it has been affirmed that the choicest animals can be procured by the favor of some official only. In several instances it has been brought to this country, and here, as in England, its intermixture with the old established breeds has been of great benefit. It is itself not so well adapted to the farmer's wants, as the crosses thus produced.

Which have been distinguished in England as the best breeds?

A few only of them can be named; for, to detail the origin and good qualities of the whole number, would require a volume of no mean size. The most conspicuous within the last few years, are the Leicestershire, the Kenilworth, the Woburn, the Essex, the Improved Suffolk, and the Berkshire. The last named is generally considered the best on account of their smallness of bone, hardihood, fecundity, aptitude to fatten upon a little food, together with their early maturity. Some animals have attained the weight of over eight hundred and fifty pounds.

When was the hog introduced into America?

Probably in one of the voyages of Columbus, but the first reliable account of its introduction upon the continent, is of its having been brought from Cuba to Florida, in the year 1538, by Ferdinand de Soto. In 1553, the Portuguese took it to Nova Scotia and Newfoundland. The French carried it with them to Acadia, or New France, in 1604; and, five years afterward, six hundred swine were brought to Jamestown in Virginia. From its domestic habits and wonderful fecundity, it is an animal that was every way suited to the wants of the colonists, in their first years of toil and hardship. The hogs were turned into the woods to procure their own subsistence, and they multiplied with such rapidity, that in 1617, it became necessary to erect a palisade around James-town, in order to prevent their troublesome incursions.

What was the character of the swine thus introduced?

Of that nothing certain is known. The animal brought by the French and Spaniards was, probably, of the black Spanish breed, which possessed many good points, and which, it is thought, has had a perceptible influence upon the whole stock in the southern part of the United States. The old English breed was, without a doubt, the one introduced into the English settlements, and shows traces of its blood in the awkward, long-legged, long-nosed, small-bodied, unprofitable brutes that were more common a few years since, than at the present time.

What was the first improved breed brought to the United States?

In the latter part of the eighteenth century, a pair of Woburn pigs was sent to General Washington by the Duke of Bedford, under whose care they had originated. This breed is supposed to have sprung from a cross of the Chinese upon the best native stock of England. The messenger dishonestly sold the pigs in Maryland, where they laid the foundation of an extensive family, that has exercised a beneficial influence upon the old breeds throughout the northern states. The character of the Woburn breed stands high, but it is thought to have lost its original purity, in this country as well as in England.

Where did the Byfield, or Grass, breed originate?

In the town of Byfield, Massachusetts. About thirty years ago, a farmer one day picked up in market a fine looking pig, the progeny of which were distinguished for their handsome shape, and other good qualities. The breed became very popular, and were widely disseminated; but, in the course of time they were superseded in the public favor by other varieties of greater pretensions.

To whom does the Mackay breed owe its name?

To Capt. John Mackay of Boston, Massachusetts. It originated somewhere about the year 1825, and was brought to its present perfection by successive crosses with some fine animals that, in his different voyages, he obtained in several parts of the world. It is now highly esteemed, and is bred by numbers of our best farmers—the Hon. Daniel Webster among others.

When was the Berkshire breed introduced into the United States?

As early as 1823, by the late John Brentnall, an Englishman settled in Canterbury, N. Y. Other importations have been made from time to time, and no reason exists why the blood should not be as pure as in England. The Berkshires were favorably received, and at one time sold at very high prices. Without doubt, it is one of the best—if not the very best,—breeds in the country, but its merits now appear to be greatly undervalued.

Are these all the breeds that have attracted attention in the United States?

By no means, and to enumerate them, as well as describe their respective characteristics, is altogether different from our plan in the present condensed summary. They have been produced by frequent, and, in many cases, injudicious crossings of the old varieties; hence a large proportion of the whole number find little to recommend them, excepting in the eyes of their original breeders. Among the foreign breeds, the Improved Suffolk, imported and bred by William Stickney, Esq., of Boston, stands conspicuous; while the Medley breed of C. N. Bement, of Albany, N. Y., deserves a favorable consideration among those that have originated on our own soil.

How is the hog esteemed in the United States?

It is raised extensively in every section—few persons being so poor that each one cannot maintain his pig—but the annual number produced in the western states is almost incredible. According to the census of 1850, the whole number in the country at the first of June was thus showing that they form an important item of our national wealth.

What is the general character of the hog?

It is generally considered stupid, intractable, rapacious, and filthy. But, with some persons it has been a question whether this character be not mainly owing to the mismanagement of the keeper, instead of being the animal's natural disposition. In its native state, it is not destitute of affection for others of its kind, herding with them for mutual safety.

Instances are not wanting of its tractability, for it has been taught the tricks of a hunting dog, and in Italy to discover truffles that are buried a few inches below the surface of the ground. Who has not seen or heard of various learned pigs, that show a degree of sagacity quite unusual in the brute creation? Its voracity may be attributable to the excellence of its digestive organs furnished by nature; and may not its fierceness and bad temper be frequently the result of rough treatment? Surely it is not so filthy as has been averred, for it loves a clean bed, and its habit of rolling in the mud is merely to cool its skin, as well as to protect itself against the attacks of insects.

What are the good qualities of the hog?

Although naturally the inhabitant of a warm climate, yet it adapts itself to almost every country. It comes soon to maturity, and increases rapidly, being more prolific than any other domestic animal, except the rabbit. It is easily susceptible of improvement, for no kind of live stock can be so quickly moulded to the fancy of the breeder. It thrives with comparatively little attention, and will live—nay, grow—upon the refuse of the field and kitchen, consuming what would otherwise be wasted, or prove a nuisance upon the premises. We add, the value of its dung as a fertilizer.

For what purpose is the hog valuable?

The skin is used by the saddler, and the bristles are employed by the shoemaker and the manufacturer of brushes. The bones, when reduced to charcoal, are used by the sugar refiner. The excellence of the flesh is well known, and will be attested by every nation excepting the Jews, Egyptians, and Mahommedans. It takes salt more kindly than any other meat, and is, therefore, better capable of preservation for household use, naval and military stores, or exportation. The lard is highly esteemed by the cook in culinary preparations; it moreover yields, when subjected to pressure, a pure limpid oil, called oleine, which is used for illumination, wool-dressing, and upon machinery, leaving a substance called stearine that resembles spermaceti, and is valuable for making candles. At the west, when pork is cheap, it has become the practice to extract the lard from the whole carcase, after the hams have been taken out, by means of a steam bath. In fact, every part of the animal—even the head, feet, and intestines—is found useful, and contributing as well to social comfort as to public wealth.

What is said of the animal senses in the hog?

The sense of touch is very imperfect, owing probably to the thickness of the skin; but their other senses are good, and the huntsmen know that wild boars both see, hear, and smell, at a great distance; since, in order to surprise them, they wait in silence during the night, and place themselves under the wind, to prevent the boars perceiving their smell, of which they are sensible at a great distance, and which always immediately makes them change their road.

What is said of the taste of the hog?

In their taste hogs discover a strange degree of caprice; for whilst they are singularly delicate in their choice of herbs, they will devour with voracity various kinds of animal substances, even the most nauseous and putrid carrion.

How is the hog affected by metereological influences?

He seems to be peculiarly affected by the approach of stormy weather. On such occasions, he runs about in a restless and excited manner, sometimes uttering loud cries or taking up sticks and carrying them in its mouth?

What anecdote does Mr. Craven relate of an American sow?

This animal, says he, passed her days in the woods, with a numerous litter of pigs, but returned regularly to the house in the evening, to share with the family a substantial supper. One of her pigs was, however, slipped away to be roasted; in a day or two afterwards another; and then a third. It would appear that this careful mother knew the number of her offspring, and missed those that were taken from her, for after this she came alone to her evening meal. This occurring repeatedly, she was watched out of the wood, and observed to drive back her pigs from its extremity, grunting, with much earnestness, in a manner so intelligible, that they retired at her command, and waited patiently for her return.

What anecdote is related of a pig in the Naturalist's Library?

Early in the month, it is said, a pig that had been kept several days a close prisoner to his sty, was let out for the purpose of its being cleaned and his bed replenished. On opening the sty-door he anticipated the purpose of his liberation by running to the stable, from which he carried several sheaves of straw to his sty, holding them in his mouth by the band. The straw being intended for another purpose, it was carried back to the stable; but our porter, seizing a more favorable opportunity, regained it to the amazement of several persons, who were pleased to observe the extraordinary instinct of this wonderful pig.

What anecdote is related of a pig belonging to Mr. Kinghorn?

He says the pig became so attached to a bull-dog that it would follow and sport with him, and follow her master, when he was accompanied by this dog, for five or six miles. The dog was fond of swimming, and the pig imitated this propensity; and if any thing was thrown into the water for the dog to fetch out, the pig would follow and dispute the prize with him very cleverly and energetically. These two animals invariably slept together.

What anecdote is related by Mr. Henderson of a sow?

I have a sow, says he, of a good breed, so docile that she will suffer my youngest son, three years of age, to climb upon her back and ride her about for an half hour at a time, and more; when she is tired of the fun, she lays herself down, carefully avoiding hurting her young jockey. He often shares his bread and meat with her.

INFLUENCE OF SCIENCE ON AGRICULTURE.

Famines have depopulated whole districts, and millions of the human race have died of starvation, and yet we have no evidence that all this suffering and all the evils necessarily connected with them, have ever operated to the improvement of agriculture, or have been instrumental in causing two blades of grass to grow where only one grew before. The agricultural world has jogged along as if nothing had happened, and as if nothing could be done to save men from these wide spreading calamities. When, however, the mind has been awakened by the light of science, when discoveries are announced which, if they illuminate only a small part of his field of labor, it usually happens that an impulse is given to his dormant powers which propels him forward in a career of improvement. What calamity, therefore, fails to produce, what the strongest incentives fail to do, is in truth effected by an agency the least expected, the gentle light of discovery beaming from a kindred department of knowledge. The same things happen in morals; earthquakes swallow up their thousands, and their continual shocks day by day startle the living, but they have never created or even improved the religious sentiment; their frequent alarms and the exposure to such imminent dangers and continual sufferings, have produced rather a recklessness of conduct than a life of religion and charity.

It is not my purpose to stop here and inquire into the cause of such seeming anomalies in the human constitution; it is sufficient to allude to the facts. I pass on to say that agriculture had made only a feeble effort to improve its mechanical modes of tillage until the period when chemistry had so far advanced that it was an established truth that its principles stood in very intimate relationship to it. So botany and geology which had been cultivated as independent systems, about the same time with chemistry began also to be studied in their relations to other sciences, and hence these, together with physiology and other collateral branches, implanted clearer views of the wants of agriculture, as well as to furnish striking illustrations of the true nature and import of the principles which lie at the foundation of its system.

It is true that practical agriculture is not deeply interested in questions relating to life in the abstract or essence; but certainly much more so to those powers which modify or control its developments. These powers belong to the deep and profound inquiries which in later times are destined to achieve triumphs for her, of a still more decided character than the world has yet witnessed. It is the peculiar province of the sciences to improve the outward condition of men. Literature had attained its highest state of excellence, and yet men were not discontented in hovels, nor with straw beds nor coarse food spread upon rough boards. Literature was brilliant as well as solid in Queen Elizabeth's day, and yet laboring men were more poorly fed and cared for, than the cattle in the period in which we are permitted to live.

Times have therefore changed; the necessities of men have increased—the value of time is felt—the supremacy of mind is acknowledged—the schemes of life are of a more exalted character—the destiny of the race begins to assume its importance; and now awakened from slumber, man tames the wildest elements and compels them to speed his progress towards an universal dominion over the powers of matter. Light paints for him pictures true to life. Lightning bears his commands. He imprisons the steam and compels it to roll his car over mountains and through vallies, and transport his products to the most distant parts, over water and over land. The mind once aroused, turns itself to find where it may still have something more to do. Agriculture could not be overlooked, the art which makes all other arts possible, and which perfected is civilization itself.

Agriculture *is civilization*, and hence its progress is linked with the highest destiny of the race. But regarded in a subordinate light and in following out the practical requirements of the age, that of drawing from the earth greater supplies of bread, it was soon found that it might be overtaxed. Such a result could not fail to open the whole field of inquiry relating to production and exhaustion, and the relation in which they stood to each other. From exhaustion originated the analysis of soils and the more modern analysis of productions in which are locked up the elements they have drawn from this store-house; the first leads to a knowledge of what and how much the soil contains; the latter, of what and how much has been taken from it. So also the fact is brought out by inference what must be returned to maintain it at least in its present state of fertility, or increase it to an indefinite extent.

The state of agricultural knowledge at the present time, is characterized by an accumulation of facts which are unclassified and unarranged. They are like the brick and stone piled before and around the site of a great edifice about to be founded, and which are ready to be arranged in the walls of a spacious building. Many of these facts it is true, have a definite signification, or in other words their relations are well known, but a great majority of them have no known collocation, although they clearly belong to the edifice. So too, to keep up the simile, I may with truth remark that the master builder is yet to be found, whose sagacity and skill is equal to the task of putting together the discordant parts, and to construct from them a symmetrical whole.

Notwithstanding the illustration I have employed to show the view which I entertain of the state of agricultural science, it is still true, that it requires only a moderate amount of information of chemistry and the collateral sciences to understand many of the applications of the principles upon which the practices of husbandry are based. When I speak, therefore, of the accumulation of facts, I mean to be understood, that it is their relation to a system and not to the meaning which they may have as individual facts. For example, the good effects of draining may be explained on philosophical principles, though the theory of agriculture is yet to be put into form and shape. Draining operates beneficially in many ways; it may merely remove superfluous water by the construction of artificial underground channels, or it may, in addition to this, carry off water charged with astringent salts which are poisonous to the more valuable plants. In either case, the principal result upon which the good effects depend is, the permanent elevation of the temperature of the soil. Surfaces constantly bathed in water, and which are supplied with this element from living springs, cannot attain the temperature required for the better grasses, cereals, or esculents, so long as it is in this condition. The principles of draining then are perfectly understood, and this is the case with many other agricultural practices. The practice of hoeing or stirring the soil is far more general than draining, but the principles upon which the practice is founded are not so well understood. Generally farmers suppose that the object is to kill the weeds; so far it is good; but the effect of hoeing is not confined to this single result; for hoeing, when all the weeds are already extirpated is followed by the most decided advantage to the crop; hence something more than the destruction of weeds comes to pass. One result undoubtedly arises from the absorbent powers of a fresh surface. Nutritive matters, such as carbonic acid and ammonia dissolved in atmospheric air, are readily taken up in this state of the surface, but an old and indurated surface becomes inert and inactive. The power of surface alone is effectual in promoting absorption and decomposition of the most active bodies. The perfect combustion of vegetable and animal matter takes place first upon the surface, upon which they rest.

There is probably no substance in use as a manure which as frequently disappoints the farmer as plaster. In the first place it may operate far more effectually than is expected, and again it may have no effect whatever; and, finally, when it has operated very beneficially for a time, it ceases to do so. This is what is called plaster sickness. Now these facts ought to be explained. On what principle does plaster ever promote vegetation? Liebig says that it is by the absorption of ammonia; sulphate of ammonia being the product of change. Were this always true, I can see in it reasons why it should always benefit crops. Sulphate of ammonia always does, but plaster does not. But there is another reason why plaster is useful. Its sulphur is wanted in the nitrogenous bodies—the protein compounds. It may, too, operate well in virtue of its lime, which is an element of the highest importance to vegetables. There may be, therefore, three reasons why plaster promotes vegetation—the supply of ammonia for the nitrogenous bodies, the supply of sulphur for the same, and finally, the supply of lime. But why it should cease to do good, is a question which has been answered only hypothetically.

We may suppose that in the first place the soil requires, at the time, no additional matter which plaster itself can furnish; it is in this case a negative. When it ceases to do good at the end of a few years, it may be from exhaustion; that is, the soil originally light, may be deprived of phosphoric acid, of chlorine, of magnesia or soluble silica and the alkalies particularly, at a much earlier period than if plaster had not been used. It has aided in the removal of a larger quantity of inorganic matter, different from itself, in a less time than if it had not been employed. If a crop is increased one-third, it has taken up one-third more of the potash of the soil than would have been obtained without it. If this is true, we may see that the further use of plaster will be worse than useless.

There is nothing plainer than this, that every element which is found in a plant in analysis, is necessary to its constitution, and is liable to be removed in a series of croppings. This leads to the necessity of supplying it directly; but what element or elements may be wanting, can be known for a certainty only by analysis. In plaster sickness, therefore, our remedies need not be hypothetical, if we pursue the method proposed; analysis will reveal the cause of plaster sickness, and probably any other sickness which follows from constant cultivation.

The application of science to agriculture, appears of the

highest importance when viewed in this light; as pointing out first, the composition of productive and barren soils, and afterward, the true method of maintaining and restoring them to fertility at the least possible expense in labor or cash. In the same line of investigation lies the business of determining the composition of the inorganic matter which vegetables remove from the soil; indeed, in one sense, this work should precede the other, for it is by the composition of the inorganic matter of plants that all that "is essential to a fertile soil is determined." But chemists went to work the other way, and determined first, the composition of the soil; and inferred from their results what they supposed on the one hand constituted its fertility, or what on the other its barrenness. This method was unquestionably defective, and probably for that cause alone gave a doubtful importance to the value of the analysis of soils.

The analysis of soils, and of the inorganic matter of plants, stood in very singular relations to each other; the elements of the former, which are in the smallest quantities, formed by far the largest in the latter; thus the alkalies and phosphates of soils are always inconsiderable in amount, and hence were not sought for, while in the parts of plants they formed by far the largest proportion. Fertility depends upon those elements of which only traces appear, where only one hundred grains of the soil are employed in analysis. When, therefore, an analysis of two soils, one a fertile one and the other known to be barren from experience, were left unfinished, that is, those elements which were small in amount, were not sought for, it was impossible to see an essential difference in their composition; the barren soil looked as well on paper as the fertile one, and so it was said that no benefit could arise from the analysis of soils. This I believe is a fair statement of the case. I have now I believe said enough upon the points to enable you to form correct views of the subjects in question.—*From an Address before the New-York State Agricultural Society, at Albany*, 1849, *by* PROF. E. EMMONS, M. D.

DISTRICT SCHOOLS IN THE COUNTRY.

Education is one of those indefinite terms which admit of almost any latitude. In its real and true signification, it is a progressive, and never ending work. The whole life time of men is but a movement onward, and it is perhaps safest to believe the elevated and beautiful idea that, throughout all eternity, man will continue in knowledge and advance in wisdom. But it is not in this broad view that I now propose to regard the term education. I define it for present purposes as a disciplinary process, fitting the mind for the business of life; not only the accumulation of knowledge and intelligence, but the acquisition of habits of order, industry and economy in proportion for the active duties and responsibilities of life. This work belongs to the school room; there the boy is to be prepared for manhood.

In the process of time nature will develop the full capacity of the physical system, but the mind is not made of the same material, and cannot alone come to its full strength and capacity. Its food and nourishment are made of different matter than that which feeds and invigorates the body; it must have the aid of other minds—must have facts and figures, arbitrary rules and distinct principles, and obtain them, not by instinct, but by hard study, severe thinking, and the rigid application of the mental faculties.

The school-room, the school-book, and the school-master are these important requisites in training the mind and bringing out its powers and energy. Mind, like the body, is the work of the great Architect, it is the gift of God, and may and does exist in all its glory and majesty in the poor man as in the rich; it knows no distinction, only in its means of development and in its educational polish. Then how glorious to educate all the people—how high and solemn the duty to give all the advantages of mental culture. The district school belongs emphatically to the masses; they are the peoples' school; they know no caste, nor recognize any distinction, but broadly unfold their beautiful panoply and cover all alike, and say, without respect to person or condition, "come and partake of my benefits." God has given the mind; ours is the duty to unfold the power and prepare for systematic and useful action, this richest and mightiest of God's gifts.

It is the highest and proudest boast of the Empire State that she has thus provided a system for the education of her children. Rightly does she judge, and wisely act, when she thus provides for the safety of herself and the elevation of her people. And have the farmers no interest in this matter? Yes, they have most of all; for they are more numerous than all other classes. The district school is truly almost exclusively their own; it is to most of them their only school, and it behoves them to look well to these seminaries, so peculiarly their own. Their children, nine out of ten, if not ninety-nine out of every hundred, will be educated in them, for they have nowhere else to go. Then let the district school be elevated, improved and made what it should and may be. As one improvement almost indispensably necessary to the farmers, there should be, and must be, a department devoted to agriculture. I can discover no reason why it should not form a regular branch of common school education; nor why every college and academy in the state should not have their professorship department devoted to agriculture as a distinct branch of study and education.

Is there any thing in the subject which precludes this? Is there any difficulty in reducing to a regular science, and of so managing and classifying its different branches, as to permit it being made a part of the educational process of the young? I think not; but on the contrary, agriculture is a science, possessing, in all its ramifications, distinctive features; is governed by fixed facts and unerring principles, which the young farmer should learn by study and close application of his mental faculties. They should be engraven on his mind when it is young and plastic, and capable of receiving and retaining impressions, and this subject may, I imagine, be introduced into every district school in the state, without any detriment to the branches now taught in those schools, and without interfering with the regular course of common school education.

Much reflection has satisfied my own mind of the great importance of the subject. I regard it as an essential step towards the elevation of the farming interests—a necessary ingredient in lifting up to their real position the farmer of this country. The state has been beneficent in her school funds; but the farmer has not yet had his full share of the benefits accruing from them. He has been content to look on listlessly, and let other classes reap the harvest which his own industry has provided. Let him now arise from his lethargy, and begin to cast about and see it there be no place where his sons can go and learn to become farmers, as well as doctors, lawyers, and divines. It seems to me that the farmers have a right to use a portion of the money which belongs to them to advance their own calling—not, indeed, to tear down, or prejudice others, but to elevate their own business to the dignity of a science, to be taught and learned in the schools of the state.—D. A. Ogden, *in the Genesee Farmer.*

DOMESTIC EDUCATION OF FEMALES.

The greatest danger to females, at the present time, is the neglect of domestic education. Not only to themselves, but to husbands, families and the community at large, does this danger impend. By far the greatest amount of happiness in civilized life is found in the domestic relations, and most of this depends on the domestic culture and habits of the wife and mother. Let her be intellectually educated as highly as possible; let her moral and social nature receive the highest graces of vigor and refinement; but along with these let the domestic virtues find ample place.

We cannot say much to our daughters about their being hereafter wives and mothers, but we ought to *think* much of it, and to give the thought prominence in our plans for their education. Good wives they cannot be, at least for men of intelligence, without mental culture; good mothers they certainly cannot be without it; and more than this, they cannot be such wives as men need, unless they are good housekeepers; they cannot be good housekeepers without a thorough and practical teaching to that end. Our daughters should be practically taught to bake, wash, sweep, cook, set table, make up beds, sew, knit, darn stockings, take care of children, nurse, and do every thing pertaining to the order, neatness, economy and happiness of the household.

All this they can learn as well as not, and better than not. It need not interfere in the least with their intellectual education, nor with the highest style of refinement. On the contrary, it shall greatly contribute thereto. Only let that time, or even a portion of it, which is worse than wasted in idleness, sauntering, gossip, frivolous reading, and the various modern female dissipations which kill time and health, be devoted to domestic duties and domestic education, and our daughters would soon be all that can be desired. A benign, regenerating influence would go forth through all the families of the land. Health and joy would sparkle in many a now lustreless eye; the bloom would return to grace many a faded cheek; and doctor's bills would fast give way to bills of wholesome fare.—REV. E. H. WINSLOW.

THE BEAUTY OF TREES.

What can be more beautiful than the trees?—their lofty trunks, august in their simplicity, asserting to the most inexperienced eye their infinite superiority over the imitative pillars of man's pride—their graceful play of wide-spreading branches, and all the delicate and glorious machinery of buds, leaves, flowers, and fruit, that with more than magical efforts burst from the naked and rigid twigs with all the rich and heaven-breathing, delectable odors, pure and animating essences, pouring out spices and medicinals, under brilliant and unimaginably varied colors, and making music from the softest and most melancholy under-tones to the full organ-peal of the tempest!

We wonder not that trees have been the admiration of men in all periods and nations of the world. What is the richest country without trees? What barren and monotonous spot can they not convert into paradise? Xerxes, in the midst of his most ambitious enterprise, stopped his vast army to contemplate the beauty of a tree. Cicero, from the throng, and exertion, and anxiety of the forum, was accustomed, Pliny tells us, to steal forth to a grove of plane-trees to refresh and invigorate his spirits. In the Scaptan Grove, the same author adds, Thucydides was supposed to have composed his noble histories. The Greek and Roman classics, indeed, abound with expressions of admiration; but above all, as the Bible surpasses, in the splendor and majesty of its poetry, all books in the world, so is its sylvan arborescent imagery the most bold and beautiful.

Beneath some spreading tree is the ancient patriarch revealed to us, sitting in contemplation, or receiving the visit of angels; and what a calm and dignified picture of primeval life is presented to our imagination at the mention of Deborah, the wife of Lapidoth, judging the twelve tribes of Israel, between Ramah and Bethel, in Mount Ephraim, beneath the palm-tree of Deborah! The oak of Bashan and the cedar of Lebanon are but other and better names for glory and power. The vine, the olive, and the fig-tree are imperishable emblems of peace, plenty, and festivity. David, in his Psalms; Solomon, in his Songs and Proverbs; the prophets, in the sublime outpourings of their awful inspiration; and Christ, in his parables—those most beautiful and perfect of all allegories—luxuriate in signs and similies drawn from the fair trees of the East.—Wm. Howitt.

OUR WONDROUS ATMOSPHERE.

The atmosphere rises above us with his cathedral dome, arching towards the heavens, of which it is the most familiar synonym and symbol. It floats about us like that grand object which the apostle John saw in his vision—"a sea of glass like unto crystal." So massive is it, that when it begins to stir, it tosses about great ships like playthings, and sweeps cities and forests, like snow-flakes, to destruction before it. And yet it is so mobile, that we have lived years in it before we can be persuaded that it exists at all, and the great bulk of mankind never realised the truth that they are bathed in an ocean of air. Its weight is so enormous that iron shivers before it like glass; yet a soap bubble sails through it with impunity, and the tiniest insect waves it aside with its wing.

The atmosphere ministers lavishly to all the senses. We touch it not, but it touches us. Its warm south winds bring back color to the pale face of the invalid—its cool west winds refresh the fevered brow, and make the blood mantle in our cheeks; even its north blasts brace into new vigor the hardened children of our rugged climate. The eye is indebted to it for all the magnificence of sun-rise, the full brightness of mid-day, the chastened radiance of the gloaming, and the clouds that cradle near the setting sun. But for it the rainbow would want its "triumphal arch," and the winds would not send their fleecy messengers on errands round the heavens. The cold would not either shed snow feathers on the earth, nor would drops of dew gather on the flowers. The kindly rain would never fall, nor hail storm, nor fog diversify the face of the sky. Our naked globe would turn its tanned and unshadowed forehead to the sun, and one dreary, monotonous blaze of light and heat dazzle and burn up all things.

Were there no atmosphere, the evening sun would in a moment set, and, without warning, plunge the earth in darkness. But the air keeps in her hand a sheaf of his rays, and lets them slip but slowly through her fingers; so that the shadows of evening are gathered by degrees, and the flowers have time to bow their heads, and each creature space to find a place of rest, and to nestle to repose. In the morning, the garish sun would at one bound burst from the bosom of night, and blaze above the horizon; but the air watches for his coming, and sends at first but one little ray to announce his approach, and then another, and by and by a handful, and so gently draws aside the curtain of night, and slowly lets the light fall on the face of the sleeping earth, till her eyelids open, and like man, she goeth forth again to her labor till the evening.—QUARTERLY REVIEW.

THE RETURN OF SPRING.

Dear as the dove, whose waiting wing
  The green leaf ransomed from the main,
Thy genial glow, returning Spring,
  Comes to our shore again;
For thou hast been a wanderer long,
  On many a fair and foreign strand,
In balm and beauty, sun and song,
  Passing from land to land.

Thou bring'st the blossoms to the bee,
  To earth a robe of emerald dye,
The leaflet to the naked tree,
  And rainbow in the sky:
I feel thy blest benign control
  The pulses of my youth restore;
Opening the spring of sense and soul,
  To love and joy once more.

I will not people thy green bowers,
With sorrow's pale and spectre band;
Or blend with thine the faded flowers
Of memory's distant land;
For thou wert surely never given
To wake regret from pleasures gone;
But like an angel sent from Heaven,
To soothe creation's groan.

Then, while the groves thy garlands twine
Thy spirit breathes in flower and tree,
My heart shall kindle at thy shrine,
And worship God in thee:
And in some calm, sequestered spot,
While listening to thy choral strain,
Past griefs shall be a while forgot,
And pleasures bloom again.

SELECTED

THE KINGS OF THE SOIL.

Black sin may nestle below a crest,
And crimes below a crown;
As good hearts beat 'neath a fustian vest,
As under a silken gown.
Shall tales be told of the chiefs who sold
Their sinews to crush and kill,
And never a word be sung or heard
Of the men who reap and till?
I bow in thanks to the sturdy throng
Who greet the young morn with toil;
And the burden I give my earnest song
Shall be this—The Kings of the Soil!
Then sing for the Kings who have no crown
But the blue sky o'er their head—
Never Sultan nor Dey had such power as they,
To withhold or offer bread.

Proud ships may hold both silver and gold,
The wealth of a distant strand;

But ships would rot, and be valued not,
Were there none to till the land.
The wildest heath, and the wildest brake,
Are rich as the richest fleet;
For they gladden the wild birds when they wake,
And give them food to eat.
And with willing hand, and spade and plough,
The gladdening hour shall come,
When that which is called the "waste land" now,
Shall ring with the "Harvest Home!"
Then sing for the Kings who have no crown
But the blue sky o'er their head—
Never Sultan nor Dey had such power as they
To withhold or to offer bread.

SELECTED.

A MOTHER'S GRAVE.

The trembling dew-drops fall
Upon the shutting flowers—like souls at rest—
The stars shine gloriously—and all,
Save me, is blest.

Mother!—I love thy grave!
The violet, with its blossom blue and mild,
Waves o'er thy head—when shall it wave
Above thy child?

'Tis a sweet flower—yet must
Its bright leaves to the coming tempest bow,
Dear mother—'tis thine emblem—dust
Is on thy brow!

And I could love to die—
To leave untasted life's dark, bitter streams,
By thee, as erst in childhood, lie,
And share thy dreams.

And must I linger here
To stain the plumage of my sinless years,

And mourn the hopes of childhood dear
With bitter tears.

Aye—must I linger here,
A lonely branch upon a blasted tree,
Whose last frail leaf, untimely sere,
Went down with thee?

Oft from life's withered bower,
In still communion with the past I turn,
And muse on thee the only flower
In memory's urn.

And, when the evening pale,
Bows like a mourner on the dim, blue wave,
I stray to hear the night-winds wail,
Around thy grave.

Where is thy spirit flown?
I gaze above—thy look is imaged there—
I listen, and thy gentle tone
Is on the air.

Oh come—whilst here I press
My brow upon the grave—and, in those mild
And thrilling tones of tenderness,
Bless, bless thy child!

Yes, bless thy weeping child,
And o'er thy urn—religion's holiest shrine—
Oh give his spirit undefiled
To blend with thine.

By George D. Prentice.

THEORY OF FEEDING ANIMALS.

In the feeding of animals, is a knowledge of animal and vegetable physiology of importance?

It is; for without this knowledge, the food may not be adapted to the wants of the animal; it may consist more of

A DEVON BULL, owned by AMBROSE STEVENS, ESQ., New York.

some particular elements and less of others than is required for animal growth and development. It may indeed contain little or nothing of the elements most needed by the animal, and consist almost wholly of elements which are needed only in a very small measure. The constituents of animal organizations can no more be produced without the materials entering into their construction, than any piece of human mechanism can be formed without the wood and the metallic substances used in making the several parts of it.

How is this made manifest?

The idea would be absurd to make a steam engine out of wood; or to make the sails of a ship out of iron; or to make a clock out of the materials that form cloth. But this is no more absurd than it would be when desiring to promote the growth and hardness of bones in young animals, to furnish food to them which contain none of the elements that enter into the formation of bones; or if we desired to fatten animals, to furnish them with food containing none of the elements entering into the formation of fat. It is evident, therefore, that economy in providing food for animals, and the best results from it, are intimately connected with, or are dependent on a knowledge of animal and vegetable physiology.

But, can no one feed animals properly unless he has made animal and vegetable physiology a study?

This is by no means to be affirmed. Nor is it the fact. A person may, from seeing what others have done, or from his own experience, in a course of time, produce the best results in rearing and fattening farm animals. He may produce the largest hogs and the fattest oxen found in the country, and at the cheapest rates possible, from having seen how his neighbors have done it, although he never saw a book or heard a lecture on the subject—or although he does not even know how to read. This he does in the way of imitation, as another person may learn to shoe a horse or make a watch without the slightest knowledge of mechanical philosophy—or as a person may learn to sing by rote without a knowledge of musical science—or as a parrot may learn in sundry cases to imitate the human voice without a knowledge of philology.

Why is not this as good as any other way?

In the first place, if a man is to learn every thing on the

subject by his own experience, or by making successive experiments, a long period is required in perfecting himself, he may lose much in making experiments, and the probability is he may not at the last arrive at the best mode of practice. In the second place, if he is to learn from observing what others do, he may not have those around him who practice the best mode of feeding animals—it is only the few whose example may safely be followed. And, in the third place, if no difficulty arises from either of the above means of informing himself, he becomes a mere automaton, without that mental satisfaction and pleasure experienced by the person who knows the reason for every thing he does; it is reducing a man of rational reasoning powers to the level of a machine.

Does the man of these scientific attainments derive any other advantage in feeding animals, besides knowing what articles of feed to give them?

He does. The growth and fattening of animals depends comparatively as much on the manner in which the food is prepared as on the food itself. The food, to answer its legitimate and destined purposes, on going into the stomach, is to be decomposed—that is, to be reduced to its original elements, as well as to be reduced to minute particles—and, then assimilated to the different parts of the animal frame. Without such decomposition, it cannot be assimilated, and hence will impart no nourishment. For instance, the kernels of Indian corn, frequently seen to pass through an animal whole, as when taken into the mouth, can have done no good. This may be an extreme case. But between this and a due decomposition of the food, there may be numerous degrees or stages of decomposition and animal assimilation. In some cases from a defect of decomposition, a fourth of the food may be lost—in others half—in others three fourths—and, again, the whole, as in the case of the kernels of Indian corn, may be lost.

What may be affirmed generally in regard to a due preparation of food for animals on a farm?

There are persons who have in a most commendable manner made various experiments with different kinds of animals, and the results have been highly satisfactory. They testify that from one fourth to one half of the food may be saved, by adopting the most approved modes of preparing feed. Some of these modes are the cutting, or steaming, or both, of grass, straw, and corn-stalks for horses, cattle and sheep—the grinding and cooking of grain—and the cooking of tubers, when used for feeding the stock of a farm.

What is said of cutting hay, straw, and corn-stalks for cattle?

It has been estimated that this will in some instances save one half of the feed, particularly with straw and corn-stalks. Corn-stalks thrown out without being cut will be as much trodden under foot and wasted as eaten. The same is true with straw and inferior hay; whereas, if well cut, the most of it will be eaten. Besides, if thus cut, what remains will readily be converted into manure, so that there will be no real loss. Nor is this all; stock will fill themselves with cut food in less than half the time they will do it when the feed is not cut. Then they may lie down and rest, so that the same feed will do them more good than though they were kept constantly on their feet. The labor of being continually employed in masticating food produces weariness, and exhausts a portion of the nutriment in what is eaten, the same as any other labor. With working oxen and horses this is an important circumstance.

How else may this description of feed be improved?

Let the cut hay and straw be thoroughly moistened by scattering water upon it, and then scattering upon, and mixing ground feed with it, more or less, as the condition of the animals may require, and they will be kept economically and in good condition. The steaming cut hay and straw is a further improvement. Cut corn-stalks well cooked, for which large kettles may be had, and a few quarts of ground feed for each animal, well mixed and cooked with the cereals, will sustain laboring animals in the very best condition, and will fatten others. This is perhaps the cheapest mode of doing it.

What is said of the difference between giving cattle and horses ground feed, or the grain without being ground?

If ground and well prepared in meshes, the whole of it will be well digested and assimilated to animal products. If not so prepared, and especially Indian corn, a portion of it will not be digested and assimilated, and will pass off in the form of excrements, without doing the animal any benefit. What is thus saved will far overbalance the cost and labor of thus preparing feed. This is true in all cases; but, with animals whose teeth are much impaired, the gain is greatest.

What does Stephens say of boiling or steaming dry food for farm animals?

He recommends it; also the same practice with potatoes and turnips. He says the practice is extensive in England, and is generally considered economical, although some have had less confidence in its utility, when given to horses for heavy draught and fast work. Notwithstanding this exception, steamed food is getting into considerable use for coach horses in that country.

How is the value of tubers affected by age?

It is said that the potato, by being kept, loses much of its nutritive value, even before it begins to sprout; and afterwards it deteriorates much more rapidly. Turnips, that have shot into flower, are found to add much less to the weight of fattening animals, than they did before. This is a fact well known to all who have made observation on the subject.

How does the absence of light operate on the thrift of animals and the quantity of food required by them?

It is found that a diminished quantity of feed will answer for animals confined in the dark; for, whatever excites their attention and causes them to be restless, lessens the effect of food in enlarging the body. It is said that fowls in the dark will fatten with astonishing rapidity. It is so with calves. In India, the eyelids of some animals are sewed up when being fattened. It was found that some sheep shut up in the dark from the 18th of November till the 9th of March following eat 886 pounds of turnips, while an equal number unsheltered eat 1912 pounds; and, that in the former case, 100 pounds of turnips added but a pound to the weight of the animals, while in the latter case, the 100 pounds added to the weight of the animals three pounds. This experiment was made by Mr. Morton in England.

What is said of cleanliness and ventilation when animals are shut up for feed or being fattened?

Both are highly important, being helps to economy in the feeding of all animals. Shelter and warmth may be detrimental to the thrift and health of them, unless kept clean and are amply supplied with fresh air. The cleaner their houses and skins are kept, the more they thrive under any given form of treatment and allowance of food.

How should calves designed for the butcher be fed?

While the calf is very young, during the first two weeks, or so, its growth is mostly in the bones and muscles. Of course during this period of its existence materials for bones and muscles are mostly required, and less for fat, and hence, half of the milk given it may be skimmed, and a little bean meal may be mixed with it, which is favorable to the growth of the muscles. Afterwards the creation of fat being the object of desire, the food should be rich in substances which yield fat. The cream of the milk is its most natural element; but if the calf requires more of this than what is furnished by its dam, as cream is expensive, it is usual to provide it daily with linseed meal in such quantities as required. This is the most economical food a calf can have. A few eggs each day will contribute much to its thrift. Calves designed for the butcher should not be allowed to run at large; if they do, it will be difficult to fatten them; by running, much of the fat from the food will be wasted. If shut up in clean, well ventilated, dark pens, they will thrive best.

How should young animals be fed, which are to be raised?

With animals generally it is desirable to increase their size; to enlarge the frame, increasing the dimensions of the bones, and covering them amply with muscles. The food should have constant reference to this design. This food, therefore, should contain a large supply of the phosphates, from which bone is formed, and of gluten or fibrin, by which the muscles are enlarged. Some kinds of fodder contain a larger proportion of these phosphates. Such are corn seeds in general, and the red clover among the grasses. Some again contain more of the materials of the muscles. Such are beans and peas. When young animals receive an insufficient allowance of food they become stinted, and never can be made to attain their full natural size and weight.

What quantity of food does an ox daily require?

It has been ascertained that an ox or a cow kept in the stall will daily eat about one fifth of its own weight in turnips, or one fiftieth of its weight in dry food, such as hay and straw. With this allowance of food the animal would probably, in some measure, add to its weight; but from actual experiment it is known, that one sixtieth of the weight of the animal in dry hay is necessary to sustain itself. Cows with calf need more, because it is apparent, that they have not only to sustain their own frame, but to nourish the fœtus. Cows also that give milk must receive additional food and of appropriate quality for the production of milk, or they will become lean, and the supply of milk will fail.

How should the food of cattle be regulated in reference to the production of the several parts of which the animal frame consists?

It is advisable that they should not always be fed on the same kind of food, but that it be mixed, or they may not retain a healthy condition. Food containing starch or sugar, alone, or pure fibrin or gelatine alone, will not sustain the animal body, because these substances do not contain all that is requisite to build up all its parts, or supply what is daily given off during respiration and in the secretions. Hence, it is evident that there should be such a variety of food as will furnish all the materials which the body requires. Occasionally, therefore, there should be a change from one kind of food to another, or the proportions of the different kinds should from time to time be varied.

On what food will cattle fatten best?

On Indian corn, or oil cake, or oil mixed with cut straw and hay, or upon oily seeds and nuts. Poultry will fatten quickest on a mixture of meal and suet, or on the scraps from which lard or tallow has been pressed. The reason of this is very obvious; for these articles of food contain a large proportion of fatty matter ready formed, which the animal can easily extract, and after a slight chemical change can convert into a portion of its own substance.

For what purposes in the living animal does fat subserve?

It lubricates the joints, answering the same end that grease does on machinery generally and the axles of wheels in particular, by preventing friction and causing a free action. It covers and protects the internal viscera, keeps the muscles separate, and enables them to play freely among each other, makes the hair and skin soft and flexible, and, by filling up the hollows, contributes to the roundness and plumpness of the parts, and defends the extremity of the bones from external injury. When exercise is taken, a portion of the fat of the body passes away in perspiration, and even in the solid excrements. On the latter account, the dung of a fat animal is better for manure than that of a lean one. To supply this waste of fat, additional feed is necessary to prevent leanness.

What is said of the time when cattle should be fed?

Thaër says, what every person should know, that the utmost regularity must be observed in the hours of feeding, and the quantity of food given at each meal. Animals acquire a most exact knowledge of time. This is remarkably conspicuous in old beasts of draught, which, when the proper hour of rest arrives, refuse to work and want to go to stall, or the field in which they graze. Stalled cattle become restless when their feeding time is not punctually observed, though very quiet before the time arrives. If the provender is not given to a horse at the customary season, he commences pawing and ceases not till his keeper is reminded of his neglect. Cattle also know pretty well the quantity usually given them—when they have received and eaten it, they go to rest; but if they have not had their full allowance they continue restless.

At what age will cattle fatten best?

It is supposed that horned cattle will fatten most readily at the age of seven and eight years. Younger animals not arrived at their full growth may indeed grow very fat, and furnish delicate and savory meat; but they always require more fodder and longer time. Old beasts do not fatten so easily; if, however, their organs of mastication are strong enough to chew the fodder well, they will ultimately attain to a good condition.

What is said of the expediency of keeping swine?

If all things be taken into the account, it will be found that in rearing swine where they must be fed wholly on grain, they will cost more than they are worth. But in all rural establishments, both great and small, the keeping of pigs is almost indispensable, because the various kinds of refuse of the dairy, or kitchen, and garden can scarcely be used in any other manner. In addition to this, should proper pains be taken, hogs will convert different kinds of dying vegetables and loam into manure, being mixed by them, with their excrements, nearly if not quite sufficient to pay for the grain they eat.

What experiments in fattening hogs have been made?

Different experiments have been made, but the one most

deserving notice was in the state of Maryland. The Agricultural Society of that state instituted an inquiry into the relative merits of two modes of feeding, and the following is one of the results. On the first day of December, four shoats of the same breed, nearly of a size, and as much alike in every respect as could be selected from a herd, were made choice of, each being carefully weighed, and placed in a single stye where their food could be exactly regulated. Two of them weighed together one hundred and eighty-five pounds. These were fed on one gallon each of shelled Indian corn, the gallon weighing seven pounds. This was the allowance for twenty-four hours, and as much water as they needed. The other two were fed on half the amount, by weight, of Indian meal, made into hasty pudding with a little salt. The seven pounds of meal when cooked weighed thirty pounds and measured three gallons. Before the experiment had progressed a fortnight, it was perceived that the two fed on hasty pudding were outstripping the two fed on whole corn; and on the thirty-fourth day they were again weighed; the corn-fed ones together weighing twenty-five pounds more than they did on the first of December, while the two fed on mush—half the quantity—had gained forty-four pounds!

AGE OF ANIMALS.

A bear rarely exceeds twenty years; a dog lives twenty years; a wolf, twenty; a fox, fourteen or sixteen. The average age of cats is fifteen years; of a squirrel or hare, seven or eight years. Elephants have been known to have lived to the great age of 400 years. When Alexander had conquered Porus, king of India, he took a great elephant, which had fought valiantly for the king, and named him Ajax, dedicated him to the sun, and let him go with this inscription—"Alexander, the son of Jupiter, hath dedicated Ajax to the sun." This elephant was found with this inscription 350 years after. Pigs have been known to live to the age of thirty years; the rhinoceros to fifty. A horse has been known to live to the age of seventy-two, but avarages twenty-five to thirty. Camels sometimes live to the age of 100. Stags are long-lived. Sheep seldom exceed

"Siberian Crab Apple. The tree of the crab apple is small, but is beautiful, having a rich foliage. The fruit being fair and pendent, exhibits a fine appearance. Nor is it simply ornamental; it is highly esteemed for preserves, tarts, and sauces."

the age of ten. Cows live about fifteen years. An eagle died at Vienna of the age of 104 years; ravens frequently reach the age of 100. Swans have been known to live 300 years; pelicans are long-lived. A tortoise has been known to live much above 190 years.

ANIMAL AND VEGETABLE FOOD.

Joseph Bentley, in his "Health and Wealth," adduces the following singular facts in relation to animal and vegetable food:

It is a fact worthy of remark, and one that seems never to have been noticed, that throughout the whole animal creation, in every country, and clime of the earth, the most useful animals cost nature the least waste to sustain them with food. For instance, all animals that work, live on vegetable food; and no animal that eats flesh works. The all-powerful elephant and the patient, untiring camel, in the torrid zone; the horse, the ox, the donkey, in the temperate; and the reindeer, in the frigid zone, obtain all their muscular power for enduring labor from nature's simplest productions, the vegetable kingdom. But all the flesh eating animals keep the rest of the animated creation in constant dread of them. They seldom eat vegetable food until some other animal has eaten it first, and made it into flesh. Their only use seems to be to destroy life—their own flesh is unfit for other animals to eat, having been itself made out of flesh, and is most foul and offensive. Great strength, fleetness of foot, usefulness, cleanliness, and docility, are then always characteristic of vegetable eating animals; while all the world dreads flesh eaters.

POETRY OF LABOR.

Pause not to dream of the future before us,
Pause not to weep of the wild cares that come o'er us;
Hark, how creation's deep, musical chorus
  Unintermitting, goes up into Heaven!

Never the ocean-wave falters in flowing;
Never the little seed stops in its growing;
More and more richly the rose heart keeps glowing,
  Till from its nourishing stem it is riven.

"Labor is worship!"—the robin is singing;
"Labor is worship!"—the wild bee is ringing;
Listen! that elegant whisper upspringing,
  Speaks to thy soul from out Nature's great heart.
From the dark cloud flows the life-giving shower;
From the rough sod blows the soft breathing flower;
From the small insect the rich coral bower;
  Only man, in the plan, shrinks from his part.

Labor is life!—'Tis the still water faileth;
Idleness ever despaireth, bewaileth;
Keep the watch wound, for the dark rust assaileth;
  Flowers droop and die in the stillness of noon.
Labor is glory!—the flying cloud lightens;
Only the waving wing changes and brightens;
Idle hearts only the dark future frightens;
  Play the sweet keys, would'st thou have them in tune

Labor is rest from the sorrows that greet us;
Rest from all petty vexations that meet us.
Rest from sin-promptings that ever entreat us,
  Rest from the world-syrens that lure us to ill.
Work—and pure slumbers wait on thy pillow;
Work—thou shalt ride over Care's coming billow;
Lie not down wearied 'neath Wo's weeping willow!
  Work with a stout heart and resolute will!

Droop not, tho' shame, sin, and anguish are round thee!
Bravely fling off the cold chain that hath bound thee!
Look to yon pure Heavens smiling beyond thee!
  Rest not content in thy darkness—a clod!
Work!—for some good—be it ever so slowly:
Cherish some flower—be it ever so lowly!
Labor! All is noble and holy;
  Let thy great deeds be thy prayer to thy God!

Mrs. Frances S. Osgood

THE MIND AND THE SOIL.

The practical farmer may and should be continually improving his mind by study and reflection; reading and study in his leisure hours, reflection and observation in his daily toil. The field of his labor is unbounded, and his mind should be continually employed in searching out the nature of his soils, and what they require to make them productive. If a piece of land is wet, how shall it be drained and made productive, at a reasonable expense? If it is cold, with what shall it be mingled, to give it genial warmth? If the soil is light and thin, how shall it be deepened and made more stringent? If dry and porous, what retentive mixture will give it the power of retaining moisture and manures? These are questions to be solved by every farmer, in regard to his own fields, and they must be, to make him a successful agriculturist.

First, he must learn the general character of his soils; the general classes to which they belong, the proper treatment of these classes of soils; and then he must patiently ascertain, by trial and experiment, the peculiar modifications of that treatment which his particular soil may require. By so doing, by this practical application of his knowledge, he can enliven dead and unproductive lands, and make them teem with the choicest productions of the earth. He can, in short, give to soils just such a combination as he pleases, and can make a good soil where nature has denied it both mechanically and chemically, by adding earths, and by adding manures. Different soils require different treatment, to dispose them to production.

The intelligent, practical farmer knows the character of lands when he sees them; he knows what treatment they want. His common sense and observation will teach him how they may be made productive. He is many times surprised, when he sees sterile and unproductive fields near very flourishing villages, where unemployed lie most of the materials necessary to enrich them. A few shillings and dollars, with some labor and energy, will apply the fossil manures, and apply them to produce fertility. The limestone beds are reposing all around, waiting for the attention of the owner, or the capital of some one who deals in stocks, to bring them to the furnace, and prepare them to correct and fertilize the soil, making those sterile fields not only productive, but ornamental.

There may be, and there are, portions of our land, where the soil is itself so rich and inexhaustible in fertility, that a knowledge of making an application of manures is of no consequence. But we all know that our soils generally are not of that character. Here the science of preparing manure lies at the very foundation of agricultural success. The variety of manures is much greater, and their application to the variety of soils and crops, is much better understood than heretofore. The farmer has learned much by his own experience, by his own experiments, patiently and perseveringly carried out, modified by his own good sense and careful observation. He is no longer content with the small amount of manures which his farm naturally supplies. Nature, at home and abroad, furnishes manures, and the materials from which manures can be made, in almost endless variety; and if the farmer is diligent in inquiry and experiment, he can create in his barn-yard and compost-heap, a bank, from which he can draw at pleasure, and without an endorser; for the manure-yard, the compost-heap, is the farmer's bank,—it is his mine of wealth, and, without it, he is poor indeed.

But, besides all that the farmer can learn upon this subject from his practical experience, he can undoubtedly derive great assistance from chemical science, and to this he should by no means shut his eyes. If he would elevate his profession, if he would ensure to his labor the reward of wealth and competence, he must make the discoveries of science tributary to it. The true course of the farmer is, not to spend his time in complaining of the advantages which knowledge and science give to other pursuits, but rather to make knowledge and science subservient to his profession. Instead of decrying the importance of these, let us use them for our own purposes. Let us use them with good sense, care and economy. We see the vapor fall in rain, and vegetation receiving it and rejoicing. The winds of heaven are continually passing over our fields, and the air is purified, and made the proper food for vegetable life. The sun is continually warming and urging the plant to grow; it also gives its light, that it may mature. The air carries with it oxygen to the seed, that causes it to germinate; and furnishes to the leaves, the lungs of the plant, carbon, without which the plant must cease to be.

These are facts well understood by the farmer. But should he stop here? Should he rest content with but this amount of knowledge, which nature has vouchsafed to him almost unasked for? Surely not, if he would be master of his profession, and maintain that professional dignity which becomes him. He should call to his aid the knowledge of scientific men, and by this, learn to mix and manage the elements of earth and air with more than magic skill. He should take care that the mechanical and manufacturing arts do not enjoy all the benefits of the discoveries of chemical science. He should claim for agriculture its just proportion. In his experiments, especially upon the subject of manures, he can thus make much greater progress, and attain much more important results.

The discoveries of scientific men are of no use, until practically applied, and tested by practical farmers. And who are to test and apply them, if we are not? How shall all these mixtures be applied? What does each particular soil require? Shall they be applied in a fermented, or in an unfermented, state? On the surface, or turned under? Put on, as a top-dressing, in spring, or in autumn? What particular dressing does each particular crop require? These are questions involving the laws of nature, the principles of art, and careful experimental labor and observation at the compost-heap, and in the field Our vocation is to render the earth productive. One great and important agent in production, is manure. Its abundant production, and its economical application, require the combination of scientific knowledge and practical industry. Science is doing her part: let agriculture be ever ready to do hers. The farmer is never too old to learn, until he is too old to labor.

Those who succeed well in any business or profession are usually found to conduct their business with system, order and neatness. System, order and neatness, should ever be the study of the farmer. When the fields of the farm are found to be badly laid out, bordered with hedge-rows; his fences crooked and half down; the stable used for everything, and not fit for anything; the outbuildings awkwardly situated, with few or no conveniences; his stock one third more than can be made comfortable, and be kept in a thrifty and growing state upon his farm; his tools and implements of husbandry left some in one place and some in another, scattered over his farm, and many of them unfit for use;—surely the want of success of such a farmer needs no explanation. These appearances fully explain it. They show why he has been kept poor when he should have been made rich. Let our young farmers, then, when they begin life for themselves, study system, order, and a continual regard for practical convenience, and their industry will be sure to meet its just reward.

The effects of system, order and care, appear to striking advantage in the rearing and management of domestic animals. The ox, the cow, the horse, sheep and swine, are animals of practical use, and some of them are parties with us in the labors of the field. It is but a part of civilization and Christianity to treat them with kindness and humanity. But more than this can be done, as the experience of many will attest. What a wonderful triumph of human reason and power, that the animals, given and subjected to man by the Author of Agriculture, can be moulded and fashioned to his liking, as the clay is fashioned by the potter. Form and figure, size and power of bone and muscle, adaptedness to the peculiar service required, kindness and docility of disposition, and even agreeable countenance and expression, by patient and skillful breeding, can be successfully attained. What can be asked more? Modern discoveries are said to enable the farmer to decide, with unerring certainty, which of his heifer calves should be raised for the dairy, and which should be prepared for the shambles; and this by marks and indications readily perceived by the common observer.—*Address before the Berkshire Agricultural Society, by* JOSHUA R. LAWTON, ESQ.

TRANSFORMATIONS IN NATURE.

Numerous transformations take place in nature; indeed it may be said, that everything in the physical world at one period or another is metamorphosed. The figure of objects continually varies; certain bodies pass successively through the three kingdoms of nature; and there are compound substances which gradually become minerals, plants, insects, reptiles, fish, birds, quadrupeds, and man. Every year millions of bodies blend together, and are reduced to dust. Where are the flowers which, during the spring and the summer, ornamented our fields and our gardens ? One species has appeared, withered, and given place to others. The flowers of March, and the modest violet, after announcing by their presence the arrival of spring, have yielded their place to the tulip and the rose. In the room of these we have seen others, till all the flowers have fulfilled their design.

The same holds good with regard to man. One generation shows itself, and another disappears. Every year thousands of human bodies return to the dust from whence they were taken; and of these evanescent bodies, others more beautiful are formed. The salts and the oils of which they were composed, dissolve in the earth; the more subtile particles are raised into the atmosphere by the sun's heat, and mixing there with other matters, are dispersed in different directions by the winds, and fall down in rain and dew, sometimes in one place, and sometimes in another; whilst the grosser particles mix with the earth. The grass which is nourished by them grows up into long blades; and it is thus that the flesh of men, transformed into grass, serves as aliment to the flocks, whose wholesome milk is again converted to our own subsistence.

These continual transformations, thus operating in nature, are so many certain proofs that the Creator has designed that nothing should perish or be useless. The dust of flowers, used in the fecundation of plants, is only a very small part of what each flower contains; and that the superabundant portion may not be lost, bees are created, which make use of it to form their honey. The earth daily presents us with new presents, and it would in the end be exhausted, if what it gives was not in some way or other returned again.

All organized bodies suffer decomposition, and are at last converted into earth. During this dissolution, their volatile parts rise into the air, and are dispersed in every direction. Thus the remains of animals are diffused through the air, as well as through the earth and the water. All these particles so dispersed, unite together again in new organic bodies, which in their turn will undergo similar revolutions. And this circulation, and these continual metamorphoses, which commenced with the world, will only terminate with its dis solution.

The most remarkable transformation, or at least that which interests us the most, is that in which we are immediately concerned. We know that our body was not once composed, and will not be so in the end, of the same number of parts as it is when in its greatest perfection. Our body was at first extremely small; it became much larger when we were brought into the world, and since then has increased to fifteen or twenty times the bulk it then had; consequently blood, flesh, and other matters, supplied by the vegetable or animal kingdom, and which formerly did not belong to our body, have been since assimilated to it, and are become parts of ourselves.

The daily necessity of eating, proves that there is a continual waste of the parts of which we are composed, and that this loss must be repaired by alimentary matter. Many parts insensibly evaporate; for since the experiments which a certain great physician made upon himself, it is ascertained, that of eight pounds of nourishment necessary to support a healthy man in one day, only the fiftieth part is converted into his own substance; all the rest passing off by perspiration and other excretions. Hence also it may be inferred, that in ten years there will not remain many of the same particles that now constitute our bodies. And at length, when they shall have passed through all their different changes, they will be converted into dust, till the blessed day of the resurrection, when they will undergo that happy and final revolution that will place them in a state of eternal rest.—STURM.

AGRICULTURAL EDUCATION.

The wholesome habits of society have been so broken up by the civil and political convulsions of the age, and the inordinate thirst for acquiring wealth and fashionable consequence, through mercantile and other speculations, that honest productive labor has been thrown entirely into the background, and considered not only ungenteel, but menial and servile. Yet I venture to lay down this proposition, that he who provides for the wants and comforts of himself and family, and renders some service to society at large, by his mental and physical industry, performs one of the highest duties of life; and will ultimately be rewarded in the conscious rectitude of his life, by a greater measure of substantial happiness, than he who makes millions by fraud and speculation, to be squandered in extravagance, or wasted in folly, by his children or grandchildren. The revolutions which are constantly taking place in families, sufficiently admonish us that it is not the wealth we leave to our children, but the industrious and moral habits in which we educate them, that secures to them worldly prosperity, and the treasure of an approving conscience.

What class in society have within their reach so many of the elements of human enjoyments—so many facilities for dispensing benefits to others,—one of the first duties and richest pleasures of life—as the independent tillers of the soil? "The farmer," says Franklin, "has no need of popular favor; the success of his crops depends only on the blessing of God upon his honest industry." If discreetly conducted on the improved principles of husbandry, agriculture offers the certain means of acquiring wealth, and as rapidly as is consistent with the pure enjoyments of life, or with the good order and prosperous condition of society. Agriculture is the golden mean, secure alike from the mushroom of opulence, and the craven sycophancy and dependence of poverty. "Give me neither poverty nor riches," was the prayer of the wise man of Scripture; "lest," he added, "I be full and deny thee, and say, who is the Lord? or lest I be poor, and steal, and take the name of my God in vain."

When we consider that agriculture is the great business of the nation—of mankind; that its successful prosecution depends upon a knowledge, in the cultivators of the soil, of the principles of natural science,—and that our agriculture stands in special need of this auxiliary aid,—we cannot withhold our surprise and regret, that we have not long since established professional schools, in which our youth, or such of them as are designed to manage this branch of national labor, may be taught, simultaneously, the principles and practice of their future business of life,—on which, more than any other branch of business, the fortunes of our country, moral, political, and national, essentially depend. We require an initiatory study of years in the principles of law and medicine, before we permit the pupil to practice in these professions. We require a like preliminary study in our military and naval schools, in the sciences of war and navigation, ere the student is deemed qualified to command.

And yet, in agriculture, by which, under the blessing of Providence, we virtually "live, and move, and have our being," and which truly embraces a wider range of useful science than either law, medicine, war, or navigation, we have no schools, we give no instruction, we bestow no govermental patronage. Scientific knowledge is deemed indispensable in many minor employments of life; but in this great business, in which its influence would be most potent and useful, we consider it, judging from our practice, of less consequence than the fictions of the novelist. We regard mind as the efficient power in most other pursuits; while we forget, that in agriculture, it is the Archimedean lever, which, though it does not *move*, tends to *fill* a world with plenty, with moral health, and human happiness. Can it excite surprise, that under these circumstances of gross neglect, agriculture should have become among us, in popular estimation, a clownish and ignoble employment?

In the absence of agricultural professional schools, could we not do much to enlighten and raise the character of American husbandry, by making its principles a branch of study in our district schools? This knowledge would seldom come amiss, and it would often prove a ready help under misfortune, to those who had failed in other business. What man is there, who may not expect, at some time of life, to profit directly by a knowledge of these principles? Who does not hope to become the owner, or cultivator, of a garden, or a farm? And what man, enjoying the blessing of health, would be at a loss for the means of an honest livelihood, whose mind had been early imbued with the philosophy of rural culture—and who would not rather work than beg?

An early acquaintance with natural science is calculated to beget a taste for rural life and rural labors, as a source of pleasure, profit, and honor. It will stimulate to the improvement of the mind—to elevate and to purify it—to self-respect; to moral deportment. And it will tend to deter from the formation of bad habits, which steal upon the ignorant and the idle unawares, and which consign thousands of young men to poverty and disgrace, if not to premature graves. A knowledge of these principles, to a very useful extent, can be acquired with as much facility in the school, or upon the farm, as other branches of learning. Why, then, shall they not be taught? Why shall we withhold from our agricultural population that knowledge which is so indispensable to their profit, to their independence, and to their correct bearing as farmers? Why, while we boast of our superior privileges, keep in comparative ignorance of their business, that class of our citizens who are truly the conservators of our freedom? I know of but one objection,—the want of teachers. A few years ago, civil engineers were not to be found among us. The demand for them created a supply. We have demonstrated that we have the materials for civil engineers, and that we can work them up. We have materials for teachers of agricultural science, which we can also work up. Demand will always insure a supply.

We have professional schools in almost every business of life, except in the cultivation of the soil, one of the most important and essential of them all, and one embracing a larger scope of useful study in natural science, and in usefulness to the temporal wants of the human family, than any other. The policy of monarchs, and of privileged orders, has been to repress intelligence in the agricultural masses, in order to keep them in a subordinate station. But, neither the policy nor the practice should be countenanced by us. Our agriculturists are our privileged class, if we have such. They are our sovereigns, because, from their superior numbers, they must ever control our political destinies, for good or for evil. And the more intelligent and independent we can render them, the more safe we make our country from the convulsions of internal feuds, and the danger of foreign war.

I put the question to fathers,—Would you esteem that son less, or think him less likely to fulfill the great duties of life, who had been educated in a professional school of agriculture, with all the high qualifications which it would confer for public and domestic influences, than him who had been educated for the counter, the bar, or other high professional callings? On which could you best rely for support and comfort in the decline of life? Nay, I will venture to carry the appeal further—to the discriminating judgment of the unmarried lady;—would you reject, as a partner for life, the student of such a college, coming forth with a sound mind, deeply imbued with useful knowledge, and a hale constitution, invigorated by manly exercise, whose cares and affections were likely to be concentrated upon home and country, and whose precepts and examples would tend to diffuse industry, prosperity and rural happiness around him? The father's response would be, I think, an unhesitating "*No*," to the first question; and the lady's, after due deliberation, I verily suspect, would be a half-articulate "*Amen!*"—HON. JESSE BUEL'S *Address before the Agricultural and Horticultural Societies of New Haven county, Conn.*, 1839.

MY OWN PLACE.

Whoever I am, wherever my lot,
  Whatever I happen to be,
Contentment and duty shall hallow the spot,
  That Providence ordains for me;
Not covetous striving and straining to gain
  One feverish step in advance,—
I know my own place, and you tempt me in vain,
  To hazard a change and a chance.

I care for no riches that are not my right,
  No honor that is not my due;
But stand in my station by day or by night,
  The will of my Master to do;
He lent me my lot, be it humble or high,
  And set me my business here,
And whether I live in his service, or die,
  My heart shall be found in my sphere.

If wealthy, I stand as the steward of my King,
  If poor, as the friend of my Lord,
If feeble, my prayers and my praises I bring;
  If stalwart, my pen or my sword;
If wisdom be mine, I will cherish his gift,
  If simpleness, bask in his love,
If sorrow, his hope shall my spirit uplift,
  If joy, I will throne it above!

The good that it pleases my God to bestow,
  I gratefully gather and prize;

The evil—it can be no evil, I know,
  But only a good in disguise;
And whether my station be lowly or great,
  No duty can ever be mean,
The factory cripple is fixed in his fate,
  As well as a king or a queen!

For Duty's bright livery glorifies all
  With brotherhood, equal and free,
Obeying, as children, the heavenly call,
  That places us where we should be;
A servant—the badge of my servitude shines
  As a jewel invested by heaven;
A monarch—remember that justice assigns
  Much service where so much is given!

Away then with "helpings" that humble and harm,
  Though "bettering" trips from your tongue;
Away! for your folly would scatter the charm
  That round my proud poverty hung;
I felt that I stood like a man at my post,
  Though peril and hardship were there,—
And all that your wisdom would counsel me most,
  Is—"Leave it—do better elsewhere."

If "better" were better indeed, and not "worse,"
  I might go ahead with the rest,
But many a gain and a joy is a curse,
  And many a grief for the best;
No!—duties are all the "advantage" I use;
  I pine not for praise or for pelf,
As to ambition, I care not to choose
  My better or worse for myself!

I will not, I dare not, I cannot!—I stand
  Where God has ordained me to be,
An honest mechanic—or lord in the land—
  He fitted my calling for me:
Whatever my state, be it weak, be it strong,
  With honor, or sweat on my face,
This, this is my glory, my strength, and my song,
  I stand, like a star, in my PLACE.

MARTIN F. TUPPER.

ENDURING MONUMENTS.

The tomb of Moses is unknown; but the traveller slakes his thirst at the well of Jacob. The gorgeous palace of the wisest and wealthiest of monarchs, with the cedar, and gold, and ivory, and even the great temple of Jerusalem, hallowed by the visible glory of the Deity himself—are gone; but Solomon's reservoirs are as perfect as ever. Of the ancient architecture of the Holy City, not one stone is left upon another; but the pool of Bethesda commands the pilgrim's reverence at the present day. The columns of Persepolis are mouldering into dust; but its cisterns and aqueducts remain to challenge our admiration. The golden house of Nero is a mass of ruins; but the Aqua Claudia still pours into Rome its limpid stream. The temple of the sun at Tadmor, in the wilderness, has fallen; but its fountain sparkles as freshly in his rays, as when thousands of worshippers thronged its lofty colonnades.

It may be that London will share the fate of Babylon, and nothing be left to mark its site, save mounds of crumbling brick work. The Thames will continue to flow as it does now. And if any work of art should still rise over the deep ocean of time, we may well believe that it will be neither a palace nor a temple, but some vast aqueduct or reservoir; and if any name should still flash through the mist of antiquity, it will probably be that of a man who in his day sought the happiness of his fellow-men, rather than their glory, and linked his memory to some great work of national utility and benevolence. This is the true glory which outlives all others, and shines with undying lustre from generation to generation—imparting to works something of its own immortality, and in some degree rescuing them from the ruin which overtakes the ordinary monuments of historical tradition or mere magnificence.—Edinburgh Review.

FREE SCHOOL EDUCATION.

I now allude to an enlarged system of popular education. By education I do not mean merely instruction in the arts of reading and writing, but that whole system of moral, intellectual, and religious training and cultivation which is necessary to develop the nobler faculties of our nature, and give to the character of man the impress and likeness of Him in whose image he was created. In Europe popular education is, in a political point of view, of comparatively little importance. The great mass of the people have few political privileges. They exert no influence on public opinion. They give no impress to national character. Indeed, it may well be doubted whether an ignorant people do not make the most loyal and obedient subjects.

But in America the case is reversed. Our institutions rest upon the virtue and intelligence of the people. The wise administration of our government requires the constant exercise of both these qualities, not only by the magistracy, but by the constituent body. The only hope of preserving our freedom is by diffusing knowledge and sound principles amongst the people, and by keeping them up to the level of our institutions, and of their duties under them. If this cannot be done, the government must sink to their level. Let the people become ignorant and debased, and the laws must be adapted to their capacity, and the constitution brought down to their standard of morality and intelligence. Public sentiment will become vitiated, and a spirit of licentiousness and disorganization pervade the whole body of society. It requires no spirit of prophecy to foretell the consequences of such a state of things.

If the foundation of our political edifice becomes rotten, the superstructure must inevitably fall. Disguise the fact as we may, under declarations of rights, constitutional guaranties, legislative sanctions, and parchment muniments of title, it is nevertheless true, that in all popular governments, the only security for life, liberty and property, is an enlightened public opinion. Our lives, our fortunes, and our freedom are all held by that tenure. The law is but an embodiment of public sentiment. If our rights are infringed, the mode and measure of redress must be ascertained by the opinions of judicial tribunals, which consist of the agents of the people, or the people themselves. If the title to our property be questioned, or our characters defamed, or our lives or liberty put in jeopardy by a criminal accusation, the shield of our defence is in the concurrent opinion and verdict of twelve honest and enlightened workingmen. Let the character of those whose voice is the law, whose agents are its judicial expounders, and who are themselves, in the jury box, its administrators, be debased by ignorance or vice, and what becomes of this bulwark of our defence in the hour of danger?

If my time permitted, it would be easy to show that the danger to liberty from the encroachments of executive power upon popular privilege, is always in proportion to the decline in the standard of virtue and intelligence. The pages of history abound with admonitions on this subject, which are no less frequent than impressive. An ignorant populace has always been the instrument by which ambition and treason have accomplished their unhallowed purposes. And if, in the progress of events, the day shall ever arrive in which some artful demagogue, or bold military chieftain, shall erect a throne upon the ruins of the constitution of his country, his pathway to power will be strewed with the fragments of the school-houses, the pulpits and the printing-presses, which now sow the seeds of virtue and knowledge broad-cast through the land! I say, then, to the people and the government of the United States, let the work of education go on; let the schoolmaster be sent abroad; let primary schools, and academies, and colleges, spring up in all parts of our confederacy, until the whole continent shall be dotted over with them, as the heavens are bespangled with stars.

Let this system of policy be adopted, and these primary duties of the government be faithfully performed, and who shall assign a limit to the onward march of this giant nation! She is already the wonder of the world—towns, cities, states, spring up within her borders as if by magic! The circles of her prosperity and greatness are continually becoming wider and wider, and in less than half a century she has added five-fold to her population, and doubled the number of republics which repose in security beneath her flag.

But it is not in these respects only that we witness her advance in the fulfilment of her destiny. Her institutions have made the pathway to honorable distinction as broad and as straight from the door of the humblest cottage, as from the proudest mansion in the land. Genius, and industry, and energy, find no barriers to arrest their career. The abolition of arbitrary distinctions and classes of society, has given all men an equal start in the race of preferment, and brought thousands of eager competitors into the field, whose nobler faculties would otherwise have remained forever undeveloped.

The whole talent of the country is thus forced into action, and the results are visible in every vocation of life. They are to be seen alike in the fairs of the American Institute, and in the council chambers of the nation. How often do we see men who in early youth guided the plough, or wielded the hammer or the axe, in maturer age giving direction to the policy of their country! Consult the pages of our history, or go into the capitol of the Union, and inquire how many of those who sit in the high places of the land, and shed lustre on the republic, at one time made the anvil or the lapstone ring with their vigorous blows, or plied the busy needle, or with their own brawny arms brought into action that most potent of all human agents, the printing press!

But there is an illustration of my proposition, which addresses itself to you in a still more striking manner. Visit the exhibition rooms of the Institute, and examine the specimens of American ingenuity and skill which are there displayed. There we have visible and significant monuments of the genius and industry of our people. Each annual exhibition is the index to the industrial history of the preceding year. It marks its progress in all its departments, and like a table of contents, appended to the volume, points to each specific improvement that has been made. Consult this index; inquire who have been most successful in gaining distinction for themselves, and conferring benefits on mankind, and you will soon find that they are not those who had the greatest advantages of wealth, or station, or opportunity for doing good, but the self-made, industrious, practical working-men.

What a glorious reflection it is to every citizen of America, that in our happy land the tablets of fame are unfolded to every man! The genius of our country places the pen in his hands! and it is not the fault of our institutions if he fails to inscribe his name high on the roll of those destined to a glorious immortality! Such are the institutions which are committed in trust to the working-men of America, to be carefully preserved, and handed down unimpaired to their posterity.—*Anniversary Address, before the American Institute, during the Seventeenth Annual Fair, October*, 1844, *by* HON. ALEXANDER H. H. STUART, *of Virginia.*

UTILITY AND PROFIT OF FRUITS.

In the whole routine of cultivation—and it is all delightful—there is no department more pleasing or useful than fruit growing; and our advantages in this country for its production, are varied and extensive. With due attention, we can have a great variety of the most delicious fruits; and the trees, with their beautiful bloom, luxuriant foliage, and rich and gorgeous crops, are among the most ornamental scenery.

Good fruit is a great luxury, in which we may freely indulge, not only with impunity, but with advantage as to health, as well as pleasure. It forms a wholesome sustenance, and lessens the excessive use of various articles of diet, the too free use of which tends to inflammations, fevers, dyspepsia, constipation, apolexy, gout, jaundice, and a host of other ills. In numerous instances, violent diseases, and almost hopeless cases of chronic complaints, have yielded to the constant use of fruits.

The vast amount of unhealthy meats, from the sudden change of filthy matters to slaughtered animals, and by far a too liberal consumption of those that are good; also of fine flower, and fine hot bread, of butter, cheese, fat, oils, strong tea and coffee, (all injurious in excess,) the high state of cookery; the free use of condiments and seasoning, and various rich dishes, and compounds, commingled and confused; all call aloud for more fruit to lessen their use, or palliate their effects, and save thoughtless beings from untimely graves, or from lingering out a wretched state of existence. Fruits have a cooling and greatly laxative effect, regulating the stomach and bowels, correcting bilious affections, and attenuating and purifying the blood, which is the very life of the whole system.

We have very excellent fruits. How delightful, refreshing, and salutary are strawberries and cream; or delicious cherries, ready to burst with their rich juices; the golden apricot, with its fine flavor; the plum, with its honied juice; the splendid peach, with its luscious sweetness; the melting pear, with its rich, sugary, or vinous flavor; the apple, in all its variety and excellence, and multifarious preparations, extending from one end of the year to the other; the rich, luscious grape; and others equally delicious—the currant, the raspberry, gooseberry, blackberry, whortleberry, mulberry, and cranberry, and the high scented quince in its conserved state; all excellent, and conducing largely to health, pleasure, sustenance, and happiness They add a charm to social life, affording a delightful treat to friends, and to children a constant, harmless feast.

As a social entertainment, these fruits serve as a grateful substitute for the once ruinous cup, thus having a powerful moral influence. Every fruit tree is a silent preacher in the cause of temperance, a formidable ally in morality, religion, and philanthropy; for the lusciousness of fruits, and the beauty of their attendant scenery, furnish an Eden, where one may sit under his own vine and fruit trees, with none to molest, and no serpent to beguile; but with an Eve, as God's last, best gift, and perhaps cherubs gamboling in his Elysian grounds, as so many multiplied existences, in which he lives and revels amidst the charms of nature and munificence of heaven, in the happy results of his own skill and industry, and faith in Him who gives seed time and harvest. Children taught the science of horticulture and pomology, will be greatly improved in their moral and social attributes; inasmuch as they will acquire an attachment to their early home that will probably last as long as life.

Every one who has a spot of land should raise fruits, that he may have them fresh from his trees; for in no way will it yield more profit for one's own use; and where there is a good market, they are profitable for that also. Numerous instances might be named in proof of it. Mr. Moses Jones, of Brookline, in the vicinity of Boston, set out one hundred and twelve apple trees, two rods apart, and peach trees between both ways. The eighth year, he had two hundred and twenty-eight barrels of apples, and in a few years from setting the trees, four hundred dollars worth of peaches in one year; and the best part of the story is, that large crops of vegetables were raised on the same land, nearly paying for the manure and labor. The tenth year from setting, many of the trees produced four or five barrels each, the land still yielding good crops of vegetables, the peach trees having mostly gone by old age. He grafted a tolerably large pear tree to the Bartlett, and the third year it produced thirty dollars worth.

Mr. S. Dudley, of Roxbury, adjoining Boston, sold the

crop of currants from one-eighth of an acre, for one hundred and eight dollars, the next year for one hundred and twenty-five dollars, and had good crops for several years. He picked five hundred quart boxes from one-eighth of an acre the next season, after setting the bushes in the fall. He had twenty-five dollars worth of cherries from one Mazzard tree. Mr. Job Sumner, of the same place, raised of early Virginia strawberries, at the rate of sixteen hundred dollars to the acre, at twenty-five cents per quart. Strawberries on an average yield what may be sold to from three to four hundred dollars per acre. The quantity sold in Cincinnati in 1847, was six thousand bushels; and, in the previous year, one grower picked one hundred and twenty-eight bushels daily, during the height of the season.

N. Wyeth, Esq., of Cambridge, Mass., had from a Harvard pear tree, nine barrels of fruit, which sold for forty-five dollars. Hovey states that a Dix pear tree, in the same town, produced forty-six dollars worth of fruit at one crop. We saw in Orange, N. J., one hundred bushels of apples on a Harrison tree, which would make ten barrels of cider, then selling at ten dollars a barrel in New-York. Downing mentions that the original Dubois Early Golden Apricot, produced forty-five dollars worth in 1844, fifty dollars worth in 1845, ninety dollars worth in 1846. A correspondent of the Horticulturist, says that Mr. Hill Pennell, of Darby, Pa., has a grape vine that has produced seventy-five bushels yearly, which sell at one dollar per bushel. James Laws, of Philadelphia, has a Washington plum tree that yields six bushels a year, that would sell for sixty dollars. And Judge Linn, of Carlisle, Pa., has two apricot trees that yield five bushels each, worth one hundred and twenty dollars. These may be extreme cases; but, they show what may be done, and should stimulate to greater and more general efforts in the culture of fruit.—S. W. Cole, *Editor of the New England Farmer.*

A NEW YEAR'S SERMON FOR FARMERS.

And now, farmers, have you done justice to your profession, to your families, and to your country, for the last year? To your profession—have you cultivated your grounds with all the assiduity and zeal of which you are susceptible?

Have you called to your aid all the agricultural reading within your reach, and taken advice from those of your neighbors who are competent to give it? Are your farms generally in better condition than they were one year ago? Are your fields better laid out and enclosed—your waste grounds less—your ditches opened—useless stones removed—and the general surface of the ground better adapted for the raising of crops? Has your land been made richer, to enable it to yield more, and have you collected a large amount of materials to increase your annual stock of manure? Are your houses more comfortable, besides being of a neater appearance, from the labors of the year? Have you added to the conveniences and safety of your barns, to make them better adapted to the purposes for which they were built?

Again, has your stock of cattle and horses improved, not only in numbers, but more in quality and appearance, and consequently in value? Have you selected, and do you raise the best kinds of sheep—we mean those kinds that are the most profitable to the owner? Have you the most profitable breed of hogs, and do you carry just so many through the winter as best conduces to your interest? In short, have you so farmed it in all things, that you have no cause of regret, because you have given to all a proper degree of attention and care?

If so, we congratulate you; but if not so,—if you have not done one, a part, or all, of these things, the year has been in a measure lost to you, and you have not done justice to your farm or your profession. Take another year of probation, turn it to better account, and let your diligence give evidence of a thorough reformation. But if you will not,—if experience cannot teach—and the prospect of harassing debts hereafter cannot incite you to a noble industry,—you will soon become an evil in a neighborhood, your example will be injurious to others, and your slothfulness and unthriftiness will assuredly lay your farm under a cumbersome mortgage,—and this once imposed, the next step is a disposal of it by a creditor, at auction.

We turn from such with disgust, and ask next, have you for the last year done your duty to your family? That is, have you made the labors of the farm as cheerful to all your dependants as circumstances would allow? Have you been so far kind and indulgent as was consistent with the proper management of a well-regulated household? Have you attended to the education of your children and apprentices, and, as far as one short year would allow, given them all the opportunities to acquire information that may be useful to them hereafter in their several pursuits, and that with intelligence they may support the free institutions of our country? If you have done this, you have done your duty; but let us at the same time remark, that education is on the advance; what was necessary for our generation, is not enough for them. The march of intellect is onward, and our present attainments are comparatively small, and will be still more undervalued, in the advance of the generation to come.

Have you done your duty to your country? Have you given the necessary aid that the good of society demands at your hands—to the bridges, roads, public improvements in your respective neighborhoods, to schools, seminaries of learning, public morals, and religious institutions? These are all great and important duties, and in a well-regulated community ought not to be slighted or forgotten. Society cannot flourish without them—they are the stamina that give stability and health to our country and its government, and that man is unfit for associated life, he is wanting in principle and reckless of consequences, who will not lend his aid to the attainment of these great and important objects.—ALBANY CULTIVATOR.

HORSE POWER GRAIN MILL. This cut represents the Horse Power and Grain Mill, manufactured by Emery & Co., of Albany, N.Y. Those having a good horse power would do well to have one of these Mills attached to it; in time of drought, and all times where the common Grist Mill is distant, it will be an economical investment. From it four bushels of fine meal can be made in an hour; and a greater quantity, if coarse.

PREJUDICES AGAINST AGRICULTURE.

NOTWITHSTANDING that Agriculture, as it was the first, has also always been the most general pursuit of civilized men, yet it is nevertheless true that it has been, more than all other sciences and arts, neglected. We generally plough, we sow, and we reap, not with enlightened knowledge of the processes we prosecute, but by habit, and with a blind following of customs established before knowledge had been gained. We suffer disappointments, which we might have prevented, and, charging the misfortune to accident and destiny, we perseveringly renew our culture in the same—I had almost said wilful—ignorance, and at the risk of the same ever-recurring disasters.

Permit me to say plainly and with some emphasis, that this indifference to Agricultural Science cannot be suffered to continue. While Commerce, aided by vigorous and well-sustained invention, is reducing the dangers and diminishing the cost of navigation, and thus bringing the similar productions of various nations into competition in common markets, population is crowding on subsistence in many countries, so rapidly as to oblige them to study how to increase the fruits of the earth which constitute that subsistence. The statesmen of Great Britain and Continental Europe have already employed Science to check the tide of an impoverishing and exhausting emigration. Even, therefore, if we should continue to neglect Agricultural improvement, England, Ireland, France, Spain, Italy, Germany and Russia would not. They must improve, are improving, and will continue to improve Agriculture; and if we neglect to follow,—aye, and if we fail to keep up with them in that improvement, they will not only exclude us from foreign markets, but will even, ultimately undersell us in our own. A pretty figure we should make in that case. This is what they are already doing in manufactures, and by the process I have indicated.

I think that there is no lack of schools and seminaries and professorships, adapted and qualified for advancing and disseminating Agricultural science. Our present seminaries and the teachers of natural science in them, are quite sufficient; and text books, guides to experiment and laboratories, are not wanting in the country. What then is wanting? Only pupils. The students in all our seminaries, intent on—not Agricultural pursuits, but what are called the learned or liberal professions—rush by the Agricultural chair, to attend to instructions in mathematics, rhetoric, and classical literature. Certainly the professor ceases to explore for new acquisitions, when no one will listen to his expositions of what he already has. A desire to communicate to others, is always combined with the passion for the pursuit of knowledge. Why then are there no pupils? The fault—again I pray you—pardon my boldness—the fault is chiefly with the farmers themselves. A farm, of course, is necessary to him who is to be a farmer. Generally, only farmers' sons have or expect farms, and so they are the class who must supply the candidates for the profession of farming. But the farmers' sons are generally averse from scientific study. There is a general prejudice that Agriculture is a simple, easy art or trade, which can be taken up and practiced without academic instruction or systematical apprenticeship, and that theoretic precepts serve only to mislead and bewilder.

On the contrary, Nature has left all the human faculties in one sense incomplete, to be perfected by general education and by training for special and distinct pursuits. She has left those faculties not less incomplete, and without adaptation, in the farmers' case than in any other. Her laws are general and inflexible. Brutes only have perfect instincts. Man can do nothing, and indeed can do nothing at all, but by the guidance of cultivated reason. Notwithstanding admitted differences of natural capacity, and of tastes and inclinations, it is nevertheless practically and generally true that success, and even distinction and eminence, in any vocation, is proportioned to the measure of culture, training, industry, and perseverance brought into exercise. So he will be the best farmer, and even the best woodsman or well digger, as he will be the best lawyer, the greatest hero, and the greatest statesman, who shall have studied most widely and most profoundly, and shall have labored most carefully and most assiduously.

There is another prejudice even more injurious than that which I have thus exposed. The farmer's son is averse from the farmer's calling. He does not intend to pursue it, and is always looking for some gate by which to escape from it. The prejudice is hereditary in the farm-house. The farmer himself is not content with his occupation; nor is the farmer's wife any more so. They regard it as an humble, laborious and toilsome one; they continually fret about its privations and hardships, and thus they unconsciously raise in their children a disgust towards it. Is not this frequently so? Is there a farmer here who does not desire, not to say seek, to procure for his son a cadet's or midshipman's warrant, a desk in the village lawyer's office, a chair in the physician's study, or a place behind the counter in the country store, in preference to training him to the labors of the farm? I fear that there is scarcely a farmer's son who would not fly to accept such a position, or a farmer's daughter who would not prefer almost any settlement in town or city, to the domestic cares of the farmhouse and the dairy.

Whence is this prejudice? It has come down to us from ages of barbarism. In the savage state, Agricultural labor is despised, because bravery in battle, and skill in the chase, must be encouraged; and so heroism is still requisite for the public defence in the earlier stages of civilization, and the tiller of the soil, therefore, rises slowly from the condition of a villein, a serf, or a slave. Nevertheless, ancient, and almost universal as this prejudice is, I am sure that it is unnatural to mankind, in ripened civilization, such as that at which we have arrived. Of all classes of society we practically have the least need of hunters; and we employ very few soldiers, while the whole structure of society hinges on the Agricultural interest. A taste, nay, a passion for Agriculture is inherent and universal among men. The soldier or the sailor cares little for learning, mechanics or music; but the solace of his weary watchings and his midnight dreams, are recollections and hopes of a cottage home. The merchant's anxieties and the lawyer's studies are prosecuted patiently for the ultimate end of graceful repose in a country seat; and lunatics, men and women, are won back to the sway of reason by the indulgence of labor in the harvest field, and the culture of fruits and flowers in the gardens of the Asylum. I know that frivolous persons, in what is called fashionable society, who sleep till noon, still continue to depreciate and despise rural pursuits and pleasures. But what are the opinions of such minds worth? They equally depreciate and despise all labor, all industry, all enterprise and all effort; and they reap their just reward in weariness of themselves, and in the contempt of those who value human talents, not by the depth in which they are buried, but by the extent of their employment for the benefit of mankind.

The prejudice, however, must be expelled from the farmer's fireside: and the farmer and his wife must do this themselves. It is as true in this case as in the more practical one which the rustic poet had in view:

"The wife, too, must husband, as well as the man,
Or farewell thy husbandry, do what thou can."

Let them remember that in well-constituted and highly advanced society like ours, intellectual cultivation relieves men from labor, but it does not at all exempt them from the practice of industry; on the contrary, it obliges the universal exercise of industry; and that notwithstanding the current use of the figures of speech, "wearied limbs, sweating brows, hardened sinews, and rough and blackened hands," there is no avocation in our country that rewards so liberally with health, wealth and honor a given application of well-directed industry, as does that of the farmer. If he is surpassed by persons in other pursuits, it is not because their avocations are preferable to his own, but because, while he has neglected education and training, they took care to secure both.

When these convictions shall have entered the farmhouse, its respectability and dignity will be confessed. Its occupants will regard their dwellings and grounds not as scenes of irksome and humiliating labor, but as their own permanent home, and the homestead of their children and their posterity. Affections unknown before, and new-born emulation will suggest motives to improvement, embellishment, refinement, with the introduction of useful and elegant studies and arts which will render the paternal roof, as it ought to be,—attractive to the young, and the farmer's life harmonious with their tastes, and satisfactory to their ambition. Then the farmer's sons will desire and demand education as liberal as that now chiefly conferred on candidates for professional life, and will subject themselves to discipline, in acquiring the art of agriculture, as rigorous as that endured by those who apprentice themselves to other vocations.

Then with the certain improvement of agriculture, we shall have the improvement and elevation of the agricultural class of American society. Have you considered how much that class renounce in denying themselves the self-improvements I have urged? Have you considered that in practice they widely renounce the functions of representation in the conduct of the Government in favor of other classes, no more privileged than their own! This is unnecessary, unwise, unsafe; indeed, it is not republican,—it is not American. In nearly all civilized States the farmers, or those who cultivated the soil, have constituted far the greater part of the population. The chief control of society and Government then, it would seem, should of right have been vested in them. Yet in truth, they have never since the age of the patriarchs, attained any such control, except just here, and just now. In Great Britain, they divide authority, but are overbalanced by merchants, manufacturers, and privileged classes. Notwithstanding modern constitutional concessions to them in France, they are nevertheless ruled there alternately by the city population and the army. In Germany, by the army. In parts of Italy, by the Church: and in Russia they are slaves.

It has always been otherwise here. Farmers planted these colonies—all of them, and organized their governments. They were farmers who defied the British soldiery on Bunker Hill, and drove them back from Lexington. They were farmers—aye, Vermont farmers, who captured the fortress at Ticonderoga and accepted its capitulation in the name of the "Great Jehovah and the Continental Congress," and thus gave over the first fortified post to the cause of the Revolution. They were farmers who checked British power at Saratoga, and broke it in pieces like a potter's vessel at Yorktown. They were farmers who re-organized the several States and the Federal Government, and established them all on the principles of equality and affiliation.

In every State, and in the whole Union, they constitute the broad electoral faculty, and by their preponderating suffrages the vast and complex machine is perpetually sustained and kept in regular motion and operation. That it is in the main well administered we all know by experienced security and happiness; that it might be better administered, our perpetual and intense passion for change fully proves; that it is administered no better, results from what? From the fact that the electoral body—the farmers, intelligent and patriotic as they are, may, nevertheless, become more intelligent and more patriotic than they now are. The more intelligent and patriotic they become, the more effective will be their control, and the wiser the direction of the Government. Is there not room? Nay, is there not need for more activity, energy and efficiency on their part for their own security and welfare? In the Federal Government commerce has its minister and department, the law its organ and representative, and the arts their commissioner and bureau. But the vast interest of Agriculture has only a single desk and a subordinate clerk in the basement of the Patent Office. It is scarcely better in the States. An empty charter of incorporation, with a scanty endowment, constitutes substantially all that has been anywhere done for Agriculture. Gentlemen, I like not that it should be so. Our nation is rolling forward in a high career, exposed to shocks and dangers. It needs the utmost wisdom and virtue to guide it safely; it needs the steady and enlightened direction which of all others the farmers of the United States can best exercise; because being freeholders, invested with equal power of suffrage, they are at once the most liberal and the most conservative element in the country.—*From an Address of the* HON. WM. H. SEWARD *before the Vermont State Agricultural Society, at Rutland, Sept.* 2, 1852.

BENEFITS OF AGRICULTURAL FAIRS.

THE principle of association—the practice of bringing men together bent on the same general object, pursuing the same general end, uniting their intellectual and their physical efforts to that purpose, is a great improvement in the present age. We saw it years ago,—perhaps I might say centuries ago. It began in the corporations of the cities of the old world. It began in professional associations of the old world—in the legal, the medical, and the theological. But it was long, in that country and in this, before this principle of combination came to be acted upon in the great system of Agriculture—before it was brought to that pursuit of life which is the main pursuit of life—before agriculturists were brought to act in unison. And the reason is obvious. In cities communities strive together. The merchants and ship-owners can come together at the sound of a bell. The mechanics, generally living in populous places, may do the same. They have the opportunity of interchange of sentiments every hour, and what one knows, all know, and what is the experience of one all soon become acquainted with. But the agricultural population is scattered over all the fields of the country. Their labors and their toils are, in some degree, isolated. They are in the midst of the hills and the valleys, and in the recesses of every solitary forest. There is no 'Change for them to assemble upon at noon. There are no coffee-houses, there are no Atheneums for them to meet at in the evening, and converse on their interests.

It has, therefore, become essential to the best interests of the farmers of the commonwealth, that these annual fairs should be established, and that they should be universally attended. And, as his Excellency the Governor has remarked, it is not so much on account of what is to be learned by the most eloquent discourses in the public houses, or at the other establishments, as from the meeting of men together who have the same general object, who wish for improvement in the same general pursuits of life—that they may converse with one another—that they may compare with each other their experience, and that they may keep up a constant communication. It is in this point of view—in this greatly practical point of view—that these annual fairs are of importance.

Why, gentlemen, every man obtains a very great portion of all that he knows in this world, by conversation. Conversation—intercourse with other minds is the general source of most of our knowledge. Books do something, but every man has not the opportunity to read. It is conversation that improves. If any one of us here to day, learned or unlearned, should deduct what he has learned by conversation from what he knows, he would find but very little left, and that little not of the most valuable kind. It is conversation—it is the meeting of men, face to face, and talking over what they have common in interest—it is this intercourse that makes men sharp, intelligent, ready to communicate to others, and ready to receive intimations from them, and ready to act upon those only which they receive by this communication.

Therefore, if there were not a thing exhibited—if there were not a good pair of steers, nor a fine horse, nor likely cow in the whole county, if there be society—if there be ladies—wives and daughters—if there be those connected with the tillage of land, I say that these annual meetings are highly important to progress in the art to which they refer. I come here as a poor farmer, to meet with other better farmers, ready to receive from them any intimations their experience may have taught, and desirous only of suggesting something for their reflection which, now or hereafter, may draw their attention, and draw it usefully to something in the agricultural art.

There is nothing that I know of in my mode of culture of the thin and light lands which I possess, different from the general method of cultivation in this commonwealth, except this, that I have been persuaded, by reading and observation abroad, that there is one species of cultivation almost unknown in the State of Massachusetts, which is still very well suited to the counties of Norfolk, Bristol, Plymouth, and other places where there is a great proportion of light land ; I mean the root cultivation—that of turnips and beets. And, from all flights of oratory upon agriculture, I come down to simple beets and turnips, and give you one word of recommendation upon that subject.

Now the time is coming when the light lands must yield to this culture. I argue from analogy ; I see that the cultivation of the turnip crop is the very soul and substance of English husbandry. I see that England would fail to pay interest of her national debt if turnips were excluded from her culture. It is just as certain as any thing in the world, it would be impossible for the cultivation of England to go on without the culture of turnips. It is several years since it was first introduced there. Nobody here has hardly yet attended to it, and everybody there knows all about it. It is the green crop of turnips which has rendered England so rich in agriculture.

What are the cattle and sheep of England supported on? On corn or dry wheat? It is not extravagant to say that four-fifths of the beef of England—the best in the world—are fattened upon the turnips of England. As to rearing so many sheep—I do not know the number—perhaps sixty or seventy millions of fleeces per year, without turnips, nobody thinks of doing such a thing. Now, I would say that the light lands of this part of the country are as favorable for turnips as in England, with one exception. In England, the turnip crop is usually consumed by the animals in the field. It is not a country where there is a great severity of winter. The ground is not frozen, for any considerable part of the time to prevent the sheep from being out in all weather. But this is not the case in Scotland, where the crop is especially profitable. There, they draw it, as it is called—they house it, and a very slight covering will keep off the frost there, and also here.

This is the basis of their prodigious production of animal food and wool, which those countries yield. I have compared statistics—I have looked at the products of my own small field, and I find that I have now, down there—the poorest farmer, on the poorest farm, in the poorest county—I have turnips, which I am willing to show to, and compare with any farmer in Yorkshire. We talk about the breed of cattle, and to be sure, it is very important. But allow me to say that no breed can be good, and strong, and fat, and handsome, without good keeping. We must adopt three rules, which the old Cato gave with regard to the raising of animals. Feed—Feed well—Feed high!

Now, gentlemen, we see, in the eastern part of the commonwealth cattle not such as they should be, because we see pastures so dry, so rocky, so covered with bushes, that they would be able to defy all the breeds in creation. The great point to which improvement should tend, is to *improve the means of sustaining animals—to increase the quantity and to improve the quality of the food for animals.* And I believe that in pursuit of that end, when the turnips of Norfolk will compare not unsatisfactorily with the turnips of England—when there will be numerous individuals, who have made their fortunes in this species of culture, that good breeds will necessarily arise. That is my opinion, and I record it now, and leave it recorded for future use. It may not come into fashion in my day. If you do not try it, your children will try it. And if there be any young man here, with a few acres fit for this purpose—if he will put it into proper shape, and will obtain the seed of the sweet turnip, and put it into the ground, between the fourth and the tenth of July, pay some little attention to it, thin it out well once, and if it does not turn out the best and most profitable crop that he ever had, why, I wish him to come down immediately to Marshfield.—*From an Address of the Annual Fair of the Norfolk Agricultural Society, at Dedham, Mass., by* HON. DANIEL WEBSTER, LLD. 1849.

WRONG FARMING AND RIGHT FARMING.

THERE is a current impression or opinion, that agriculture, in New England, is not and cannot be profitable. And some respect is due to current opinions, though there are many such that are mere saws of the day, and worth about as much as the new saws that prevail, every successive year, in regard to medicine. Our farmers scour over three or four times the amount of surface which they are really able to cultivate, wearing down land, teams, bones and patience, to get a bare subsistence, and then declare, with a sigh, that agriculture here is not profitable, they must sell and go west.

Then springs up a dilettanti, or gentleman farmer, who declares it is not so, as he will shortly prove. He buys a small farm, for he has some right notions, and proceeds to bring it under cultivation, on scientific principles. He gives orders to the men to make heaps of compost, that will cost him, before he gets through, at least five dollars a load; lays in a stock of patent utensils; plants orchards that are certainly to pay all his expenses in five years; purchases a herd of cows that are to give half a barrel of milk a day; sheep that will yield a fleece of ten pounds every year, besides being mutton themselves, and a choice breed of swine that will almost fatten themselves on their own reputation, and thus he begins. Then he comes out of his scientific library, in his gloves, to see how the men get on, and how the old worn out farm rejoices under his new scientific dispensation. Or perhaps he is a professional man, and only goes out occasionally to observe the working of his experiment. By-and-by he begins to think the bills come in too fast, much faster than his money. He goes into a reckoning, and finds to his great surprise, that his farm has cost him about a thousand dollars a year. Now he also concludes that agriculture in New England is certainly an unprofitable business, and the fact is proved. The old style farmer thought so; the new style farmer has made it certain. Henceforth it is an established fact.

Now I have the greatest respect for science. There is no doubt that it has added very great assistance to agriculture already, and will, in future times, add a great deal more than it has done. Still it can do nothing, separated from practical economy, personal industry, inspection and experience. Botany, by itself, will not raise potatoes; vegetable physiology will not keep them from rotting after they are raised; a knowledge of manures will not enrich a farm; a knowledge of soils, and the relations of soil and climate to the several kinds of product will not maintain good economy, or make good bargains.

The true farmer must be neither a mere theorist, nor a dull-minded drudge, following in the rut made by his father's wheels. While he takes off his coat, and wipes the sweat from his brow, he must have his wits at work too. He must dig ditches and make figures. He must throw out rocks, and what is quite as hard, must loosen the dull ideas that habit has bedden in his brain. He must be alive all over, in body and mind, in mind and body. There is no business so complex, requiring so much of steady, well digested economy, so great sharpness of judgment, so nice a balancing between theory and practice. Ten thousand chances are all at work about him, and he must have his eyes open to them all,—frosts, droughts, excess of water, good and bad seeds, insects, diseases, good successions of crops, good divisions of fields, appropriate and cheap manures, the relative yield of harvests, the markets,—all these and a thousand other distinct matters he must have in view, and it requires a mind full of intelligence and wide awake observation, to choose his way.

The time is passing away when farming at haphazard, raising any thing, any how, anywhere, can be profitable. The farmer of the coming age must be a different style of man, or he will come to naught. The future owners of the United States, and that at no very distant day, are likely to be a class of men who understand agriculture as an art; for just as the wealth of manufactures is passing into the hands of the great practical operators, so the lands will pass, ere long, by the same law, into the hands of a class who have skill to manage them profitably; while the mere drudges, who are now scrubbing over their old dilapidated farms, among rocks, and brakes, and bogs, and daisies, will descend, as they ought, to the mere rank of day laborers. If there be any class of farmers among us, who cannot awake to the necessity of improvement, cannot understand that anything is needed but to keep ploughing and planting and raising weeds, as their fathers did, I am not sure that they had not better remove to the west. The new scenes and hard trials of western life, and perhaps a good shake of ague will wake them up. If not, if they still adhere to their old vegetable habit, that is certainly a better place for the spontaneous vegetable growths of all sorts than this, and will be for at least two or three generations to come.

But the young man who has a mind awake, a sound practical judgment in a sound practical body, can do better. If he has slender means to begin with, it does not follow that he must go where land is cheapest; certainly not if that is the hardest, most uncertain way to increase his means, as in many cases it unquestionably is. Let him select for purchase, some small farm, of only twenty or thirty acres, worth perhaps thirty or forty dollars an acre, favorably situated for improvement, and of such a description, that it is capable of being easily raised in value. On this let him make his beginning. There are many such farms in the market, which, in five years, can be made worth eighty or one hundred dollars per acre, repaying, meantime, by what they produce, every expense incurred. I do not say that this generally can be done; for some kinds of land are more intractable as regards improvement than others. I only say, that a man of sharp sighted judgment, assisted by science, will select many such. For the first year or two, the land will not pay the expenses of labor and manure. But the farther you go, and the more expense you make, if wisely made, the better the return becomes, till at length, when the soil is brought up into the very highest tone possible, the income yielded is enhanced in a geometrical ratio; for the taxes, the expenses of cultivation, harvesting and fencing, are scarcely greater than before, the new manure falls into a soil that is already coming into a hearty and vigorous action, and the growths take their spring from a higher level. I doubt whether there is any method of increasing in property so certain and easy as this, or any that is more within the bounds of rigid computation. So also facts most abundantly prove.

Thus an English gentleman, who, fifty years ago, received from one of his estates five thousand pounds a year, has so increased its productiveness, by an improved agriculture, that he is now receiving forty thousand pounds a year. That is, he so managed the estate, as to make it eightfold its own former value, yielding him, all the time, a large and increasing revenue for his own expenditure: Such examples are frequent in England. The whole island, taken as a single estate, has nearly three-folded its power of production, within the last fifty years. To an American, passing through, it appears to be a vast cultivated garden, clean of weeds, covered with luxuriant growths, every hedge and field in the nicest keeping, and the highest state of production, and yet I heard them complaining in parliament of their wretched and slovenly agriculture, and declaring without scruple, and I have no doubt with truth, that the island is capable of being made to yield more than double its present product.

A similar process is going on in Prussia. Vast sterile plains of sand, that were considered worthless a few years ago, are now producing luxuriant crops of wheat. A school farm that cost two thousand dollars, was raised, in twelve years, to the value of twelve thousand dollars, by nothing but an improved method of agriculture. Similar facts are furnished in our own country. Two gentlemen, in the State of Delaware, bought a farm, at the rate of thirty dollars per acre. In a few years, the farm had paid all their expenditure, and was found producing a clear annual income, equal to the interest of five hundred dollars per acre. A worn out farm near Geneva, in the State of New York, was bought for ten dollars per acre. At the end of fifteen years, it was found to have supported a family, paid its own expenses, and was yielding, for the whole four hundred acres, the interest of one hundred and fifty dollars per acre. More conclusive still, because it is a proof on a larger scale, the current price of lands in Duchess county, New York, was, twenty years ago, only twenty or twenty-five dollars per acre. They are now selling currently at one hundred and one hundred and fifty dollars per acre. Meantime, the lands of George Washington, formerly valued at forty dollars per acre, are now selling at seven dollars.

If you desire proofs closer at hand, there is a farmer in our own State—Connecticut—who will tell you that he and his father used to sow an hundred acres of rye to get one thousand bushels of the grain, which they sold, one year, as he recollects, for six hundred and twenty-five dollars. Last year, he sold, from six acres of the same land, products to the amount of eight hundred dollars; a sum equal to the interest on two thousand two hundred dollars per acre. This gentleman began with a farm of five hundred acres. Convinced that it was too large, he commenced selling it off, and wisely reduced it to one hundred and seventy-five acres, which now, with a much smaller amount of labor, produce a much larger income than the whole five hundred, and are also worth more in the market. He lately refused two hundred dollars per acre for several acres of land, which, twenty years ago, he valued at forty dollars. Nor is the experience of this gentleman at all singular. Others have realized to a greater or less extent, the same general results, by a similar process.

I see no reason why agriculture may not be prosecuted to advantage, on a large scale, as well as any other kind of business. But the care, labor and expenditure must be proportionate, in order to carry forward the desired improvement. For the same treatment which will enrich twenty acres, will enrich a thousand. At the same time, twenty acres, well wrought, are better and more profitable than five hundred shiftlessly managed, or merely run over to catch what they will yield of their own accord, or under half cultivation. This hitherto has been the folly of our agricultural methods. There probably is no farm in Connecticut, however large, the whole amount of labor and expenditure on which might not be more profitably employed on fifty or seventy-five acres.—*From an Address of the* Rev. Horace Bushnell, DD., *before the Hartford County Connecticut Agricultural Society*, 1846.

INDUSTRY AND THE YANKEE NATION.

There is little in the past that can be useful except in admonition. It is too often but a record of woe, a beacon light flaming, with blood-red glare, on wrecks along the shore of time, that the present may be warned by the example, and not add its mournful testimony, bequeathing to the future another monument of ruin.

The industry of by-gone ages, if it could be faithfully delineated, would give a picture of national and domestic life, which might be studied with interest and instruction. It now comes to us in fragments, gathered from crumbling tombs and buried cities. It peers from the canopy of sand, wreathed over broken statue and fallen temple. It tells its tale of misery on idol and obelisk, where its own hands have written, in the everlasting granite, a memorial of its wrongs and sufferings. It lies strewed on plain and hill-side, to arrest the pilgrim, and point out for what it toiled, and starved, and died. It shows how, at the will of priest and despot, it carved out their theology in black basalt, left the Sphinx to crouch, with her unsolved riddle, by the eternal pyramids, raised the chanting Memnon to salute the morning with his now unstrung melody, and burthened the encumbered earth with huge, unmeaning monuments, that unnamed kings might moulder in forgotten graves. Poor humanity groaned out its agonized existence, that, at some distant day, when the Arab should pitch his tent on mounds covering regal palaces, the curious stranger should unfold the long hidden customs of Assyrian life, and laying open the abodes of monarchs and the altars of unknown rites, bear to far off lands the symbols of old religions, and the ensigns of ancient sovereignty.

Through the varied stages of a better known antiquity, there is no cheering sign of well employed industry, giving a character and prosperity to their times. Grecian art and Roman conquest were pre-occupied, either with sublimated visions or wholesale pillage, and had no time to watch the elements of wealth and greatness within their borders. The mighty ruins, over their widely extended dominion, bear evidence how the working masses bent to iron rule, and dragged through a life of labor, that amphitheatre, and column, and triumphal arch, might gratify the vanity or commemorate the achievements of their masters. Nor, when the Roman Empire passed away, and Christianity, mounting the throne of the Cæsars, gave hope of "peace on earth and good will toward men," was there any change to denote that the precept and example of the great founder were intended for all God's children. Then came the Feudal Ages, black as night; and the Crusades, with their prodigality of life and treasure, and war, and civil strife, and priestcraft, worse than all, to bow down honest industry, making it the passive instrument of intolerance and despotism. At last, when humanity could bend no longer, and, rebounding from the unnatural tension, burst out in Reformation and Revolution, it seemed that the time had come when outraged manhood would claim its own. But unskilled and untutored freedom scorched her wings in the flames of her own kindling, and grovelled on the earth to be chained and trampled on. Old superstitions, and old oppressions walked abroad again, under new names, and the cheated many fell back to their original position, and to the long accustomed bondage.

The desolating wars which make up history, leaving little to be told except the shock of armies and the results of conquest, have been fatal to the advancement of the human race. True progress recedes from the clash of arms. Silence reigns in shop and factory at the flaunting of martial banners, and rural toil ceases when the drum beat tells that men are to be harvested. "The thunder of the captains, and the shouting," are the death knell of industry. It flourishes where they are never known, and rejoices only in the songs of peace. The farmer's frock and the mechanic's apron, are more honorable badges than the warrior's tinselled livery, and the horse is a more respectable animal, tugging at the cart or the plough, than in saying, "ha, ha! among the trumpets."

Thirty-seven years of continued peace have done more for the useful arts, and the general welfare of the world, than all the patronage of conquerors, in the rarely lucid intervals of their madness, through all past time. War breeds war, not only in perpetuating national hatred, but in creating a tiger-taste for blood, and a love for the vagabond, lawless, and exciting life of the soldier, disqualifying for useful purposes. The pride of military glory is a country's curse, wasting her energies, and demoralizing her people. But a long period of peace brings out a better ambition, and a more civilized rivalry. Deeds of arms become matters of anecdote and story, and are heard with lessening interest as the old actors drop off the stage. New views of duty are generated, and activity is turned to productive labor, and the promotion of human happiness.

There has been, recently, a development of the spirit of the age worthy the great nation which produced it, and alike honorable to its chief projector, and those whose best interest he consulted, and who, so cordially, aided and appreciated the effort. The world has seen the novel spectacle of a gathering of the industry of all lands, and a mingling of every garb and tongue, in a commendable and peaceful contention. England looks back on many a "stricken field," and the salt seas cover many a wreck where hostile fleets were vainly arrayed against her. But her last great triumph outweighs them all, and is more glorious to her than Trafalgars and Waterloos. If the sins of nations have their retribution, may this good work arise to plead for mercy in her day of judgment.

Industry has never had such an unimpeded progress as in this country. Free from the military exactions of Europe, where standing armies drain the land of its best blood, and the liability to compelled service disheartens the exertions of the laborer, we are left to our own resources, without interference or encumbrance. However little reason we have to boast of the advance made in those portions, where local causes keep down improvement, and show no change, save exhausted fertility, and the return of the wild deer to the worn out and abandoned plantation, we can present our own New England as an evidence of what may be accomplished by diligence, when allied with determined perseverance and temperate habits.

The stern necessity which brought the fathers, has made the children what they are; and the resolute, dogmatic, conscientious, rigid Puritan, has produced the pains-taking, money-making, enterprising, working Yankee. He is an original formation. History has offered nothing like him, and you may go the wide world over, ransacking all nations, without encountering his parallel. Circumstances have created him, and although they operate on the newly landed immigrant, astonishing him, in time, with the fact, that his head can think of two things, and his hands do a dozen, they can never mould him into the think of every thing, do every thing, natural growth of the soil. The real native turns his hand to any thing, lays the foundations of a city with an axe and a jack-knife, unites all trades in his own person, whenever it is most convenient to do so, it being a matter of unconcern to him whether he shoes himself or his ox, builds a house for his family or a sty for his pigs, holds a plough or tends a spinning jenny, goes a lumbering or to the Legislature, opens tavern or keeps school. He is farmer, trader, carpenter, mason, shoemaker, blacksmith, and legislator; and is ready to give instruction in either, including most other branches of human knowledge. If he wants an article, he makes it, and if the right thing has never been made before, he invents it for the occasion, never dreaming that he is doing what is not, of course, done by all other men. He longs for foreign travel, concludes to see the world, and indulges in the episode of a four years whaling voyage; hears of California, and, in a few months, is whittling on the banks of the Sacramento, finding it more profitable to let others dig, and trade for the proceeds. He is slow to wrath, and can endure much, when it is against his interest to show fight.

The thing is, however, in him, as may be ascertained, by painful experience, by pressing him beyond his bearings. He has a reverence for wealth, and a decided inclination to convey as much of it as possible into his own keeping; thinks well of education, and has great respect for learning, when it does not cost too much, but grumbles if it increases the tax bill, and is apt, for the time, to manifest, like the spiritual rappers, a ghostly indifference for such minor matters as orthography and syntax. He has great economy of time, grudges the lost hours given to sleep and food, takes the fastest boat, the chances of being blown up being quite a secondary consideration to that of landing a few minutes earlier; wonders why the train goes so slowly, when it is tearing along its thirty miles an hour, and rises from his seat the moment its speed slackens, fearing that somebody may, somehow or other, have an advantage over him by getting out first. He has a natural inborn courtesy to woman, gives up his comfortable seat in coach or car without a murmur, considers that she has, indisputably, the best right to the best things, and has more real gallantry than existed in all the cutthroat days of chivalry. His sole relaxation is in talking politics, and it is surprising where he ever found time to pick up the details which he rolls out with such amazing volubility. He would seem almost to forget his own business, in the exciting theme; but that, though kept in the back ground, is never lost sight of, and he admits no argument, and utters no doctrine that can endanger it. He loves his home and country, and boasts of it at the furthest ends of the earth. He cares nothing for boundary lines, and drives over them with a wagon full of "notions," trading with Camanche or Mexican with as little concern as though within hail of his native village.

Let the Yankee alone, without the intermeddling aid of Government, and he will show the inland route to Patagonia. He thinks the sword a very poor instrument of conquest, and that tin is a much better agent of annexation than steel. He is the pioneer of the new world; the foremost leader in the advanced guard, as it follows the course of the sun, to herald the advent of that mighty host pressing onward to cover every hill and valley of the western wilderness. In fact, he is the man of the time, and the country; a compound of many excellent qualities, mixed with imperfections consequent to the peculiarity of his position. The character will probably change with years, conforming to the mutations incidental to old civilization; but it may be doubted if improvement will come with age, and whether posterity will not lament its decline from a race, which will have left the deeply indented marks of its energy, and the enduring and reverenced monuments of its example.

The distinguishing characteristic among us is the importance given to wealth, and its natural consequence, a desire for accumulation. It is not to be regretted, so far as it acts as an incentive to exertion, and keeps up a laudable ambition, by offering a reward to industry. But when it becomes an all absorbing passion, leaving no time for intellectual culture and rational amusement; but, engulfing every noble feeling and generous impulse, strives as though immortal life depended on the balance sheet; it may, without irreverence to popular sentiment, be questioned—whether Almighty wisdom intended this as the whole duty and final destiny of man.

Sages have differed in giving "a local habitation" to the soul. Some have supposed that it is harbored in the brain, whilst others assign the stomach as its chosen mansion, reasonably judging from the exclusive devotion often given to that organ. The philosophy of our day has settled down into a more plausible theory, and instead of imprisoning the ethereal visitor within the caverns of the head, or the integuments of the body, deposits it in that more convenient and inaccessible tenement—the pocket. In that sacred retreat it must not be disturbed. The "boundless universe" may be traversed; the range of earth, air, ocean, is left unobstructed; every science may be explored, every opinion disseminated, every policy discussed, provided that sanctuary is left untouched. But enter not that holy of holies, rend not the veil of that inner temple. Philanthropies can be tolerated that do not invade its precincts, but become alarming when they encroach on the ledger and bank book, or threaten the reduction of the semi-annual dividends. This pocket-worship makes a task of recreation, and our excursions of pleasure are races against time reluctantly abstracted from regular occupations. The work is to be got through, and the sooner it is accomplished the better. It appears to be much more important that a steamboat should make a quick passage, than that travellers should land safely.

There may be explosion, or conflagration, or collision; offering the various chances of scalding, burning, and drowning, but these trifles are not worth considering, provided a sufficient number of passengers are left alive to present a vote of thanks to the captain. A railroad train cannot abate its speed, or stop to be sure that the track is clear, without a loud swelling murmur of indignation from protruded heads, which feel defrauded of their right to be broken. Hardly a month passes without some terrible disaster, and we skim over the paragraph which narrates it, as too common place even for remark. A blow-up on the Ohio or Mississippi, or a crash-up on a Western or Southern railroad, is of no great consequence. People so far off cannot feel much, and, besides, such events are so continuous that they are presumed to become rather a natural and pleasing mode of "shuffling off this mortal coil." But when a fearful tragedy on the Sound or Hudson, or on a road terminating in our midst, comes, with familiar names and horrible details, we are aroused to the belief that there is some fault somewhere. But, after much talk and threatening, the alarm subsides, and we relapse into our usual indifference, leaving the lives of ourselves and friends to the tender mercies of grasping proprietors, and the guilty rivalship of reckless officers. There is among no civilized people such a disregard of life, or where it is so poorly protected. In the absence of war and pestilence, we resort to other modes of checking a surplus population; and, in our encouragement of the means, seem to adopt the resignation of the Chinese, who console themselves for any accident resulting in great loss of life, with the philosophical reflection that there is more room in the world for the survivors.

Although the universality of the New-Englander, and the readiness with which he adapts himself to every situation, usually overcome most obstacles, he, in common with all mankind, is often mistaken, and finds himself in a position for which he was not predestined. In the old world, the occupation of the father becomes that of the son, and generation after generation works on without a thought of extending itself beyond the area where its ancestors lived and died. But here, the first movement is to turn from the old homestead, and seek by new craft, and in strange places, the fortune, which is supposed to increase in bounty, as the distance lengthens. The whole world is open to enterprise, and, on entering it, the only reflection not considered, is whether taste and talent are suited to the calling which is so readily adopted.

It is not an uncommon notion that there is incompatibility between the labor of the head and hands—an established aristocracy, by which the one disdains a plebeian alliance with the other, and sets up as a privileged order, rejoicing in its own isolated grandeur. This may be comfortable doctrine to those whose manipulations are principally confined to the stretching of kid leather, but, like many a newspaper article, "wants confirmation." It is more generally entertained by such as have least claim to the usurpation, and who, by the incomprehensible caprices of fortune, are exempted from the only vocation for which nature has qualified them. The intimate relation between intellectual and manual power, is no where more strongly exemplified than here. Examine the industry dotted about this State; go to the working places within this city. Wherever the clink of the hammer is heard, the hand of labor is shaping out the creations of thought. The elegancies, which wealth and taste collect around them, are not the mere results of mechanical operation. The beautiful forms, which the cunning hand fashions, by slow degrees, are children of the brain, wrought out in "the mind's eye," in clear and distinct vision, while the metal or the wood is an unshaped block. There must be much study before the pattern can go to the loom, or the model to the foundry; many a hard thought, as well as hard thump, ere the splendid plate service can glitter on sideboard or tea-table.

The ornament, which decorates the ball-room beauty, comes from the aching head of invention, with its curious combination of gems and gold, while the one still nestles with its fellows in the drawer of the shop, and the other has not yet entered the smelting furnace. Every article of luxury is a consequence of intellectual application—a perpetual diploma testifying to the mental ability of its originator. The self-styled educated classes do not monopolize the treasures of thought. There are other claimants to contest the honors, who never entered college, and rarely crossed the threshold of a school-house. Take heed, O student! with thy pale countenance and delicate hands, for thy brother of the smeared face and blackened fingers is hard upon thee. Thy rare subtleties and unproductive abstractions will not be cared for by thy opponent, for he comes with knowledge gained in the smoke of real battle, and the tactics of thy sham-fight will not stand before this veteran of the actual and available. Shun not, however, the encounter, but grasp that palm, hardened among tangibilities. The contact will honor thee, and thou wilt find that it is directed by a power which will strengthen thy weakness, and profit thy uncertain speculations.

In turning to the first great employment of man, the parent of every branch of industry, it is surprising that in a country especially adapted to the pursuit, it should be less a favorite than most others. All other occupations are crowded. The streets of cities swarm with busy men eager in the struggle of life. Every variety of employment is overflowing. The roar of restless humanity swells up, amidst dust and smoke, from the rising to the setting of the sun. Trade blocks every thoroughfare. Competition, in protean shapes, invites attention, alluring by gilded sign and gaudy shop window. Mechanical arts blazon their merits in newspapers, and learned professions cover whole blocks of buildings with intimations—that, within, may be found panaceas for bodily ailment, and redress for injuries of person or estate. Even divinity sends out uncounted fledgelings, the shepherds outnumbering the flocks. Men stand idly at corners, hoping to exchange work for bread. Wave after wave of immigration, in continual succession, dash on our shores, and the broken fragments huddle together, in squalid masses, in narrow lanes and impure cellars, ripening harvests to be garnered into hospitals and almshouses.

While well patronized industry, or fortunate speculation, or a combination of luck and merit, gives out a few examples of marked success, barely sufficient to dazzle and lead on the hosts of candidates in the great lottery, which has a woful quantity of blanks, and an amazing scarcity of prizes, the majorities which congregate in towns, with a devotedness to herding worthy a better share in the prosperities of society, work hard, breathe a contaminated atmosphere, contend with adverse circumstances, live doubtfully, and die poor. Yet, beyond the strife and turmoil of these human hives, beginning where the pavement ends, and stretching in illimitable extent, there is room and opportunity where the superabundance of cities might be lost in space, and the whole population of Europe could be poured without filling the mighty void. Even, within sight and sound of our densest towns, there is land calling for cultivation. Look around us, where civilization, for two centuries, has been reclaiming the wilderness. Is nothing left to be done? Have science and labor exhausted themselves? Are there no barren hills, no undrained swamps, no desolate wastes, abandoned to the rankness of exhausting and profitless vegetation? Why, the Sabbath bells, which ring out multitudes to crowded churches, peal over solitudes almost untrodden by man; where the wild flower blossoms unnoticed, and the vine drops its ripened clusters to continue the never ending circle of reproduction. The tangled thicket, shutting out the sun, is yet to be laid open, and the ploughshare is to turn the accumulated mould of centuries, that noxious luxuriance may give place to growing cornfields, and, driving out the snake and turtle, convert their long appropriated homes to the promised bounty of seed time and harvest. We have not yet begun a frugal husbandry. Necessity has not compelled an economical use of the wealth of nature, nor can it be expected when it multiplies as we recede from home, and offers a reward even for niggard toil and indifferent management.

Playing farmer is very interesting, but somewhat expensive luxury. It is not only a harmless, but praiseworthy recreation to those who need not stop to count the cost. If it is a most effectual, it is also a refreshing and exemplary relief for a plethora of the pocket—a bleeding useful to the patient, and of essential service to those to whom prudence dictates the propriety of obtaining experience vicariously. Unlike most methods of spending money, for mere amusement, there is virtue in excess, and merit is due in direct ratio to the extent of outlay. The practitioner has little danger of injuring the constitution of his subject by repeated experiment. Mother Earth is a tough old lady, and stands dosing and cutting up with admirable fortitude. Indeed, the more she gets, the better she looks, being decidedly in favor of the allopathic mode of treatment, and despising all infinitesimal application.

A gentleman farmer is usually understood, in this country, to mean one who possesses some capital in money, and very little, if any, in agricultural knowledge. He pays for his information as he gets it, and, if endowed with a moderate share of prudence, abstains from being lavish of his opinions before his practical hired laborers. When he assumes the direction of things, his orders have very much the appearance of a declaration of hostility against first principles, being often irreconcilable with each other, and somewhat at variance with the laws of nature. Like a newly made General at a militia muster, he is apt to get the rank and file into a hard knot without knowing by what earthly process he shall disentangle them, putting them as they were. He can sympathize with the sailor's embarrassment in ploughing, who managed tolerably well before the wind, but, in going about, missed stays, and involved the whole team in inextricable confusion. He fills his barn and corn crib at an expense, which may entitle the contents of the latter to the graphic appellation of "golden grain."

The gentleman farmer talks learnedly of crops, and buys his vegetables; has the most wonderful cows, and often wants milk; is well supplied with newly invented churns, and is furnished with butter from a passing market wagon, although, occasionally, the product of his dairy enables him to exult over, what seems to be, a lump of white tallow. He is strong on poultry, mixing the ornamental with the useful, gives them crystal palaces with many curious devices to induce hens to become perpetual laying machines, and is lucky if he can eat a few of his own eggs at a dollar a piece, depending for his family supply on his poor neighbors, who can at any time sell him an apron-full hurriedly collected from old sheds and rickety hay-mows. He turns for relief to his fruit, as a perennial source of consolation, there being, at least, one element of unadulterated enjoyment, demanding care from no hand but that of nature. Her operatives are at work for him. The borer riddles his apple trees; the curculio anticipates him in tasting his plums; his peach orchard gives him a crop or two, and then surrenders to the leaf-curl or the yellows, and becomes poor fire-wood; and after waiting for years, for the rich harvest from his dearly bought imported pear stocks, he finds, dependent thereon, a few gnarled, wrinkled, warty excrescences, being apparently a cross from a stringy turnip and a third rate potato.—*From an Address before the Rhode Island Society for the Encouragement of Domestic Industry, and the Rhode Island Horticultural Society, at Providence, Sept.* 17, 1852, *by the* Hon. George R. Russell, LL.D.

INSTINCTS OF THE HONEY BEE.

The knowledge that the bee possesses, as displayed in her architecture, and general economy, is not acquired by habit, or taught her by those older than herself. She comes into the world as perfect as she goes out of it. Many are the astonishing instances of foresight and knowledge, of adapting means to ends, that have come under my personal observation, but I can give only a few of them. The skill and mathematical knowledge exhibited by the bee in her architecture has astonished philosophers and scientific men of every age. It has been fully demonstrated, that the same space occupied by their cells cannot possibly be filled with any shaped vessels, that will either be of greater capacity, or take less material in the construction. There are but three ways in which cells can be built, and have the sides of all equal, namely, square, triangular, and hexagonal; a fourth is utterly impossible. The hexagonal form is superior to either of the other modes, in strength, capacity, and saving of materials in building; and this form the bee has chosen!

The honey bee will seldom use her sting against any one when not molested, and children, in particular, are exempt. When a bee is aroused to anger, she gives immediate notice of it, and no person was ever stung, unless in the midst of hundreds, excited to vengeance, without having timely warning given him. Every bee keeper is familiar with the shrill sound emitted, when the bee approaches in a threatening attitude. It is quite unlike the soft song of contentment, that is sung as the bees return from the fields laden with honey. I have never heard of any fatal consequences arising from the stings of bees, except in animals. If a horse or a cow, or any other animal upset a hive [illegible]s generally certain death. In case of being dangerously stung in many places, tobacco is worth more than all other remedies in the world. The duration of the anger of bees is from three days to a week; and any operation disturbing them much, will not be entirely forgotten short of that time. Private injuries are seldom resented by them; that is, when molested in the fields.

The following facts will show the sagacity of the bee. On

COUNTRY HOUSE, copied from WHEELER's RURAL HOMES.

a certain occasion, I attached a large sheet of comb in a hive, for the use of a family, that I was about driving into it. Some two or three days after the bees had been placed therein, I discoved that a lateral brace had been constructed from the side of the hive to the lower end of the comb. This brace was built, in consequence of my getting the comb out of its perpendicular position several times, while turning over the hive to examine the bees. The bees, it may be, reasoned thus—He is turning our hive over every day, and our comb bends, and leans over; by-and-by it will break off, so we will put a brace across to hold it! On another occasion, I laid a sheet of comb, filled with honey, on the floor of the chamber of the hive, covering several of the holes of communication with the family below. I placed it there for the purpose of feeding the bees. A few days thereafter, I was surprised to find this sheet raised three-eighths of an inch, and supported on four pillars built of wax! This was done to give the bees an opportunity to pass through with facility. The honey had all been taken away.

The most extraordinary performance, however, that was ever put on record, occurred in the following manner:—Having an entrance to one of my hives about two inches long and half an inch wide, that was covered with a thin strip of wood, with a nail at one end to hold it in the proper place, I was accustomed to turn up the door or cover, perpendicularly, as I passed the hive and found it closed. The bees had no particular use of this passage-way, as they had abundant egress below; yet it being warm weather, I kept the cover up as much as possible. It got so loose by turning it up, that it would often fall down of its own gravity; and not thinking the matter of sufficient importance to secure it at once, I turned it up daily, for about a week, and every morning I would find it down again. At last I turned it up, and out rushed about an hundred bees, and commenced clustering around it in a very singular manner, and I left them and went to town. Not returning until evening, I could not see what the result was before the next morning, when I went to the hive, and found the cover to the opening so deeply imbedded in propolis, that it could not be easily removed! It appeared that the bees wished to have this hole open, and finding that it was down one day and up the next one, they thought that they would put a stop to it at once, and they did so.

The drones appear to be a superfluous legion, of no use at all, but rather a disadvantage. This class of the honey bee derive their name from their general lazy habits, spending their time in luxury, and feeding upon the stores gathered by the ever industrious workers. They collect no honey at all, for the reason, that nature has not provided them with honey-bags, such as the workers possess, to contain collected sweets; neither have they any cavities, or baskets upon their legs, as workers have, to hold pollen or farina. This insect is the only creature known to exist in the animal creation, unprovided with the means of supplying itself with food from the boundless store-house of nature. A drone could not exist a day, were it deprived of the privilege of feeding on the stores of the hive already gathered. They are never seen to alight on any flower, or doing anything to aid the prosperity of the colony. In one respect they differ entirely from the workers, having the liberty of entering different hives with perfect impunity, while a worker enters any hive but its own at the peril of its life.—*From the American Bee Keeper's Manual, by* T. B. MINER.

INDEX.

The articles in capitals are prepared particularly in reference to recitation in school. Those in lower case Roman letters were selected, modified, and arranged for reading exercises. Those of the Reading Lessons with ¶ prefixed, contain very interesting scientific facts and illustrations, and hence are also adapted to study and recitation. Those in Italics are not heads of chapters, but incidental topics happily illustrated in connection with the main subjects.

LIST OF POPULAR BOOKS.

	Retail Price.
FERN LEAVES,	
From Fanny's Portfolio, a new work by "Fanny Fern,"..........	1 25
THE SAME—muslin, gilt edge and full gilt sides, ...	2 00
LIFE OF NAPOLEON BONAPARTE,	
Emperor of France, by J. G. LOCKHART, fine portrait, 12mo. muslin,	1 25
THE SAME—muslin, gilt edge and full gilt sides,......................	2 25
LIFE OF THE EMPRESS JOSEPHINE,	
First Wife of Napoleon, by P. C. HEADLEY, with portrait, 12mo. muslin,..	1 25
THE SAME—muslin, gilt edge and full gilt sides,......................	2 25
LIFE OF GEORGE WASHINGTON,	
First President of the United States, by JARED SPARKS, LL. D., 12mo. muslin, new and fine edition, with portrait,......................	1 50
THE SAME—muslin, gilt edge and full gilt sides,......................	2 50
LIVES OF MARY AND MARTHA WASHINGTON,	
Mother and Wife of George Washington, by MARGARET C. CONKLING, with a portrait, 16mo. muslin,...	75
THE SAME—muslin, gilt edge and full gilt sides, ...	1 50
LIFE OF REV. ADONIRAM JUDSON,	
Of the Burman Mission, by J. CLEMENT, with portrait, 12mo. muslin,	1 00
THE SAME—muslin, gilt edge and full gilt sides, ...	2 00
LIVES OF THE THREE MRS. JUDSONS,	
By ARABELLA M. WILLSON, with portraits, 12mo. muslin,.........	1 00
THE SAME—muslin, gilt edge and full gilt sides,......................	2 00
LIFE OF LADY JANE GREY,	
Steel portrait, 1 vol. 16mo. muslin, by D. W. BARTLETT,...........	1 00
THE SAME—gilt edge and full gilt sides,............................	2 00
LIFE OF HENRY CLAY,	
Of Kentucky, by HORACE GREELEY and EPES SERGEANT, 12mo, muslin,	1 25
THE SAME—muslin, gilt edge and full gilt sides,......................	2 25
LIFE OF CHRIST AND HIS APOSTLES,	
A new edition, pica type, by Rev. J. FLEETWOOD, 12mo. muslin,	1 25
THE SAME—gilt edge and full gilt sides,............................	2 25
JOHN BUNYAN'S PILGRIM'S PROGRESS,	
Pica type, seven illustrations, 12mo. muslin, (a new and fine edition,)	1 25
THE SAME—gilt edge and full gilt sides,............................	2 25
NOBLE DEEDS OF AMERICAN WOMEN,	
Edited by J. CLEMENT and Mrs. L. H. SIGOURNEY, 12mo. muslin,	1 25
THE SAME—gilt edge and full gilt sides,............................	2 25

	Retail Price.
WOMEN OF THE BIBLE,	
Being historical and descriptive Sketches of the Women of the Bible, from Eve of the Old, to the Marys of the New Testament, by Rev. P. C. HEADLEY, illustrated, 16mo. muslin,	1 00
THE SAME—gilt edge and full gilt sides,	2 00
POETS AND POETRY OF THE BIBLE,	
By GEORGE GILFILLIAN, 12mo. muslin,	1 00
THE SAME—gilt edge and full gilt sides,	2 00
GIFT BOOK FOR YOUNG MEN,	
Or Familiar Letters on Self-knowledge, Self-education, Female Society, Marriage, &c., By Dr. WM. A. ALCOTT, 12mo. muslin	75
THE SAME—gilt edge, and full gilt sides,	1 50
GIFT BOOK FOR YOUNG LADIES.	
Or Woman's Mission; being Familiar Letters to a Young Lady on her Amusements, Employments, Studies, Acquaintances, male and female, Friendships, &c., by Dr. WILLIAM A. ALCOTT, 12mo. muslin,	75
THE SAME—gilt edge and full gilt sides,	1 50
YOUNG MAN'S BOOK,	
Or, Self-Education, by Rev. WM. HOSMER, 12mo. muslin	75
THE SAME—muslin, gilt edge and full gilt sides,	1 50
YOUNG LADY'S BOOK,	
Or, principles of Female Education, by Rev. WM. HOSMER, 12mo. muslin,	75
THE SAME—muslin, gilt edge and full gilt sides,	1 50
WESLEY OFFERING,	
By Rev. D. HOLMES, 16mo.	75
THE SAME—muslin, gilt edge and full gilt sides,	1 50
SUMMERFIELD, OR LIFE ON THE FARM,	
By Rev. DAY KELLOGG LEE, 16mo.	1 00
THE SAME—muslin, gilt edge and full gilt sides	2 00
GOLDEN STEPS FOR THE YOUNG,	
To Usefulness, Respectabilty and Happiness, by JOHN MATHER AUSTIN, author of "Voice to Youth," &c., 12mo. muslin, steel frontispiece,	75
THE SAME—gilt edge and full gilt sides,	1 50
POEMS OF JOHN QUINCY ADAMS,	
On Religion and Society, with notices of his life and character, 12mo.. muslin,	50
THE SAME—gilt edge,	75
SILVER CUP OF SPARKLING DROPS,	
From many Fountains, by Miss C. B. PORTER, 12mo. muslin	75
THE SAME—gilt edge and full gilt sides,	1 50
THE AMERICAN ORATOR'S OWN BOOK,	
Compiled by JOHN AGAR, 12mo.	1 00
FRESH LEAVES FROM WESTERN WOODS,	
By METTA V. FULLER, 12mo. muslin,	1 00
LIFE OF BENJAMIN FRANKLIN,	
Written by himself, fine portrait, 12mo. muslin,	1 25

	Retail Price.
LIFE OF GENERAL LAFAYETTE, By P. C. HEADLEY, 12mo. with a portrait, uniform with Headley's Josephine, muslin,...	1 25
LIFE OF JOHN QUINCY ADAMS, Sixth President of the United States, by WILLIAM H. SEWARD, 12mo. muslin,...	1 25
LIFE OF LOUIS KOSSUTH, Governor of Hungary, by P. C. HEADLEY, with an Introduction by HORACE GREELEY, one vol. 12mo. muslin,........................	1 25
LIFE OF GENERAL ZACHARY TAYLOR, Twelfth President of the United States, by H. MONTGOMERY, 12mo. muslin, ...	1 00
LIFE OF WINFIELD SCOTT, By E. D. MANSFIELD, illustrated, 12mo. muslin,..................	1 25
LIFE OF GENERAL FRANK PIERCE, Fourteenth President of the United States, by D. W. BARTLETT, 12mo. muslin,...	75
GENERALS OF THE LAST WAR With Great Britain, with portraits of Generals Brown, Macomb, Scott, Jackson, Harrison and Gaines, by JOHN S. JENKINS, muslin, 12mo...	1 00
LIVES OF JAMES MADISON AND JAMES MONROE, By J. Q. ADAMS, 12mo. muslin,..............................	1 00
LIFE OF ANDREW JACKSON, President of the United States, by JOHN S. JENKINS, muslin 12mo.	1 00
VOICE TO THE YOUNG, Or, Lectures for the Times, by W. W. PATTON. 12mo. muslin, ...	63
MISSIONARY OFFERING, A Memorial of Christ's Messengers in Heathen Lands, dedicated to Dr. Judson, 8 engravings, 12mo. muslin,............................	1 00
PURE GOLD, Or, Truth in its Native Loveliness, by Rev. D. HOLMES, muslin,	1 00
METHODIST PREACHER, Containing 28 Sermons upon Doctrinal Subjects, by Bishop Hedding, Dr. Fisk, Dr. Bangs, Dr. Durbin, and other eminent preachers of the M. E. Church, octavo,..............................	1 00
EPISCOPAL METHODISM AS IT WAS AND IS, Being a History of the M. E. Church in the United States, &c., by Rev. P. DOUGLASS GORRIE, 12mo. muslin,........................	1 00
LIVES OF EMINENT METHODIST MINISTERS, By Rev. P. DOUGLASS GORRIE, 12mo. muslin,....................	1 25
INCIDENTS AND NARRATIVES, For Christian Families, by Rev. A. RUSSEL BELDEN, 12mo. muslin,	1 00
HISTORY AND CONDITION OF OREGON, Including a Voyage round the World, by Rev. G. HINES, of the Oregon Mission, 12mo. muslin,....................................	1 25

	Retail Price.
FREMONT'S EXPLORING EXPEDITION, Through the Rocky Mountains, Oregon and California, with additional "El Dorado" matter, with several portraits and illustrations, 435 pages, 12mo. muslin,	1 25
SIR JOHN FRANKLIN, And the Arctic Expeditions, by P. L. SIMMONDS, 12mo. muslin,	1 25
FRONTIER LIFE, Or, Scenes and Adventures in the South-west, by F. HARDMAN, 12mo muslin,	1 25
INDIAN CAPTIVES, Or, Life in the Wigwam, being true narratives of Captives who have been carried away by the Indians from the frontier settlements of the United States, from the earliest period, by SAMUEL G. DRAKE, 12mo. muslin,	1 25
HISTORY OF THE MORMONS, Or, Latter Day Saints, with Memoirs of Joe Smith, the "American Mahomet," illustrated, 1 vol. 12mo. muslin,	1 00
HISTORY OF THE WAR WITH MEXICO, From the commencement of hostilities with the United States, to the ratification of peace—embracing detailed accounts of the brilliant achievements of Generals Tyalor, Scott, Worth, Twiggs, Kearney and others, by JOHN S. JENKINS, 514 pp. large 12mo. 20 illustrations, gilt back, muslin,	1 25
FROST'S PICTORIAL HISTORY OF CALIFORNIA, Illustrated, 12mo. muslin,	1 50
THRILLING ADVENTURES, By Land and by Sea, 12mo. muslin, illustrated by J. O. BRAYMAN,	1 25
DARING DEEDS OF AMERICAN HEROES, 12mo. illustrated, by J. O. BRAYMAN,	1 25
WESTERN SCENES, And Reminiscences, together with thrilling Legends and Traditions of the Red Man of the Forest, octavo, muslin,	2 00
WILD SCENES OF A HUNTER'S LIFE, Including Cummings' Adventures among the Lions, Elephants and other wild Animals of Africa, with 300 illustrations, by JOHN FROST, LL. D., two colored plates,	1 50
BORDER WARS OF THE WEST, By PROFESSOR FROST, 300 illustrations, octavo, muslin,	2 00
YOUNG'S SCIENCE OF GOVERNMENT, New and improved edition, 12mo. muslin	1 00
AMERICAN LADY'S SYSTEM OF COOKERY, Comprising every variety of information for ordinary and holiday occasions, by Mrs. T. J. CROWEN, 12mo. muslin,	1 25
WHAT I SAW IN LONDON, Or, Men and Things in the Great Metropolis, by D. W. BARTLETT, 12mo. muslin,	1 00

	Retail Price.
WHAT I SAW IN NEW-YORK, Or, a Bird's-eye View of City Life, by J. H. Ross, M. D., 12mo. muslin,	1 00
HINTS AND HELPS TO HEALTH AND HAPPINESS, By J. H. Ross, M. D,. 1 2mo. muslin,	1 00
YOUATT ON THE DISEASES OF THE HORSE, With their Remedies, also practical Rules to Buyers, Breeders, Breakers, Smiths, &c. Notes by SPOONER. An account of breeds in the United States, by H. S. RANDALL—with sixty illustrations, muslin, 12mo.	1 25
THE AMERICAN FRUIT CULTURIST, By J. J. THOMAS, revised and enlarged, 12mo. muslin,	1 25
THE DAIRYMAN'S MANUAL, By G. EVANS, octavo, muslin,	1 00
THE AMERICAN FARMER, Or, Home in the Country, for rainy days and winter evenings, by J. L. BLAKE, D. D., 12mo. muslin,	1 25
THE YOUTH'S BOOK OF GEMS, For the Head and Heart, (with many illustrations) by F. C. WOODWORTH, octavo, muslin,	1 25
THE STRING OF PEARLS, For Boys and Girls, by T. S. ARTHUR, 16mo. muslin,	75
STORIES ABOUT BIRDS, With pictures to match, by F. C. WOODWORTH, 16mo. muslin,	75
STORIES ABOUT ANIMALS, With pictures to match, by F. C. WOODWORTH, 16mo. muslin,	75
THE HOME STORY BOOK, Illustrated, 16mo. muslin,	38
STORIES FOR GOOD CHILDREN, By Mrs. E. OAKES SMITH, 32mo. Dandelion, Moss Cup, Rose Bud, } per set	75
DOYLE'S READY RECKONER,	25
ROBBINS' PRODUCE RECKONER,	75
THE POCKET ANATOMIST,	38
LIFE AND ESSAYS OF BENJAMIN FRANKLIN, 18mo. muslin,	38
THE AUSTRALIAN CAPTIVE, Or, Adventures of William Jackman, with plates, 12mo. muslin, Edited by Rev. I. CHAMBERLAIN,	1 25
DICK WILSON, The Rumseller's Victim, or Humanity Pleading for the Maine Law, by J. K. CORNYN, 12mo. muslin,	1 25

	Retail Price
LIFE AT THE SOUTH, Or, "Uncle Tom's Cabin as it is," by W. L. G. SMITH, 12mo. muslin,	1 25
WHY I AM A TEMPERANCE MAN, With Tales and Reveries, by T. W. BROWN, 12mo................	1 25
LIFE OF MARY QUEEN OF SCOTS, By P. C. HEADLEY, 12mo. (ready in August)....................	1 25
LIFE ON THE PLAINS, And among the Diggings of California, by A. DELANO, (ready in August)..	1 25
FROST'S PICTORIAL FAMILY MUSEUM, With 300 illustrations, ...	2 50
TWELVE YEARS A SLAVE, Narrative of SOLOMON NORTHUP, 12mo. muslin,..................	1 00
LIFE OF DANIEL WEBSTER, By B. F. TEFFT, D. D., President of Genesee College, 12mo. (in press,)	1 25
DANIEL BOONE AND THE HUNTERS OF KENTUCKY, By W. H. BOGERT, 12mo. (in press)............................	1 25
THE MORNING STARS OF THE NEW WORLD, By HELEN F. PARKER, (in press,)................................	1 25
THE SUNRISE AND SUNSET OF LIFE, By HELEN F. PARKER, (in press,)................................	50
MINNIE HERMON, Or, the Night and its Morning, a Tale of the Times, by T. W. BROWN, (in press)..	1 25
WHITE SLAVES OF ENGLAND, By J. C. COBDEN, 12mo. illustrated, (in press)....................	1 50
THE WRITINGS OF JAMES ARMINIUS, Translated by J. NICHOLS and W. R. BAGNALL, 3 vols.,............	6 00
PHILOSOPHY OF CHRISTIAN PERFECTION, By PROF. CALDWELL,..	50
THE NEW-YORK CIVIL AND CRIMINAL JUSTICE, By MORGAN and SEWARD, ...	4 00
THE NEW CLERK'S ASSISTANT, A book of 1071 practical forms for business men, octavo,..........	2 25
WRIGHT'S EXECUTOR'S, ADMINISTRATOR'S AND GUARDIAN'S GUIDE,..	1 50
THE NEW CONSTABLE'S GUIDE, ...	1 00
THE STATUTES OF NEW-YORK STATE, Of a General Nature, 2 vols., ...	5 50
THE NEW ROAD ACT AND HIGHWAY LAWS,................	50

www.ingramcontent.com/pod-product-compliance
Lightning Source LLC
LaVergne TN
LVHW021315110826
845150LV00003B/578

* 9 7 8 1 4 2 5 5 5 7 5 6 0 *